Electromagnetic Compatibility for Space Systems Design

Christos D. Nikolopoulos
National Technical University of Athens, Greece

A volume in the Advances in Computer and Electrical Engineering (ACEE) Book Series

IGI Global
DISSEMINATOR OF KNOWLEDGE

Published in the United States of America by
IGI Global
Engineering Science Reference (an imprint of IGI Global)
701 E. Chocolate Avenue
Hershey PA, USA 17033
Tel: 717-533-8845
Fax: 717-533-8661
E-mail: cust@igi-global.com
Web site: http://www.igi-global.com

Library of Congress Cataloging-in-Publication Data

Names: Nikolopoulos, Christos D., 1981- editor.
Title: Electromagnetic compatibility for space systems design / Christos D. Nikolopoulos, editor.
Description: Hershey, PA : Engineering Science Reference, [2018] | Includes bibliographical references.
Identifiers: LCCN 2017043499| ISBN 9781522554158 (hardcover) | ISBN 9781522554165 (eISBN)
Subjects: LCSH: Space vehicles--Electronic equipment--Protection. | Electromagnetic compatibility. | Extraterrestrial radiation.
Classification: LCC TL3000 .E44 2018 | DDC 629.47/4--dc23 LC record available at https://lccn.loc.gov/2017043499

This book is published in the IGI Global book series Advances in Computer and Electrical Engineering (ACEE) (ISSN: 2327-039X; eISSN: 2327-0403)

British Cataloguing in Publication Data
A Cataloguing in Publication record for this book is available from the British Library.

Advances in Computer and Electrical Engineering (ACEE) Book Series

ISSN:2327-039X
EISSN:2327-0403

Editor-in-Chief: Srikanta Patnaik, SOA University, India

MISSION

The fields of computer engineering and electrical engineering encompass a broad range of interdisciplinary topics allowing for expansive research developments across multiple fields. Research in these areas continues to develop and become increasingly important as computer and electrical systems have become an integral part of everyday life.

The **Advances in Computer and Electrical Engineering (ACEE) Book Series** aims to publish research on diverse topics pertaining to computer engineering and electrical engineering. **ACEE** encourages scholarly discourse on the latest applications, tools, and methodologies being implemented in the field for the design and development of computer and electrical systems.

COVERAGE

- Computer science
- Electrical Power Conversion
- VLSI Design
- VLSI Fabrication
- Programming
- Optical Electronics
- Chip Design
- Microprocessor Design
- Digital Electronics
- Algorithms

IGI Global is currently accepting manuscripts for publication within this series. To submit a proposal for a volume in this series, please contact our Acquisition Editors at Acquisitions@igi-global.com or visit: http://www.igi-global.com/publish/.

The Advances in Computer and Electrical Engineering (ACEE) Book Series (ISSN 2327-039X) is published by IGI Global, 701 E. Chocolate Avenue, Hershey, PA 17033-1240, USA, www.igi-global.com. This series is composed of titles available for purchase individually; each title is edited to be contextually exclusive from any other title within the series. For pricing and ordering information please visit http://www.igi-global.com/book-series/advances-computer-electrical-engineering/73675. Postmaster: Send all address changes to above address.

Titles in this Series

For a list of additional titles in this series, please visit:
https://www.igi-global.com/book-series/advances-computer-electrical-engineering/73675

Design and Use of Virtualization Technology in Cloud Computing
Prashanta Kumar Das (Government Industrial Training Institute Dhansiri, India) and Ganesh Chandra Deka (Government of India,India)
Engineering Science Reference • ©2018 • 315pp • H/C (ISBN: 9781522527855) • US $235.00

Smart Grid Test Bed Using OPNET and Power Line Communication
Jun-Ho Huh (Catholic University of Pusan, South Korea)
Engineering Science Reference • ©2018 • 425pp • H/C (ISBN: 9781522527763) • US $225.00

Transport of Information-Carriers in Semiconductors and Nanodevices
Muhammad El-Saba (Ain-Shams University, Egypt)
Engineering Science Reference • ©2017 • 677pp • H/C (ISBN: 9781522523123) • US $225.00

Accelerating the Discovery of New Dielectric Properties in Polymer Insulation
Boxue Du (Tianjin University, China)
Engineering Science Reference • ©2017 • 388pp • H/C (ISBN: 9781522523093) • US $210.00

Handbook of Research on Nanoelectronic Sensor Modeling and Applications
Mohammad Taghi Ahmadi (Urmia University, Iran) Razali Ismail (Universiti Teknologi Malaysia, Malaysia) and Sohail Anwar (Penn State University, USA)
Engineering Science Reference • ©2017 • 579pp • H/C (ISBN: 9781522507369) • US $245.00

Field-Programmable Gate Array (FPGA) Technologies for High Performance...
Julio Daniel Dondo Gazzano (University of Castilla-La Mancha, Spain) Maria Liz Crespo (International Centre for Theoretical Physics, Italy) Andres Cicuttin (International Centre for Theoretical Physics, Italy) and Fernando Rincon Calle (University of Castilla-La Mancha, Spain)
Engineering Science Reference • ©2016 • 306pp • H/C (ISBN: 9781522502999) • US $185.00

For an entire list of titles in this series, please visit:
https://www.igi-global.com/book-series/advances-computer-electrical-engineering/73675

IGI Global
DISSEMINATOR OF KNOWLEDGE

701 East Chocolate Avenue, Hershey, PA 17033, USA
Tel: 717-533-8845 x100 • Fax: 717-533-8661
E-Mail: cust@igi-global.com • www.igi-global.com

Editorial Advisory Board

List of Reviewers

Table of Contents

Detailed Table of Contents

SpaceWire is a point-to-point bit shipping protocol for high-speed data communication links and networks providing equipment compatibility and seamless component reusability. It has found great application in many space missions reducing the development cost, offering architectural flexibility and improving reliability. This chapter delves into the standard describing the SpaceWire, focusing on the lower levels that play a key role in the electromagnetic behavior of the system and concern cable assemblies, shielding, bonding, and grounding. Findings regarding emissions affecting spacecraft components are presented as well as other EMC issues that have an impact on the system performance. Recent developments and upcoming updates to the standard are also presented and discussed.

In this chapter, modeling methods of static and slowly varying magnetic field emissions that are generated by spacecraft equipment are analyzed and discussed. In particular, specific issues on the established methods for multiple magnetic dipole modeling (MDM) are investigated and validated via near field measurements of well-behaved magnetic sources. Moreover, a software-based calibration technique for measuring facilities, dedicated to magnetic characterization of spacecraft units, is described and implemented on a configuration consisting of 12 sensors. Due to increasingly strict magnetic cleanliness demands, the modeling of units' induced DC magnetic behavior has become a necessary requirement for various space missions. Therefore, a baseline methodology regarding the measurements and modeling of induced magnetic fields is presented. Finally, DC methods are complemented to cover AC magnetic cleanliness requirements for on-ground verification of low-frequency magnetic fields, including AC induced magnetization effects.

In a space system or equipment, electromagnetic compatibility is a critical aspect affecting its operation, function, and performance. In the beginning of this chapter, an overview of the electromagnetic compatibility main considerations is provided. Then the main electromagnetic compatibility standards and relative useful handbooks for space applications are briefly presented. Testing methods and procedures to assess electromagnetic compatibility of space systems and units are discussed briefly covering all kinds of measurements. Moreover, information regarding the instrumentation and the testing facilities is also provided. Mission testing requirements and limits are also presented as examples for every electromagnetic compatibility test. Recent developments in test methods are also discussed where applicable.

Miguel David Ruiz-Cabello Nuñez, University of Granada, Spain
Sergio Fernández Romero, National Institute of Aerospace Technology (INTA), Spain
Marc Pous, Universitat Politècnica de Catalunya (UPC), Spain
Enrique Pascual Gil, AIRBUS, Spain
Luis M. Diaz Angulo, University of Granada, Spain
David Poyatos Martínez, National Institute of Aerospace Technology (INTA), Spain
Mireya Fernández Chimeno, Universitat Politècnica de Catalunya (UPC), Spain
Guadalupe G. Gutierrez, Airbus Defence and Space, Spain
Amelia Rubio Bretones, University of Granada, Spain
Manuel Añon Cancela, National Institute of Aerospace Technology (INTA), Spain
Ferran Silva, Universitat Politècnica de Catalunya (UPC), Spain
Jesus Alvarez, AIRBUS, Spain
Mario Fernández Pantoja, University of Granada, Spain
Borja Plaza Gallardo, Ingeniería de Sistemas para la Defensa de España S.A. (ISDEFE), Spain
Luis Nuño, Polytechnic University of Valencia, Spain
Rafael Gómez Martín, University of Granada, Spain
David Escot Bocanegra, National Institute of Aerospace Technology (INTA), Spain
Pere J. Riu, Universitat Politècnica de Catalunya (UPC), Spain
Rafael Trallero, National Institute of Aerospace Technology (INTA), Spain
Ricardo Jauregui Telleria, Universitat Politècnica de Catalunya (UPC), Spain
Salvador G. Garcia, University of Granada, Spain

Electrical and electronic systems on board air vehicles are susceptible to electromagnetic interference (EMI). This has made the topic of electromagnetic compatibility (EMC), a major concern for aircraft safety. The use of composite materials worsens this situation, for their poor shielding and low conductive capabilities. Some of the main experimental E3 certification scenarios used in aeronautics are revisited in this chapter. Guidelines to achieve simple, yet accurate, numerical models of them are provided, with appropriate tradeoffs between computational simplicity and accuracy. The numerical method, endowed with extended capabilities, has been chosen for this task for its ability and efficiency to deal with complex problems of arbitrary materials. The feature selective validation (FSV) IEEE standard procedure, commonly used to quantify the comparison of data in electromagnetic problems, is also revisited. The simulation of three different air vehicles in several certification scenarios is finally described and the numerical results compared to experimental data.

In this chapter, the spacecraft charging effects that lead to ESD events, as well as the associated immunity testing required, are presented in detail and compared to the human-induced discharges, mainly simulated by most commercial ESD generators. Additionally, design guidelines for mitigating the effects of these ESD occurrences, essential for every space application designer, are discussed. Finally, the measurement requirements, as well as the test setups, regarding the immunity testing to power leads transients are described thoroughly. This chapter aims at reviewing the current international standards and the recent bibliography, summarizing and commenting on the EMC requirements, regarding the transient phenomena, pinpointing the differentiations of these standards and their effects on the design procedure and testing creditability.

This chapter is composed of three parts. The first is an introductory part, providing general information about magnetism and related phenomena. Magnetic materials are also discussed and presented. Afterwards, the magnetic field and various measurement techniques are discussed. In the second part, different magnetic sensors used in a laboratory or space are presented. Magnetic sensors that are discussed include anisotropic magneto-resistance (AMR), giant magneto-resistance (GMR), giant magneto-impedance (GMI), flux-gate and superconducting quantum interference

Foreword

The high and constantly increasing complexity of the on-board electronic systems in modern spacecraft structures has made the topic of electromagnetic compatibility (EMC) a major concern for space missions. EMC assessments include the capability of systems and all associated subsystems and instruments to perform within the design limits without performance degradation due to electromagnetic effects encountered during accomplishment of the assigned mission.

The study of a system with regard to Electromagnetic Compatibility is a critical process to define both the probability and impact of undesired events concerning the system. The objective is to provide the policy and methodology by which risks could be managed, minimized or ideally, completely eliminated. EMC activities, many times, address new problems and demand innovative solutions tailored to the task at hand. However, some of them regard recurrent problems that can be efficiently coped with, enlisting the help of a common repository of up-to-date technical information, specifications and guidelines. This book aims to contribute to exactly that; providing practical and helpful information for Electromagnetic Compatibility in the development, measuring and modeling of spacecraft equipment and systems. Moreover, it suggests design techniques, analyses and test methods in various open issues of EMC for space systems design.

The challenges in Electromagnetic Compatibility for Space Systems are both difficult and interesting. People working on them, do so with enthusiasm, tenacity, and dedication to develop new methods of analysis, modeling, and design and to provide new solutions to keep up with the ever-changing threats. In this new age of global interconnectivity and interdependence, it is necessary to provide EMC engineers, both professionals and students, with state-of-the art knowledge on the frontiers in Electromagnetic Compatibility. This book is a pivotal step in that direction drawing greatly from the extended and diverse expertise of the contributing authors in many industry, flight and space projects as well as important theoretical contributions in their respective topics.

Nikolaos Livanos
EMTech Space P.C., Greece

Foreword

Nikolaos Livanos *is the founder and managing director of EMTECH SPACE P.C in Greece, and the subsidiary companies EMTECH GMBH in Germany and EMTECH SPACE P.C. (CY) LTD in Cyprus. He combines a strong technical background with managerial abilities, being responsible for building and executing the company's business development strategy. Since 2009, he has been involved in Space Engineering projects for the European Space Agency (ESA), as well as prime contractors in the European space industry, such as AIRBUS and Thales Alenia Space. He acts as a project manager and quality assurance manager, having proven experience in simulations, software engineering, EMC and quality assurance space standards. He acquires the qualification of MSc. degree in Electronics and BSc. degree in Physics. His working experience includes several R&D positions in high technology companies, as well as tutor and lecturer in educational institutes.*

Preface

Electromagnetic Compatibility, as well as regulatory compliance issues are subjects of great importance in electronics engineering. Avoiding problems regarding an electronic system's operation is very important, in space missions and in satellite structures however is critical. The vast majority of these problems can be traced to electromagnetic field disturbances, as interference from unintended sources and other electromagnetic phenomena. As a result, stringent requirements are to be met in terms of electromagnetic emissions level. The inclusion of this electromagnetic environment in the design of a multi - millions mission can lead to a system able to withstand whatever challenge the harsh environment throws at it. Failure to do so, may lead to important data corruption or loss, destruction of expensive instruments, waste of resources and even a total mission failure. The main scope of this book is to address the possible major emissions contributors and provide theoretical and practical background in order to model and design space systems with regard to immunity of spacecraft components and the elimination of electromagnetic interference.

Nowadays circuits for communication and automation, instruments and electronic components in general operate in close proximity and consequently affect each other adversely. As a main system design objective is to provide to the extent possible acceptable emissions and susceptibility levels. Thereafter electromagnetic compatibility has two aspects, the ability of a device to react to unwanted electromagnetic energy (susceptibility to noise) and emissions that can cause potential interference to other products (radiated and conducted). The opposite of susceptibility is immunity which defines the electromagnetic environment where the device can operate satisfactory without performance degradation. The majority of these acceptable levels of emissions are determined and controlled by regulations established from various governments' agencies and regulatory bodies; in some cases like space missions and although the devices are certified and they meet these EMC specifications, additional stringent requirements are mandatory for a successful system operation. Chapters 1 and 3 examine such cases providing new measurement approaches for modeling purposes.

This book aims to provide the relevant theoretical framework and the latest empirical research findings in the area of Electromagnetic Compatibility and Electromagnetic Interference (EMC/EMI) for space industry. Moreover, main EMC standards and relative handbooks are presented along with testing methodologies and procedures covering various areas of expertise. The target audience of this book is composed of professionals, spacecraft designers, science and data processing managers, electrical & mechanical engineers, EMC testing engineers and researchers working in the field of aerospace industry and they want to improve their understanding of the emissions and immunity regarding aerospace design engineering. Moreover, the book provides insights and support executives concerned with the management of expertise, knowledge, information and organizational development in different types of work communities and environments (Academia and industry).

Chapter 1 provides modeling methodology and measurement techniques for accurate representation of the electromagnetic signature of spacecraft's units in terms of extremely low frequency electric field emissions. Modeling methodology exploits custom measurement setups in order to solve the inverse electromagnetic problem, producing accurate model validated in an extrapolated distance. Furthermore measurement position uncertainty issues are discussed along with other contributing phenomena in a system level architecture.

Chapter 2 presents a short description of the SpaceWire protocol with emphasis on the physical layer and reviews the types of the SpaceWire emphasizing to the most commonly used micro-miniature D-type connector. Also in this chapter, bonding and grounding techniques are discussed with regards to EMC/ EMI aspects. Concluding, it focuses on parameters that affects the electromagnetic behavior of the cable, as cable assembly, the differential characteristic impedance, the skew, and the insertion & return losses.

Chapter 3 highlights the electromagnetic compatibility main considerations. Furthermore EMC standards, test methodologies and procedures to access electromagnetic compatibility of space systems and units are briefly discussed concerning all kinds of measurements. Moreover, information regarding the instrumentation and the testing facilities is also provided. Mission testing requirements and limits are also presented as examples for every electromagnetic compatibility test. Recent developments in test methods are also discussed where applicable.

Chapter 4 revises some of the main experimental E^3 (Electromagnetic Environmental Effects) certification scenarios used in aeronautics. The chapter also provides guidelines to achieve simple, yet accurate, numerical models, with appropriate tradeoffs between computational simplicity and accuracy. The Finite Difference Time Domain (FDTD) numerical method, endowed with extended capabilities, has been chosen for this task for its ability and efficiency to deal with complex problems of arbitrary materials. The Feature Selective Validation (FSV) IEEE standard procedure, commonly used to quantify the comparison of data in electromagnetic problems, is also revisited. The simulation of three different air vehicles in several certification scenarios is finally described and the numerical results compared to experimental data, to demonstrate the feasibility of the proposed roadmap.

Chapter 5 analyzes the technological advances that advanced materials are introducing in technical designs, to achieve a successful shielding of electromagnetic interference. In addition, a comparative study of the different materials that have been proposed is carried out in order to guide the researchers and companies in the selection of the most suitable material to achieve the shielding to the electromagnetic interference for diverse applications giving a specific emphasis to space applications such as satellites, aircraft, international space station, etc. The application of macro, micro- and nanomaterials is analyzed and discussed in detail in order to identify the role that advanced materials have had in the optimization of electromagnetic compatibility properties.

In Chapter 6, the spacecraft charging effects that lead to ESD events, as well as the associated immunity testing required, are presented in detail and compared to the human-induced discharges, mainly simulated by most commercial ESD generators. Additionally, design guidelines for mitigating the effects of these ESD occurrences, essential for every space application designer, are discussed. Finally, the measurement requirements, as well as the test setups, regarding the immunity testing to power leads transients are described thoroughly. This chapter aims at reviewing the current International Standards and the recent bibliography, summarizing and commenting on the EMC requirements, regarding the transient phenomena, pinpointing the differentiations of these Standards and their effects on the design procedure and testing creditability.

Chapter 7 revises general consideration about magnetism and magnetic field while examines different types of magnetic field sensors with their applications in measuring magnetic field. Moreover, this chapter provides active and passive techniques in order to create a clean magnetic room or to provide electromagnetic shielding at low frequencies.

Chapter 8 analyzes and discuss modeling methods and techniques of static and time varying magnetic fields generated by spacecraft equipment, regarding magnetic cleanliness requirements in space missions. The analysis includes specific issues concerning the test measurement methodology both for DC and AC magnetic fields (also including induced magnetic field measurements). The developed methods that are illustrated involve the implementation of stochastic techniques on the magnetic dipole modeling inverse problem and are validated via real and simulated test measurement results. Furthermore, a calibration method is presented and validated, targeting to accurately determine the position of the measuring sensors in a magnetic measuring facility. The present analysis aims to accurately estimate the parameters of magnetic dipole models that reproduce the DC and low frequency radiated AC magnetic signature of real spacecraft units and may be used to virtually represent them.

This book attempts to present current and emerging trends in research and modeling of EMC/EMI aspects in Space Missions. Features include:

- Electromagnetic Cleanliness for Space Missions
- Standards, Specifications and Best Practices
- Instrumentation and Measurement Techniques for Space Applications
- Measurement uncertainty/ Quality control in test laboratories
- Uncoupling Test Facilities for ELF field Measurements: Methodologies
- Electromagnetic Interference between spacecraft's components
- Transient Phenomena (emissions, susceptibility)
- EUT Electromagnetic Modeling
- SpaceWire: Types and Comparison
- SpaceWire Structure Harnesses Identification and Evaluation
- Shielding Techniques in Spacecraft/Satellite

It is also hoped that this book will serve as a comprehensive reference for graduate students who wish to enhance their knowledge of all aspects of EMC/ EMI for space systems engineering.

Christos D. Nikolopoulos
National Technical University of Athens, Greece

Acknowledgment

The editor would like to acknowledge the help of all the people involved in this project and, more specifically, the authors and reviewers that took part in the review process. Without their support, this book could not have become a reality.

First, the editor would like to extend a warm thank you to each end every one of the authors for their contribution. Sincere gratitude goes to the chapter's authors who contributed their time and expertise to this book. I would like to mention at this point that all authors were extremely cooperative during the different and difficult stages of the book development process. The editor wishes also to acknowledge the valuable contribution of the reviewers regarding the improvement of the quality, coherence, and content presentation of chapters. Most of the authors also served as referees; I highly appreciate their double task.

Second, I am also beholden to the IGI Global team and especially Miss Jordan Tepper for assisting me throughout this project. I wish to include here a word of appreciation for smoothly and efficiently handling the project and for being there ready to support me in times of crisis.

Finally, I wish to extend my sincere thanks to Prof. Christos Capsalis for the mentoring and the necessary theoretical and technical background he offered throughout my academic career, and made this editorial possible.

Christos D. Nikolopoulos
National Technical University of Athens, Greece

Chapter 1
Extremely Low Frequency Electric Field Emissions for Space Applications:
Measuring and Modeling Techniques

Christos D. Nikolopoulos
National Technical University of Athens, Greece

ABSTRACT

This chapter focuses on understanding the behavior of the extremely low frequency (ELF) electric field emissions of EUTs from spacecraft subsystems and on reviewing reliable equivalent models combining measurement techniques in order to acquire a complete system electromagnetic cleanliness. More precise time variations of the electric fields produced by spacecraft equipment have to be characterized, measured, and modeled. The proposed methodology is employing equivalent dipole modeling (EDM) to describe the ELF electric field spectral dependency. Each spectral component of the measured field is considered isolated and produced by one electric dipole. In order to validate the accuracy of the proposed methodology, various ELF signals in different distances are studied. In addition, since the validation of the proposed methodology would require "on ground" measurements and due to the low-frequency range, test facilities decoupling techniques are discussed. Moreover, early considerations including contributing phenomena for the complete spacecraft system modeling are reviewed.

DOI: 10.4018/978-1-5225-5415-8.ch001

INTRODUCTION

The need to measure electromagnetic fields and particle population in space plasmas has led to several space missions (Benkhoff, Casteren, Hayakawa, Fujimoto, Laakso, Novara, Ferri, Middleton, & Ziethe, 2010; Drinkwater, Floberghagen, Haagmans, Muzi, & Popescu, 2003; Scheeres, Marzari, Tomasella, & Vanzani, 1998; Escoubet, Schmidt, & Goldstein, 1997; Antonucci, Armano, Audley, Auger, Benedetti, Binetruy, Boatella et al., 2011; Müller, Marsden, Cyr & Gilbert, 2013; Bayle, Lorenzoni, Blancquaert, Langlois, Walloschek, Portigliotti, & Capuano, 2011), such as GOCE, EXOMARS, Rosetta, Cluster, BepiColombo, LISA and Solar Orbiter, which aim to facilitate these measurements. The payload of these missions is necessarily sensitive to electric and magnetic fields and require stringent electromagnetic cleanliness with emphasis to random and periodic AC electric and magnetic field variations in frequency and time domain. This necessity for cleanliness occurs due to the sensitivity of the measuring instruments that they are placed on the spacecraft's boom. From the perspective of extremely low frequency (ELF) electric cleanliness, these instruments are meant to measure slow time- variant electric fields, equivalent to frequencies not higher than 250 KHz (THOR mission). Test limits for radiated AC electric field emissions in terms of spectral density and amplitude level which derives from the spectral density in case of THOR mission are depicted in Figures 2 and 3. Regarding these Figures, it is understood that the unintended emissions from spacecraft's units shall not exceed these limits. An example of such an emission is depicted in Figure 1, presenting radiated emissions of a PCDU unit from GOCE mission.

In the process of ensuring compliance with the strict requirements, both characterization of the sources and the accurate prediction of emissions are vital for the design and position arrangement of the equipment inside the spacecraft structure. Equivalent dipole modeling (EDM) has been demonstrated to be very efficient and robust in characterizing electromagnetic emissions from radiating elements (Obiekezie, Thomas, Nothofer, Greedy, Arnaut & Sewell, 2014). Basic modeling requires well defined and characterized problem especially focused on the localization and identification of the equivalent dipoles. Additionally, optimization techniques offer ways of determining the best solution with limited computational resources (Obiekezie et al, 2014). For extremely low frequencies it is safe to assume that the electric field, especially when the near field is of interest, can be described by a quasi-static formulation. In this case, source reconstruction can be achieved either using equivalent current densities or electric dipoles with specific position and moments. Both methods are suitable and the differences between using current densities and electric dipole moments are negligible. Given the fact that the electromagnetic fields from a dipole moment can be expressed in detail and the exact distributions of current and voltages can be described in a completely

Figure 1. Indicative ELF Electric Emissions of PCDU (14 KHz -100 MHz part) from GOCE mission (published with permission of TAS- I, ESA)

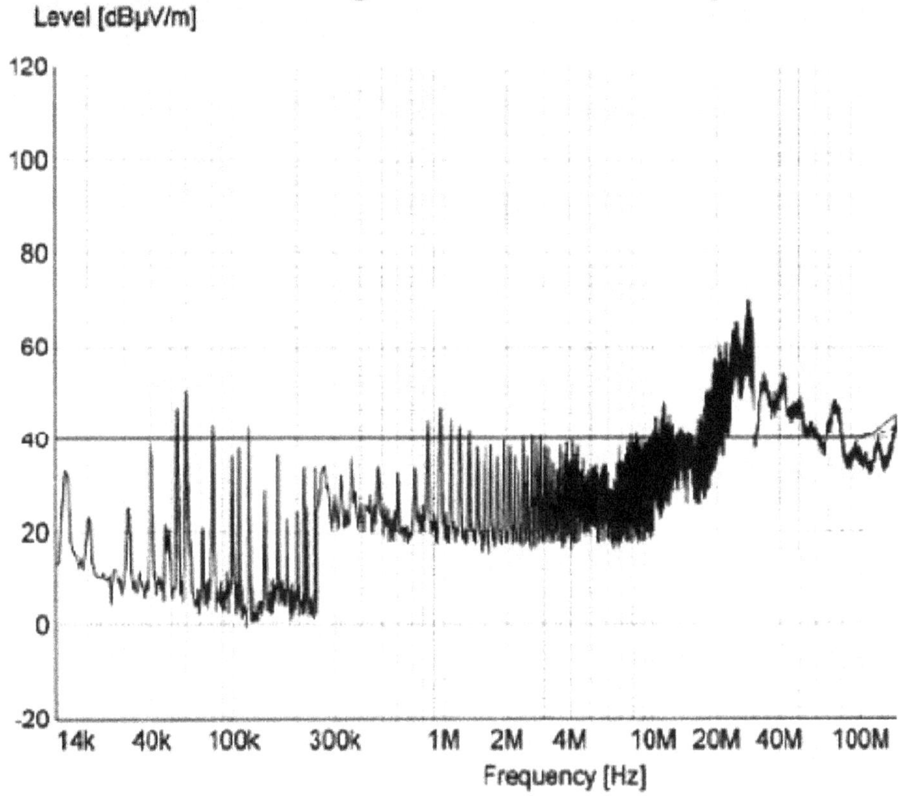

equivalent manner using dipole moments, a Multi-Frequency Electric Dipole Model methodology (MFEDM) is proposed. The novelty in the proposed MFEDM technique consists of the modeling of LF AC electric field variations assigning an equivalent dipole source for every frequency sample. The proposed methodology is able to solve the equivalent inverse problem assuming that the magnitude of each frequency component is the corresponding electrostatic field value. An iteration of this process for all frequency samples results to the full MFEDM model.

Indicative sources in the frequency range of interest (Table 1.) (Airbus Defence & Space SAS, 2014) can be modeled through a single dipole with three spatial parameters along the x, y, z coordinates and three parameters for the moment p_x, p_y, p_z, for each frequency. The mathematical formulation and background of the study will be briefly presented in the next section.

Table 1. Indicative sources for Electric Field Emissions in case of Power Control & Distribution Unit with respect to the operational frequency

Frequency	Sources
400 Hz	Tracking Frequency
65 KHz	DC/DC B push pull driver
81 KHz	DC/DC A push pull driver
130 KHz	MPPT switching
162 KHz	DC/DC A clock
250 KHz	TM data

Figure 2. Test limits for radiated AC electric field emissions in terms of spectral density (Source: ESA THOR Missions Requirements Document, 2015)

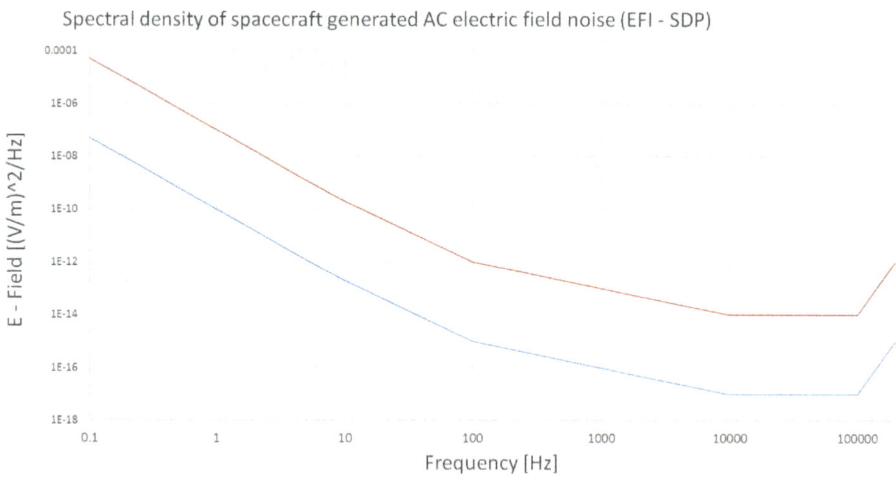

ELF E-field measurements are usually conducted with near field probes, sensors or monopole antennas where only the magnitude of the field is available. For the electric field emissions each discrete frequency corresponds to a different field magnitude value for each measurement point. In this way a multitude of quasi-static problems are formed – one for each frequency bin. In order to separate ambient noise from the useful (for modeling purposes) emitted electric field of the Equipment Under Test (EUT), a noise reference level is defined. Thus, for every distinct frequency of the spectrum above the reference level, each EUT can be modeled as one electric dipole. An iteration of this method for all frequencies of interest, results to a set of equivalent dipole models that compose the full ELF spectrum model of the EUT.

Figure 3. Test limits for radiated AC electric field emissions in terms of amplitude derived from the corresponding spectral density of Figure 2 (Source: ESA THOR Missions Requirements Document, 2015)

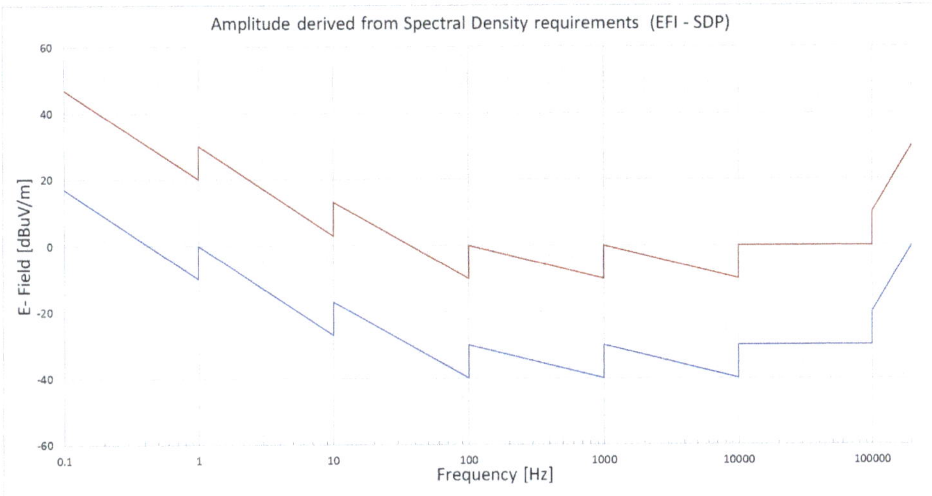

The proposed methodology is applying a stochastic method, the Differential Evolution (DE) process, in order to find a dipole (both position and moments) which produces a field that more accurately fits the electric field measurements of the EUT, on a certain set of measurement points. The source identification given the electric field values constitutes an inverse problem. After the source identification (the solution to the inverse problem) comes the validation process and in order to enable this, a second set of measurements is necessary. The correct extrapolation of the electric field on a greater distance, as compared against the second set of measurements, will validate the model. This yields the solution for a specific frequency. Repeating the aforementioned process over the whole range of frequencies that are of interest, a matrix of the dipole characteristic values for all frequencies is attained forming the EUT's Multi-Frequency Electric Dipole Model.

BACKGROUND

The electric sources inside an EUT are assumed to be represented by a small number of electric dipoles (Baklezos, Nikolopoulos & Capsalis, 2017). However, it should be noted that not all of the electric field sources inside a spacecraft are electric dipoles. For instance, electric currents flowing in the equipment and solar arrays need to be treated analytically with Biot - Savart law, and high frequency electric currents floating on spacecraft's harness can be modelled as travelling wave antennas.

It is assumed an EUT located at position $\overrightarrow{X_{EUT}}$ that produces an extremely low frequency (ELF) electric field, measured at a measurement point $\overrightarrow{X_m}$. This ELF electric field at the measurement points is calculated according to (Jackson, 1962):

$$E(\vec{n},t) = \frac{e^{-iwt}}{4\pi\varepsilon_0}\left\{k^2(\vec{n}\times\vec{p})\times\vec{n}\,\frac{e^{ikr}}{r} + [3\vec{n}(\vec{n}\cdot\vec{p}) - \vec{p}](\frac{1}{r^3} - \frac{ik}{r^2})e^{ikr}\right\} \tag{1}$$

Where $\vec{r} = \overrightarrow{X_m} - \overrightarrow{X_{EUT}}$, $r = \left|\vec{r}\right|$,$\vec{n} = \frac{\vec{r}}{r}$ and $\vec{p}$ is the electric moment vector. Moreover, it is worth mentioning that the electric field fall-off (of the dipole) follows the r^{-3} distance power law.

Wavelengths corresponding to ELF fields are in the order of kilometers and since the area of interest is in close proximity to the EUT (several meters), near field approximation is well justified. Near field approximation $(kr << 1)$ reduces Eq. 1 to:

$$\boldsymbol{E}\left(\vec{\boldsymbol{n}},\boldsymbol{t}\right) = \frac{\boldsymbol{e}^{-iwt}}{4\pi\varepsilon_0}\left\{3\vec{\boldsymbol{n}}\left(\vec{\boldsymbol{n}}\cdot\vec{\boldsymbol{p}}\right) - \vec{\boldsymbol{p}}\right\}\frac{1}{\boldsymbol{r}^3} \tag{2}$$

The electric field, with the exception of its oscillations in time, is just the static electric dipole field (Jackson, 1962). Hence the electric field can be written in the form of:

$$\boldsymbol{E}\left(\boldsymbol{t}\right) = \boldsymbol{E}_{static}\boldsymbol{cos}\left(\omega\boldsymbol{t}\right) \tag{3}$$

Fourier analysis represent a time domain signal as a sum of n sinusoidal components according to:

$$\boldsymbol{f}\left(\boldsymbol{t}\right) = \frac{1}{2}\boldsymbol{a}_0 + \sum_{n=1}^{\infty}\boldsymbol{a}_n\cos\left(\boldsymbol{nt}\right) + \sum_{n=1}^{\infty}\boldsymbol{b}_n\sin\left(\boldsymbol{nt}\right) \tag{4}$$

Based on this rationale the electric behavior of an EUT as measured in the frequency domain, can be modeled with one appropriate electric dipole per frequency yielding a Multi-Frequency Electric Dipole Model.

For each frequency an electric dipole located at point (x, y, z) produces an electric field at measurement point (x_m, y_m, z_m) with the following time invariant components in each axis:

$$Ex = \frac{1}{4\pi\varepsilon_{¿}}\left[\frac{3\left(x_m - x\right)\cdot C}{r^5} - \frac{p_x}{r^3}\right]$$
$$Ey = \frac{1}{4\pi\varepsilon_{¿}}\left[\frac{3\left(y_m - y\right)\cdot C}{r^5} - \frac{p_y}{r^3}\right] \tag{5}$$
$$Ez = \frac{1}{4\pi\varepsilon_{¿}}\left[\frac{3\left(z_m - z\right)\cdot C}{r^5} - \frac{p_z}{r^3}\right]$$

Where:

$$c = p_x(x_m - x) + p_y(y_m - y) + p_z(z_m - z)$$

And

$$r = \sqrt{\left(x_m - x\right)^2 + \left(y_m - y\right)^2 + \left(z_m - z\right)^2}$$

and the total electric field magnitude measured for frequency ω_i for every measurement point $\mathrm{N}_m\left(j\right)$ is:

$$\left|E_{static}\right|_{i,j} = \sqrt{E_x^{\,2} + E_y^{\,2} + E_z^{\,2}} \tag{6}$$

For every frequency ω_i, N_m values of the electric field are measured and comprise the adequate input dataset in order to solve the inverse problem, the solution of which is the 6 parameters (position and electric moments) of the equivalent dipole (Koutantos, Nikolopoulos, Baklezos & Capsalis, 2016).

The proposed algorithm uses a stochastic method, namely the Differential Evolution, which attempts to find the 6 variables dipole solution that produces an electric field that matches the measurements. The prososed algorthm is presented as a flowchart in Figure 4 and the DE implementation is discused in detail in the following subsection.

At this point is should be noted that an EUT can also have at the same time a magnetic behaviour. Although in the majority of the cases the magnetic moments are adequate small and the electric field that is produced due to the magnetic moment is small

Figure 4. Flowchart of MFEDM Modeling Algorithm (Koutantos, Nikolopoulos, Baklezos & Capsalis, 2016)

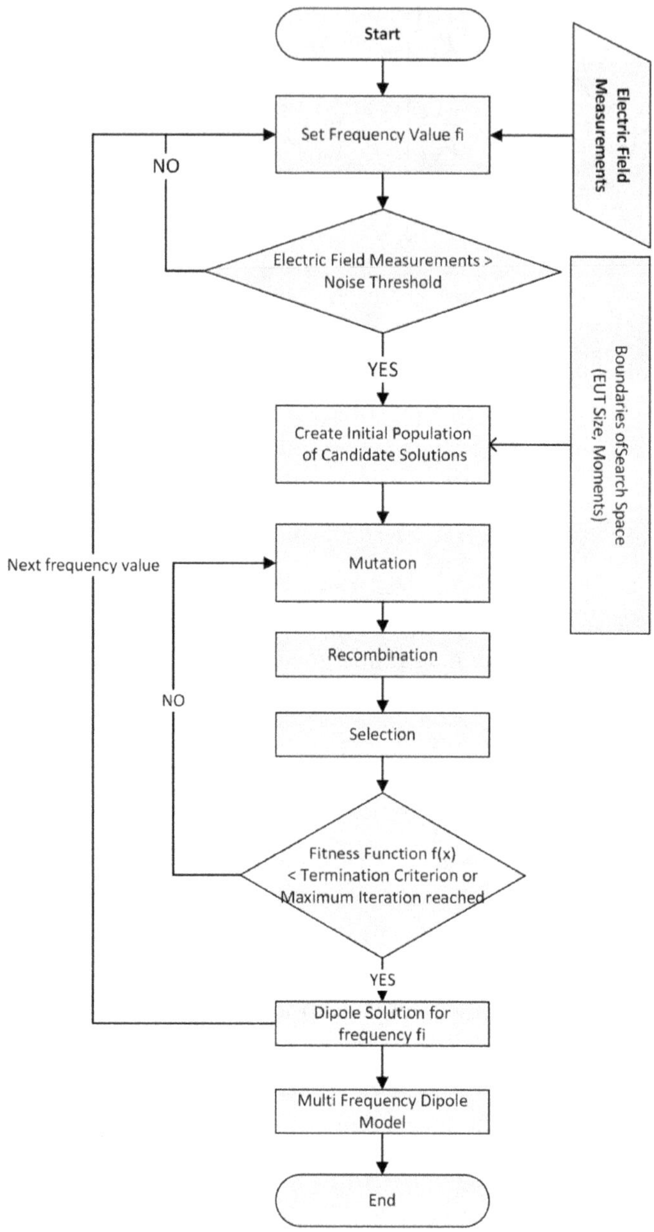

$$E\left(x,t\right)=\frac{-Z_0k^2e^{-i\omega t}}{4\pi}\left(n\times m\right)\frac{e^{ikr}}{r}\left(1-\frac{1}{ikr}\right) \tag{7}$$

That leads to a field dependence in the order of r^2:

$$E\left(x,t\right)=\frac{1}{4\pi\varepsilon_0}\left\{\frac{\dot{m}\left(t_0\right)\times n}{r^2}+\frac{\ddot{m}\left(t_0\right)\times n}{cr}\right\}$$

In should be noted that in order to produce an accurate model two steps are needed to be performed. Initially a measurement of the magnetic field shall be conducted and the EUT's magnetic signature should be captured; the magnetic moment would be estimated and evaluated. If this value can produce a substantial electric field distribution, then equation 7 should be taken also into consideration in the EDM model extraction.

DE Algorithm IMPLEMENTATION for the Inverse Problem Solution

Differential Evolution (DE) algorithm is a population based, derivative-free, Stochastic Direct Search which is using a Global Optimization procedure, first pesented by Storn and Price at 1995 (Storn & Price, 1995). DE algorithm begins with a set of initial possible solutions (initial population) that are randomly generated according to a normal or uniform distribution from within the space of valid solutions (dimensions of the EUT for the spatial parameters and indicative valid ranges for the three values of the electric moments) and are subjected to iterations of recombination, evaluation, and selection. It usually encodes decision variables as floating point numbers and manipulates them with simple arithmetic operations such as addition, subtraction, and multiplication. The recombination approach (mutation and crossover) involves the creation of new candidate solution vectror (dipole components) based on the weighted difference between two randomly selected population members added to a third population member (Brownlee, 2011).

Differential Evolution has a specialized nomenclature that describes the adopted configuration. This takes the form of $DE\,/\,x\,/\,y\,/\,z$, where x represents the solution to be perturbed (such a random or best). The y signifies the number of difference vectors used (Brownlee, 2011) in the perturbation of x, where a difference vectors is the difference between two randomly selected although distinct members of the population. Finally, z signifies the recombination operator performed such as bin for binomial and exp for exponential. The weighting factor F controls the amplification of differential variation and the crossover weight CR probabilistically controls the amount of recombination. The DE optimization algorithm is well documented and used for many problems so it will not be discussed any further. In (Drăgoi, Curteanu & Vlad, 2012) and (Oliveri, Rocca & Massa, 2012) authors have presented a wide range of DE applications in electromagnetics. For more details on the internals of the DE algorithm, the reader is directed to the rich DE literature.

The realization of DE employed to stochastically solve the inverse problem, is using the mutation strategy of $DE / rand / 1$ with the inclusion of a technique called dither (Dawar & Ludwig, 2014). Dither is the selection of factor F from the interval [0.5, 1.0] randomly for each generation or for each difference vector, a technique that is shown to improve convergence behavior significantly. Crossover weight CR is selected as 0.9 and the population size NP is set at 50. As fitness function the RMS error of the field of the solution dipole, as calculated from Eq. 7, is selected.

The proposed model is comprised of the 6 parameters (the solution variables of the DE) necessary to describe a dipole at each frequency. As depicted in Figure 4, DE is used in order to match the input data from the N_m values of the electric field measurements for each frequency with the produced field from a model dipole. For every frequency step, the DE provides the optimal set of 6 parameter's values, that are produced according to the methodology described before, that minimize the RMS error value depicted in Eq. 7. The iteration of this process for all frequency values yields the complete MFEDM model.

$$\left(RMS\right) = \sqrt{\frac{\sum_{i=1}^{N_m}\left(E_{meas}\left(i\right) - E_{\mathrm{mod}\,el}\left(i\right)\right)^2}{N_m}} \tag{8}$$

The E_{model} values are calculated according to Eq. 6.

Simulation Results & Implementation of DE on the MFEDM Method

It is assumed an arbitrary EUT with dimensions 30×30×30 cm^3 (Figure 5) placed at the origin. It should be mentioned that the corresponding equivalent source is located internally the EUT not exceeding its dimensions. As already noted, the EUT' source produces an E-field that can be modeled as an electric dipole. The emitted E-field is calculated at a number of measurement points N_m. Studying the simpler

Figure 5. EUT orientation and electric field measurement positions (Koutantos et al, 2016)

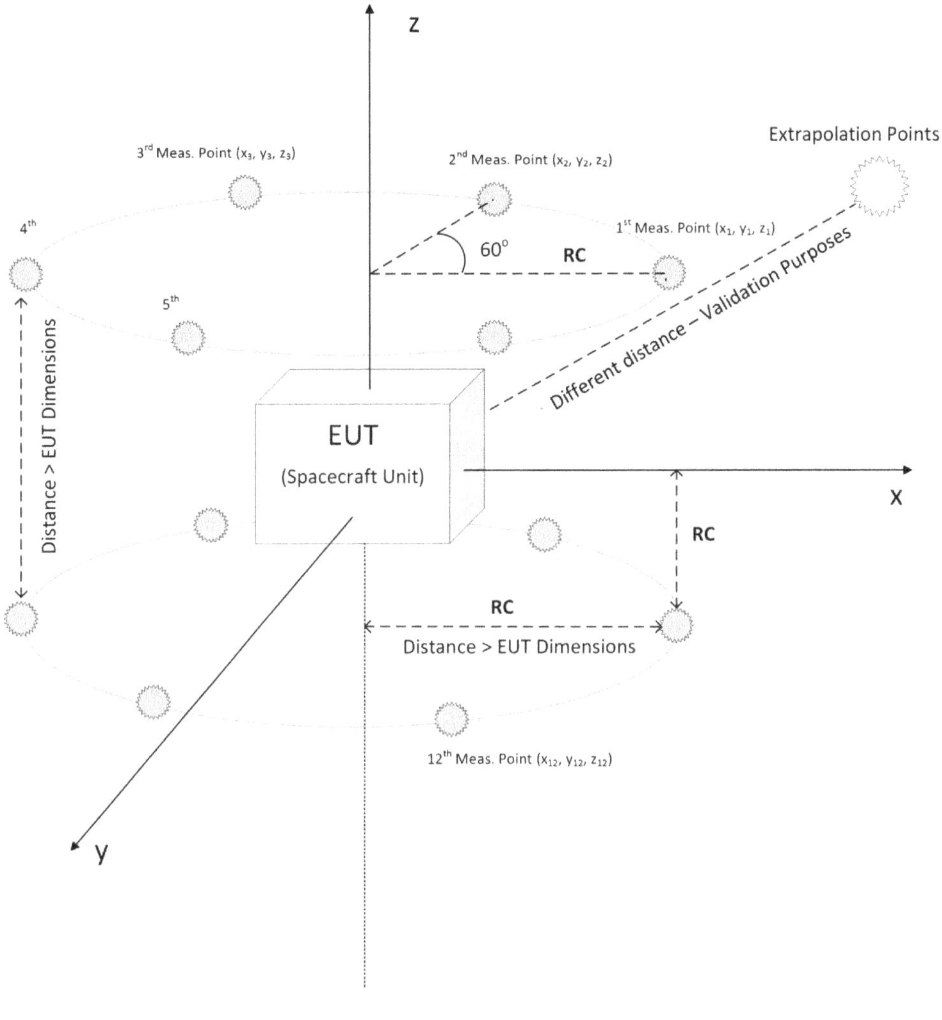

case, it is assumed that the EUT has not a magnetic dependence. In case the EUT has a magnetic dependence, equations presented at the last paragraph of background section, should be considered. The first process in order to feed the algorithm with the necessary measurement data is to populate the set of measurement points N_m. The field component magnitude is not linear with regard to the spatial parameters (x, y, and z), so the 6 parameter's value (3 for position, 3 for moment) are not adequate to accurately predict the source that produces the electric field. After extended simulations and different tested measurement setups (Baklezos et al. 2017) in order to accurately predict the equivalent model, the necessary minimum number of measurement points is found to be twelve (12) and the optimum measurement setup decided to be the one presented in Figure 5.

The number and the position of the needed measurement points in order to provide to the algorithm adequate data is depicted in Figure 5. A realistic measurement process able to acquire measurement dataset that corresponds to Figure 5, can be easily implemented in laboratory according to Figure 6.

The selected measurement points were defined in two circles of the same radius $\left(r = 40\,cm\right)$ one above and one below the EUT with six points in each circle and with the same angle φ between two points $\left(\varphi = 60^0\right)$. It is necessary that the circles be positioned outside the EUT (measurement distance bigger than the EUT dimensions). For simplicity the distance between the circles and the origin is selected equal to r. Also, another set of measurement points at a greater distance (extrapolation points at $r' = 1.5 \cdot r$) will be used for the verification of the methodology.

In order to produce simulated data at the measurement points N_m (as well at the extrapolated points), a field oscillating sinusoidally in time with frequencies

Figure 6. Implementation of measurement setup – sensor position (Koutantos et al, 2016)

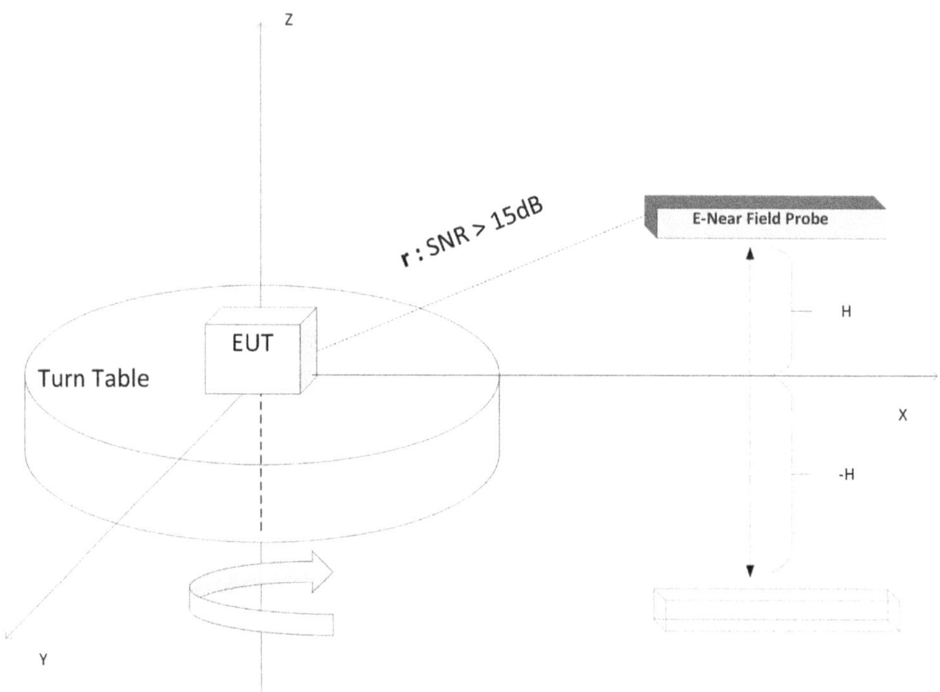

f_1=200Hz and f_2=400Hz has been produced from a dipole with spatial parameters (x, y, z) = (0.5 cm, -13 cm, 8 cm), and corresponding electric moments (P_x, P_y, P_z) = (-5 fCm,-80 fCm, 12 fCm) is considered. Dipole parameter values were selected in order for the produced field values to be similar to measurements from GOCE mission as depicted in Figure 1. So the electric field can be described by Equation 9 at each measurement point $\mathrm{N}_m\left(j\right)$.

$$E\left(t\right)\Big|_j = E_{static}\Big|_j \left\{\cos\left(\omega_1 t\right) + \cos\left(\omega_2 t\right)\right\} \tag{9}$$

Additionally, the simulated field values are mixed with Additive White Gaussian Noise (AWGN - SNR = 20dB). In the implementation of this, for each time sample of the initial signal, an AWGN value was drawn from a normal distribution with a zero mean value and standard deviation in mV equal to max signal value over SNR. A time domain snapshot of 1000 time samples (sampling rate 1 kHz) for the signal with and without AWGN for one of the measurement points is depicted in Figure 7.

Figure 7. Sinusoidally oscillating signal with AWGN in time domain (Koutantos et al, 2016)

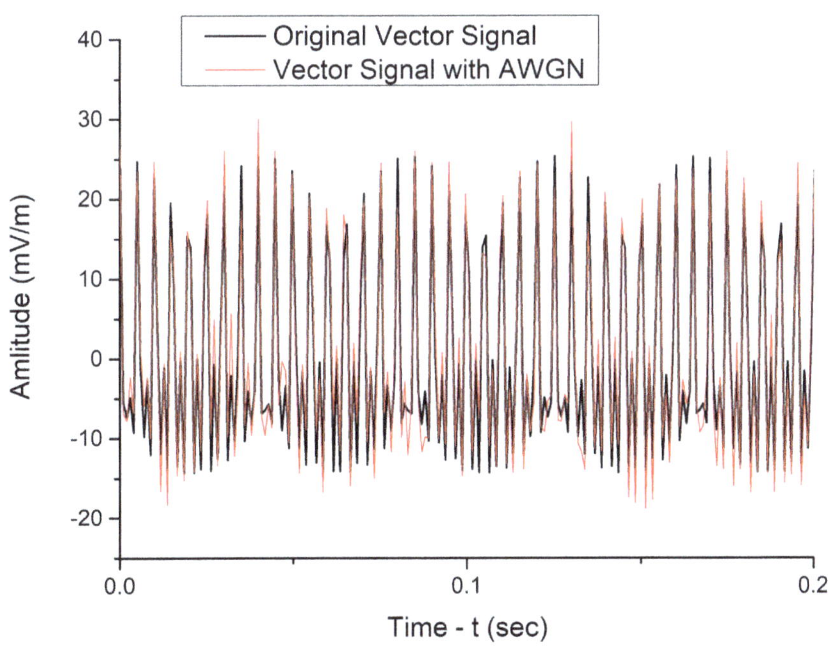

The Fast Fourier Transform (FFT) of the time domain signals (including the AWGN) is the simulated measurement data for all the measurement points. The spectrum of the sinusoidal signal of Figure 7 for one of the measurement points is depicted in Figure 8. As described before the magnitude of each spectral component across all measurement points forms the input dataset for a specific frequency. As it is depicted in Figure 8, the actual signal has spectral components at f_1=200Hz and f_2=400Hz while all the rest amplitude fluctuations is just noise. In order to distinguish the signal from the noise components and to reduce the computational time of the algorithm, a decision limit is enforced. This way the algorithm attempts to solve the inverse problem only for the frequencies that have a substantial component in at least one of the measurement points. In the simulated signal of Figure 7, the aforementioned threshold is set at 1 mV/m since the noise level is well below this value.

As described in the algorithm implementation, for the spectral magnitude of frequencies that are greater than the decision limit (for this case, in f_1=200Hz and f_2=400Hz), the inverse problem produces two dipole solutions, one for each frequency. The corresponding dipole solution vector for each frequency is tabulated in Table 2. Considering the ambient noise level, the solution produced by this MFEDM method

Figure 8. Theoretical Electric Field Values at one of the measurement points and corresponding Decision Limit in the spectral domain (Koutantos et al, 2016)

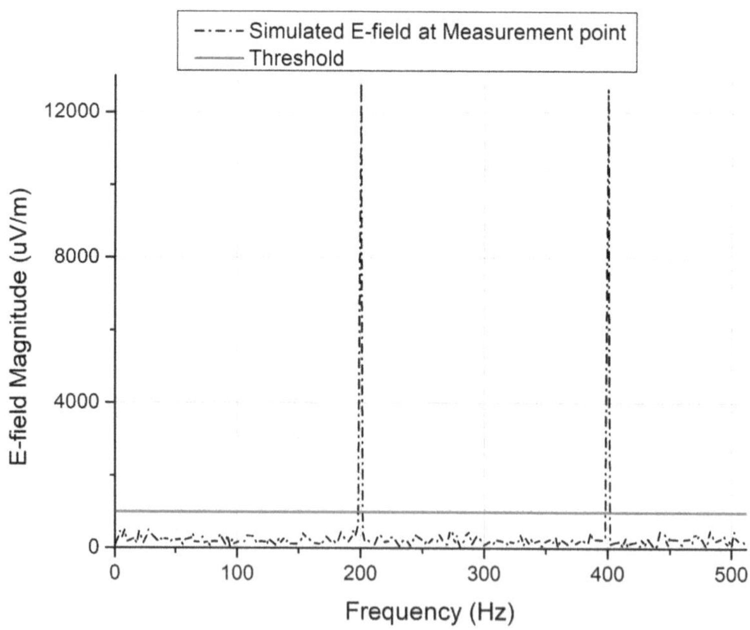

Table 2. Source Vs Model Solution Parameters

Variable	Actual Source Dipole	Model Solution $f_1 = 200Hz$	Model Solution $f_2 = 400Hz$
x	0.5	0.407	0.443
y	-13	-13.393	-13.067
z	8	8.198	7.974
p_x	-5	-6.219	5.028
p_y	-80	-80.022	80.503
p_z	12	12.437	-11.880

is found adequately accurate. Since the proposed methodology is meant to be used for the calculation of the E-field of an EUT at random distances, a comparison of the fields at the extrapolation points will increase the confidence and the accuracy of the solution. For this reason, in Figure 9 this comparison is depicted, with an RMS error across the 12 measurement points of 22uV/m.

Figure 9. Theoretical Vs Model Predicted Electric Field Values at extrapolation points in the spectral domain (Koutantos et al, 2016)

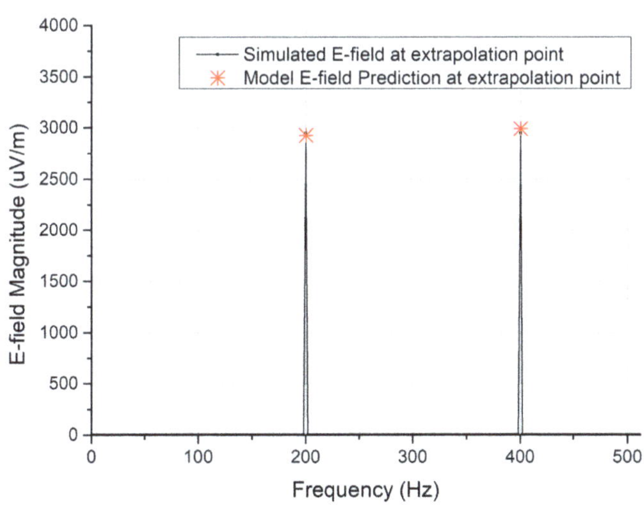

In the model solutions, spatial parameters as well as the absolute electric moment values are accurately predicted. Although signs of the electric moments in one case (for frequency 200 Hz) are correctly extracted, in the case of f_2=*400Hz* are exactly the opposite. This behavior leads to the same field amplitude values, so is also acceptable. This duality of solutions for the electric moment sign order is typical for amplitude measurements and discussed further in the next section. It should be noted that due to field dependency on the spatial parameters, they are always (with regard to the sign order) correctly extracted.

Although Table 2 demonstrates an accurate source reconstruction, this is owed completely to the simplicity of the field described by Eq. 8. In more complex time domain field forms the exact source identification would be near impossible however with the proposed methodology, which is primarily targeted at accurate extrapolated field calculation, the ELF electric spectral behavior of the EUT is well estimated.

Real behavior in actual equipment operation includes transient phenomena with a broader spectrum. To simulate such conditions a low repetition rate pulse is employed. Pulse rise and fall time are set equal to 0.25 ms, while the pulse repetition rate is 222 Hz with a duty cycle of 50%. Adding AWGN with SNR=20 dB to this pulse produces the simulated measurement for this case, as depicted in Figure 10. It should be noted that since the pulse signal has a richer spectrum compared to the

Figure 10. Pulse signal with AWGN in time domain (Koutantos et al, 2016)

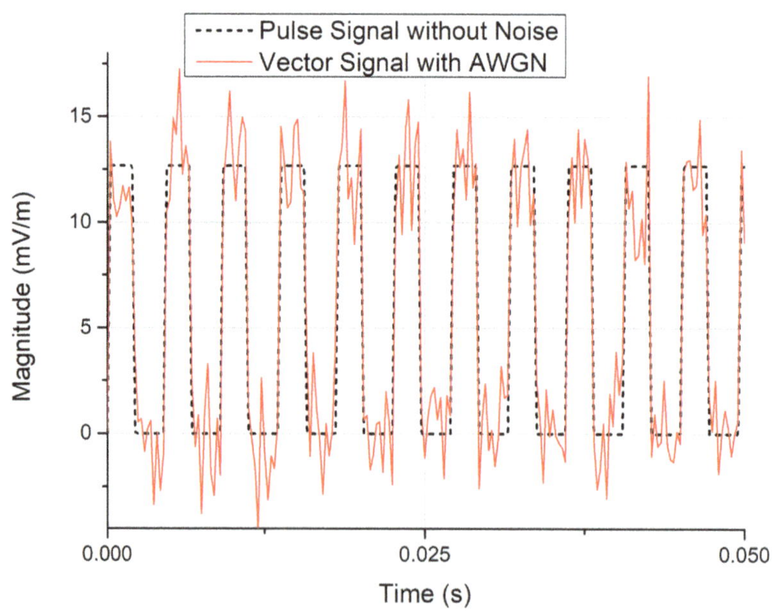

sinusoidal signal, the selected sampling frequency rises to 4 kHz in order to have a more detailed representation.

The spectral representation (FFT) of this time-domain signal is depicted in Figure 11 and the corresponding threshold is set at 0.2 mV/m. It should be noted that the decision limit value is selected roughly according to noise level per signal case. Since noise levels are different in each case, as they are expressed by SNR, the decision limit value varies accordingly. However, in real measurements, the first action taken is an ambient noise level measurement; that is with the EUT not operating. The decision limit can be safely set slightly above the maximum noise value. It is obvious that in this case the rich spectral content leads to numerous dipoles for all the frequencies above the decision limit. It should be noted that the number of solution dipoles is proportional to the sampling frequency.

The predicted values at the extrapolation points in this case are depicted in Figure 12 and are in excellent agreement with the theoretical values.

Distortion and Noise Tolerance

The field probes measuring the electric field magnitude are introducing measurement uncertainties which cannot be neglected. Furthermore, the ambient noise always

Figure 11. Theoretical Electric Field Values at one of the measurement points and Decision Limit in spectral domain (Koutantos et al, 2016)

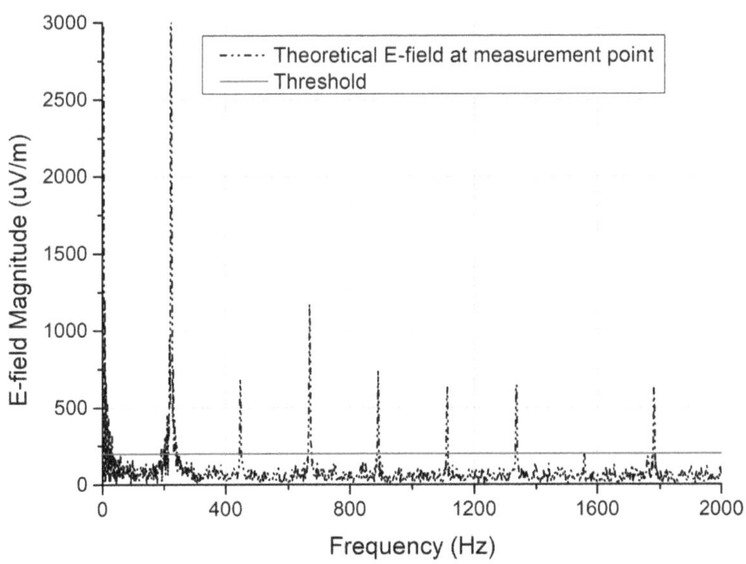

Figure 12. Theoretical Vs Model Predicted Electric Field Values at extrapolation points in spectral domain (Koutantos et al, 2016)

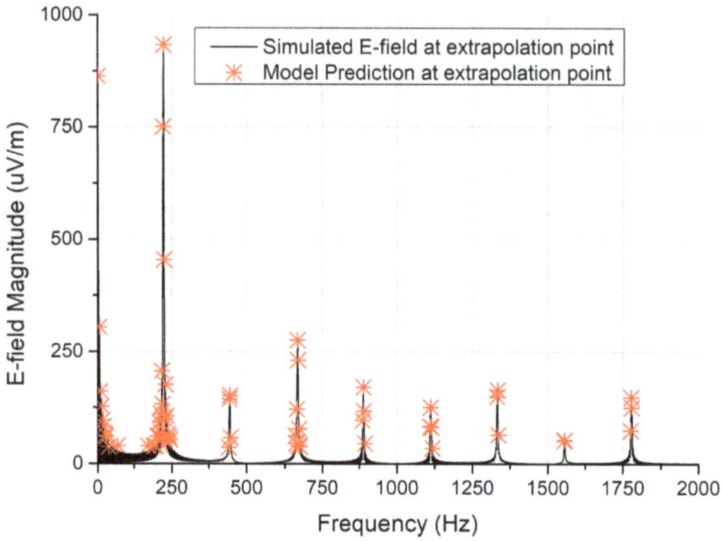

exists at any experimental procedure and can lead to models that inaccurately predict the electric field at extrapolation points.

Given these considerations, various simulations were carried out including these factors. Specifically in order to simulate measurement uncertainties (such as field probe displacement) a field distortion level of 5% is considered. Moreover, different noise levels are examined in order to determine the sensitivity of the proposed methodology to noise. With the inclusion of these factors solution produced from the algorithm are not always accurate especially for low SNR. A proposed remedy to this, is to include data from multiple measurements.

In order to simulate the multiple measurements, the theoretical electric field (E_{th}) for each measurement point is distorted for each measurement repetition according to Eq. 10 and AWGN is added:

$$E = E_{th} * \left[1 + 0.05 * \left(\boldsymbol{rand} - \boldsymbol{rand}\right)\right] + \boldsymbol{AWGN} \tag{10}$$

where **rand** is a random number between zero and one.

Afterwards, the methodology entails the solution of the reverse problem (the correct identification of the source that produces the measured field), exploiting all the previous measurements. At the next step the algorithm assigns as the solution for the spatial parameters the mean predicted values of the multiple measurements and as the moment parameters the mean of the absolute predicted values of the multiple measurements. In order to accurately predict the correct sign order of the electric moment values, four scenarios are developed for all the possible combinations or the exact opposites as explained in this section of sign order (+++ or ---,+-+ or -+-,++- or --+, -++ or +--). Every solution is evaluated according to these scenarios and a decision flag is accumulated, making the "mode" candidate the most accurate combination. These solution mean values and their standard deviations are presented in Table 3.

As it is presented in Table 3, Mean Predicted Values from the proposed methodology are in good agreement with the source dipole parameters ensuring the accurate approximation of the field distribution..

Table 4 presents the Goodness of Fit (GoF) of the proposed methodology as derived from Eq. 11 with respect to various SNR values.

$$G_{rel} = \sqrt{\frac{\sum_{i=1}^{N_m} \Delta E_i^2}{\sum_{i=1}^{N_m} E_i^2}} \tag{11}$$

Table 3. Mean Predicted Values

	Source Dipole Values	Mean Predicted Values	Standard Deviation
$x(cm)$	0.5	0.509	0.22
$y(cm)$	-13	-13.051	0.24
$z(cm)$	8	8.041	0.12
$p_x\,(fC)$	-5	-5.090	1.773
$p_y\,(fC)$	-80	-80.048	0.634
$p_z\,(fC)$	12	12.254	1.086

Table 4. Goodness of Fit for Various SNR Levels

SNR	Goodness of Fit
13.98 dB	0.1574 (15.74%)
20.00 dB 23.42 dB 26.02 dB	0.0843 (8.43%) 0.0114 (1.14%) 0.0048 (0.48%)

E_i stands for the calculated electric field amplitude from the source dipole while ΔE_i is the difference between the amplitudes of the field from the source dipole and the field produced by the predicted equivalent dipole (mean values).

For this application, values of GoF metric are considered acceptable when they are below 10%. Noise tolerance expressed as acceptable of GoF values can be reached by the proposed methodology for SNR of 20 dB or higher, which consist more than a typical noise floor in any measurement facility.

Although the discussion presented in this section concerns a single dipole hence a single frequency, the methodology is identical as presented in Figure 4 for the whole frequency range of interest. Employing multiple ELF field amplitude measurements for all the observation points and for the whole spectral range, an accurate complete ELF MFEDM model is extracted.

Ambient Conditions

In order to have a reference, the ambient levels shall be measured. The ambient measurements are defined as the background emission levels, when:

1. All the function and supervisory units of the Electrical Ground support equipment (EGSE) are powered and in operation.
2. The EUT is non-powered.
3. A resistive load is connected to the Test Equipment with a load corresponding to the power consumption of the equipment under test.
4. During the ambient measurements a LISN shall be connected to the power terminal of the resistive load.
5. The ambient electromagnetic emission levels shall be at least 6 dB below the specified limit.

In the event that the ambient measurements exceed the relevant limit-6dB every effort shall be made to reduce or eliminate those emissions, if this cannot be achieved the source of the emissions should be identified and reported in the EMC Test

Report. It should be noted that it may be necessary to address the configuration of the test equipment and its harnessing, as far as possible this should also be installed such that ground loops are minimized, harnesses may require to be over-shielded, aluminum foil can be used for this if necessary.

The impedance of power sources providing input power to the EUT shall be controlled by Line Impedance Stabilization Network(s) (LISN(s)) for all measurement. LISN's shall not be used on output power leads.

Electric Field Measurements

Considering that there are no standard methods and setups for radiated emission below 10 kHz, the author proposes a possible setup to cope with the measurement using a near field probe calibrated with the R&S AM524 (as discussed in a following section). Since the proposed antenna (R&S AM524) has a large volume, inappropriate to perform the proposed measurement setup of 12 measurement points around the EUT, a small near field probe is the best candidate to help in the acquisition. Along with the calibration of the near field probe, the prerequisites for the measurement procedure concerning the electric field acquisition are shorted in the following list:

- All ELF electric field measurements shall be performed in an anechoic chamber.
- Time domain measurements are preferable for ELF electric field modeling and ESA insists on time domain acquisition, however frequency domain measurements are sufficient. It should be noted that in any case the modeling procedure will be performed in frequency domain. In case of frequency domain measurements, transient phenomena will be lost.
- Firstly, measurements shall be performed only with power harness at EUT level → Additional information is needed concerning the topology of the harness.
- Secondly, measurements with data lines in order to obtain transient phenomena, different operational cycles of the EUT, etc. → Additional information is needed concerning the topology of the data lines and technical specifications of the data sequence on the data (e.g. LVDS).
- If electric radiated emissions are lower than instrument capability (in case of time domain) → Frequency Domain measurements.
- One measurement with and without data lines at the same operational mode of the EUT (useful for Phase 2 to investigate the contribution of the data lines).

Proposed Measurement Technique for ELF Electric Field

The proposed methodology employs electric field measurements dataset in order to extract a model corresponding to the measured EUT/component. The AC electric field measurements shall include both Time and Frequency domain datasets. In frequency domain analysis the AC electric field, either directly measured in frequency spectrum from the receiver or extracted through time domain signal processing (Fourier transform of a time – series dataset), corresponds to a number of discrete frequencies. The number of these discrete frequencies has a maximum value equal (in case of narrowband emissions) to the number of frequency bins implemented in the Fourier transform of a time series dataset or the actual bandwidth of the receiver measurements in the frequency domain. In the proposed methodology each such bin corresponds to an electric dipole composing a Multi Frequency Electric Dipole Model (MFEDM).

In order to distinguish the actual signal bin from the noise components and to reduce the computational time of the algorithm, a decision limit is enforced (Ambient Level that is measured according to the procedure described in the previous section). This way the algorithm attempts to solve the inverse problem only for the frequencies (bins) that have a substantial component (magnitude above decision margin) in at least one of the measurement points.

For every discrete frequency (even in the extended RF spectrum) each Component (An Equipment Under Test – EUT consists of a number of components) can be modelled by a number (usually 1 or 2) of small electric dipoles. The components consisting the EUT produce an electric field calculated at a number of measurement points Nm.

It should be noted that in case the EUT has both magnetic and electric behavior the AC magnetic modeling should be taken into consideration and the electric field produced from the modeled magnetic dipole should be included especially in case the frequency exceeds the 100 kHz.

Electric Field Measurement Instruments (EFI)

EFI measures components of the electric field vector in space as the potential difference between two probes. The EFI instrument measures the electric field vector in the frequency range 0-200 kHz. EFI consists of two sets of sensors: Spin-plane Double Probes (EFI-SDP) providing high sensitivity DC electric field in the spacecraft spin plane (2D), and High Frequency Antenna (EFI- HFA) providing 3D AC electric field at frequencies above ~1 kHz.

Spectral density of spacecraft generated AC electric field noise measured by EFI sensors, other than narrowband peaks, must not exceed the level depicted in Figure 2 and Figure 3.

Considering a total spacecraft electric moment magnitude |p| oriented transverse to a sensor in a distance d, the electric field magnitude in SI units can be calculated, based on EFI-SDP and EFI-HFA sensor emission spectral density requirements for the frequencies of interest, as presented in the Table 6 and Table 7. The EFI-SPD is on 50m long wire boom while the EFI-HFA is located midway on the 5 m SCM boom, so the electric field limits of Figure 2 and Figure 3 correspond to these distances. Electric field falls according to the distance law of $\sim \frac{1}{d^3}$, so the resulting maximum electric field values corresponding to the measurement points at distances 0.4 m and 1 m are presented to the Table 6 and Table 7. Measurement distances of 0.4 m and 1 m are indicative for the measurement setups that have been described before. Additionally the corresponding system level maximum electric moments according to these requirements are calculated for preliminary simulation purposes.

Antenna for On-Ground E-Field Acquisition

In order to define the test parameters (i.e. antennas) it is needed to derive the applicable limit in terms of amplitude instead of spectral density. The computed applicable limits (measured at EFI location) is shown in Table 6 and Table 7. For the frequency range of interest (frequencies below 200 KHz) it is recommended to to use specialized antennas like A.R.A. SAS-2/A or R&S AM524. In order to standardize an approach to be used at both unit and subsystem level, it is proposed to perform the measurement at 1 m from the EUT and therefore, the requirement shall be rescaled consequently.

The AC electric field emission levels, computed at 1 m from the EUT, are shown in Table 6 and Table 7.

Table 5. E-Field Instruments Accommodation and Bandwidths

	EFI - SDP	EFI-HFA
Accommodation	extended on 50 m long wire booms	located on the SCM-boom, midway to the end (SCM boom length = 5m)
Bandwidth	DC-200 kHz	~1 kHz - 200 kHz

Table 6. Spectral Density of AC Electric Field Noise– Rescaled EFI-SPD E-Field Cleanliness requirements

EFI-SDP						
f [Hz]	Spectral Density [(V/m)^2/Hz]	BW [Hz]	E-Field Limit [V/m] (@ 50 m)	E-Field limit [V/m] (@ 1 m)	E-Field limit [V/m] (@ 0.5 m)	Max. System Electric Moment [C*m]
Wideband Emissions						
1	1.00E-10	1	1.00E-05	1.25	10	1.391E-10
10	1.00E-12	1	1.00E-06	0.125	1	1.391E-11
100	1.00E-14	1	1.00E-07	0.0125	0.1	1.391E-12
1000	2.00E-17	1	4.47E-09	0.000559	0.004472136	6.22E-14
10000	2.00E-18	1	1.41E-09	0.0001768	0.001414214	1.967E-14
100000	2.00E-18	1	1.41E-09	0.0001768	0.001414214	1.967E-14
200000	1.00E-16	1	1.00E-08	0.00125	0.01	1.391E-13
Narrowband Emissions						
1	1.00E-07	0.2	1.41E-04	17.67767	141.4213562	1.967E-09
10	1.00E-09	0.2	1.41E-05	1.767767	14.14213562	1.967E-10
100	1.00E-11	0.2	1.41E-06	0.1767767	14.14213562	1.967E-11
1000	2.00E-14	2	2.00E-07	0.025	0.2	2.782E-12
10000	2.00E-15	20	2.00E-07	0.025	0.2	2.782E-12
100000	2.00E-15	200	6.32E-07	0.0790569	0.632455532	8.796E-12
200000	1.00E-13	200	4.47E-06	0.559017	4.472135955	6.22E-11

Table 7. Spectral Density of AC Electric Field Noise – Rescaled EFI-HFA E-Field Cleanliness requirements

EFI-HFA						
f (Hz)	Spectral Density (V/m)^2/Hz	BW (Hz)	E-Field Limit (V/m) (@ 2.5 m)	E-Field limit (V/m) (@ 1 m)	E-Field limit (V/m) (@ 0.5 m)	Max. System Electric Moment (C*m)
Wideband Emissions						
1	1.00E-11	1	3.16E-06	4.941E-05	0.000395285	5.498E-15
1000	1.00E-11	1	3.16E-06	4.941E-05	0.000395285	5.498E-15
200000	5.00E-14	1	2.24E-07	3.494E-06	2.79508E-05	3.888E-16
Narrowband Emissions						
1	1.00E-08	0.2	4.47E-05	0.0006988	0.00559017	7.775E-14
1000	1.00E-08	2	1.41E-04	0.0022097	0.01767767	2.459E-13
200000	5.00E-11	200	1.00E-04	0.0015625	0.0125	1.739E-13

Near-Field Probe Calibration (Correlation With R&S AM524 Field Values)

Calibration Procedure

In order to have a good correlation between the values of the measurements of Near-Field Probe and the corresponding values of the R&S AM524 antenna the following procedure is proposed:

For the frequency range of interest (lowest Available common Frequency of Signal Generator and Receiver up to 250 KHz), a Signal Generator operating in CW mode and providing sinusoidal signal at least at each measurement frequency of the actual measurements. Please consider the signal Generator's drift when choosing the receiver's resolution bandwidth in order to capture the frequency component (It is preferable to have the receiver in the mode of actual measurement settings).

A source antenna should act as the transmitter of ELF electric field, placed at a distance equal to the actual measurement distance of the antenna from the EUT.

Identification of the transfer function of the Near Field Probe from the comparison of measurements with the R&S AM524 for the whole frequency range would provide the adequate information in order to render the Near-Field Probe measurements usable for modeling purposes.

During normal measurement procedure, two (2) different measurements should be acquired:

- One measurement with the R&S AM524 antenna according to MIL-STD-461-F/G . Twelve (12) measurements with the Near-Field Probe at the

Figure 13. Near-Field Calibration Measurements

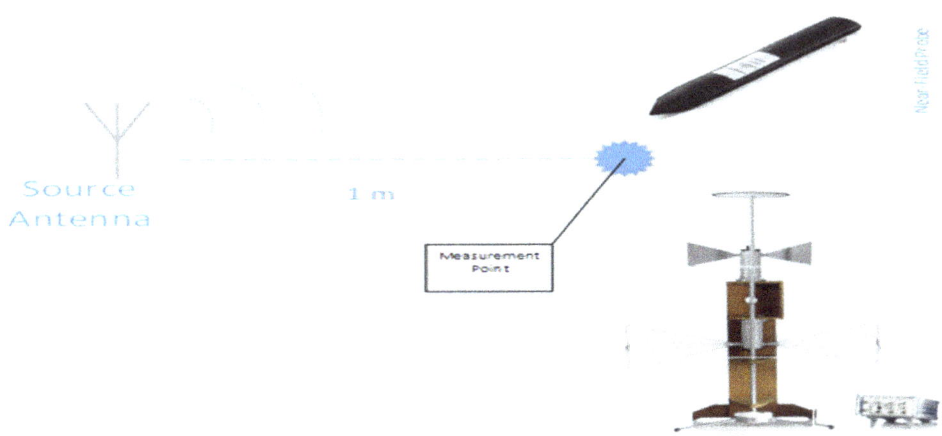

measurement points indicated in the Figure 5. It should be noted that in order to acquire significant E-Field values data for modeling purposes at the majority of the measurement points, the topology of the measurement points should be rotated at the direction of the maximum field strength.

- Another measurement set (preferably in the same setup with different distance – extrapolation distance) should be acquired at least for one EUT for model validation purposes.

Near Field Probe Positioning

It is recommended the Near-Field Probe to be positioned as close as possible to the EUT in order to have a good SNR value

As depicted in Figure 5, measurement points are placed in two planes arranged in circular geometries with a radius of RC cm. The planes are positioned at heights "RC" and "-RC" cm. Initially RC is selected as 30 cm. In case the EUT dimensions don't allow the placement of the measurement points at 30 cm (larger EUT), use as measurement radius RC 10 cm plus the half of the maximum dimension of the EUT .First Measurement point is placed at the maximum field emission direction and each of the five consecutive points of the first plane is placed at 60^0 from the previous point. The next six points are the mirrors of the six measurement points of the first plane on the second plane.

MEASUREMENT UNCERTENTIES DUE TO INACCURATE SENSOR POSITION

In the work of Koutantos et al. (2016), authors have proposed the Multi-Frequency Electric Dipole Model (MFEDM) method to describe the spectral dependence and predict extremely low frequency field emissions. Small variations in the relative distance of the source and the measurement points alter the field measurement results. This, in return, leads to different models produced by the proposed method. Consequently the precision of this method highly relies on the accuracy of the relative measurement distance. The correlation between relative measurement distance uncertainty and the electric field distribution is a key parameter to a successful model prediction.

One way to calculate the measurement uncertainty due to the sensor position inaccuracy is to perform the electric field measurement multiple times and estimate the mean value of the field with a standard deviation. In order to achieve this, the knowledge of field distribution due to distance displacement shall be known.

In this section relative measurement distance varied according to several distributions in order to evaluate this correlation. The results indicate that the electric field distribution follows always normal distribution independently with the way the sensor position is distributed.

TEST FACILITIES DECOUPLING CONSIDERATIONS

The reliability of the measurement data is the main issue when modeling an EUT. It is apparent that accurate measurement dataset can produce a valid EDM model. For that reason two substantial factors should have been taken into consideration. The first and the most critical is to provide appropriate test facilities isolated from the external- environmental noise in order to perform the electric field measurements. The second and also important issue, as discussed in the previous section, is the uncertainties of the measurement sensors position. Regarding the test facilities, the most common approach is to use full / semi - anechoic chambers or screening tents capable to provide an environmental free of extraneous signals.

Although the anechoic chamber design is often a trade-off between optimum quite zone performance and minimum system cost, the suitable type of RF absorber must be chosen to line the entire inner surface of the shielded room in order to simulate a free-space environment with no reflection from the walls, ceiling, and floor. These trade-offs are in turn influenced by the operational frequency range that the chamber is being designed for and the available space and shape of the chamber. Typical design methods that are used include empirical formulae and ray tracing methods.

Figure 14. (a) Moving sensor position with uniform distribution (b) Corresponding E-Field Distribution

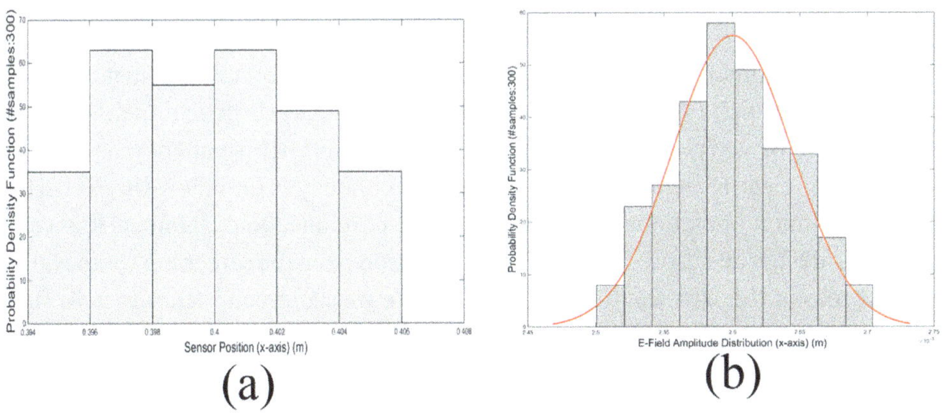

Figure 15. (a) Moving sensor position with Rayleigh distribution (b) Corresponding E-Field Distribution

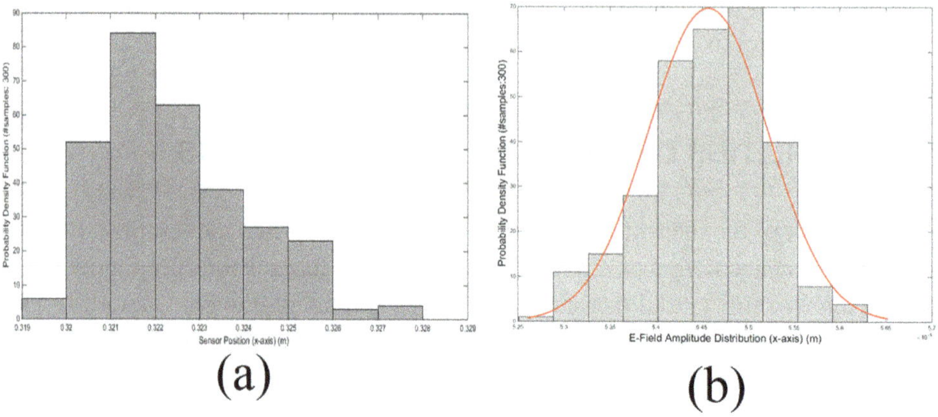

Figure 16. (a) Moving sensor position with Normal distribution, (b) Corresponding E-Field Distribution

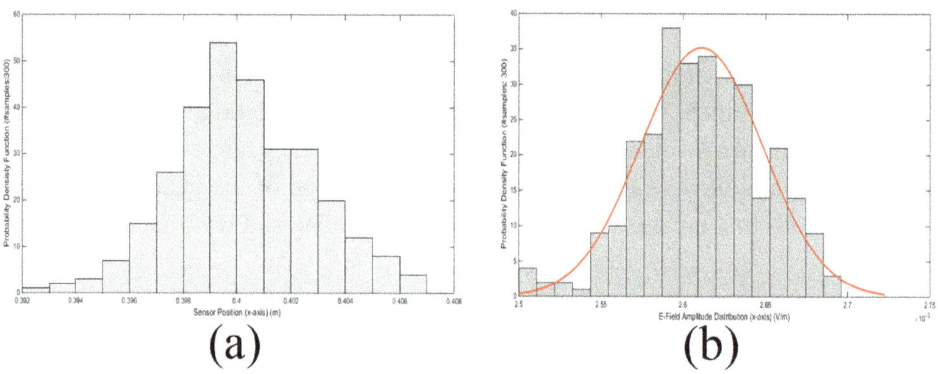

Unfortunately mathematical methods in the ray tracing family suffer from numerical inaccuracy, especially in scenarios where the room's characteristic dimensions are only a couple of wavelengths in size (i.e. typical VHF/UHF frequency bands). In the case ELF electric field, where the resonant frequency is far below the 30 MHz; typical minimum operational frequency of a standard anechoic chamber, it is very difficult to design an absorber to eliminate reflection from the screening room walls. For that reason the best way to produce accurate measurement datasets is to find a formula to decoupling the anechoic chamber contribution to the measurement. Author propose to use image theory approximation in order to calculate the screening room's conductive walls contribution. Since in this frequency (from a very few Hz

Figure 17. Image theory approximations

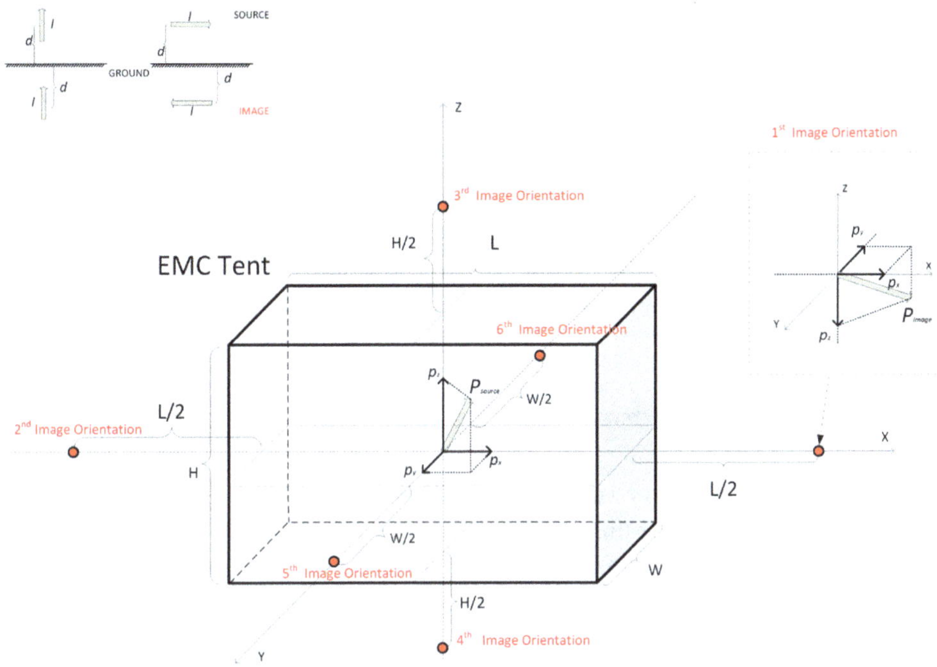

to a few KHz) of interest the chamber's walls act as a conductive plane (Figure 17), it is easy using image theory to subtract from the electric field measured value the contribution of the six images from all the cube's planes as long as the distance of the EUT from the walls is well defined. Modifying the algorithm of the proposed methodology (Figure 5) only in the section where the method is solving the inverse problem, without adding extra variables (since the images are produced from the unknown dipole of the EUT) instead of trying to identify a single dipole with six parameters, in this case the electric field is produced from six (6) dipoles oriented according to the Figure 17. In this way without increasing the algorithm's complexity, a test facility decoupling is achieved. It should be noted that the image contribution is a more complex phenomenon, particularly when two planes crossed forming a corner, but after extensive simulations the simplification of the six images; one on each cube side, founded adequate to provide a good estimation of the anechoic chamber contribution.

CONTRIBUTION OF SPACECRAFT STRUCTURE IN MODEL'S SYSTEM LEVEL PREDICTIONS

Image Theory Approximations

Spacecraft internal space is limited and possible EUT location can be near the spacecraft's conducted metallic surfaces. In that case, modelling electromagnetic behavior of the EUT should take into account the contribution of the image charges which in the majority of test cases they aren't negligible. Same considerations as in the previous section can be applied also in the case of system level contribution estimation.

Oversized Cavity Theory and Applications

The oversized cavity theory is very useful to assess electromagnetic fields inside an enclosure such as a satellite module or a launcher fairing that is much bigger than the wavelength at the frequencies of interest and geometrically complex. This theory is effectively the reverberation or stirred-mode chambers theory and is part of statistical electromagnetics. According to certain conditions concerning the size and complexity of the cavity, the field is statistically homogenous and isotropic and follows known distributions.

For a given calculated $\langle E \rangle$ the actual $\langle E \rangle_{dB\,V/m}$ may be between $\langle E \rangle_{dB\,V/m} + 6dB\,and$ - $12dB$ if one exclude the 0.1% of cases respectively higher and lower than this range. With this approximation and for the worst case of model ELF radiated emissions estimation, a level of 6 dB should be considered as a standard addition.

Contribution of Internal Sources Radiation Through Residual Apertures

According to the Bethe's theory of diffraction by small holes, an aperture smaller than λ/4 excited by an electromagnetic field can be represented by one electric dipole normal to the plane of the aperture and one magnetic dipole in the plane of the aperture.

In free space, the total power radiated by an electric dipole p_e, respectively by a magnetic dipole p_m is given by:

$$P = \frac{1}{12\pi} \frac{\omega^4 \mu_0 \left|p_e\right|^2}{c} \tag{12}$$

$$P = \frac{1}{12\pi} \frac{\omega^4 \varepsilon_0 \left|p_m\right|^2}{c} \tag{13}$$

Where $p_m = \mu_0 m$ and $\left|m\right|$ is in $A \cdot m^2$

The aperture electric and magnetic dipoles are related to the internal incident field by:

$$p_e = 2\varepsilon_0 a_e E_{sc} \tag{14}$$

$$p_m = 2\mu_0 a_m \mathrm{H}_{sc} \tag{15}$$

Where:

- E_{sc} and H_{sc} are the fields that exists on the inside of the cavity (spacecraft) at the location of the aperture if the aperture was closed (i.e shielded, short - circuited); the plane of the aperture is assumed perfectly conducting such that H_{sc} Is tangential and the E_{sc} in normal to the plane of the aperture;
- α_e is the electric polarizability of the aperture
- α_m is the magnetic polarizability of their total power through the aperture.

With the convention that it has adopted, the dipoles do not need to be imaged in the conducting plane (and therefore doubled) as this factor 2 is already included in their definition. Those dipoles radiate half of their total power through the aperture.

Electromagnetic Shielding Approximations

Electronic unit housings and spacecraft closed conductive structures naturally result in an effect of electromagnetic decoupling between the inside and the outside. In the literature, shielding effectiveness (SE) is introduced as a figure of merit describing the performance of a shield and defined as:

$$SE_{dB} = 20log\left(\frac{E_b}{E_a}\right) for\, electric\, fields \tag{16}$$

$$SE_{dB} = 20log\left(\frac{H_b}{H_a}\right) for\, magnetic\, fields \tag{17}$$

Where:

E_b and H_b are the fields before shield is installed; E_a and H_a are the field after shield is installed. The above formulas result in identical results in the far field, but not in the near field where the wave impedance is different. The classical transmission model of shielding introduced by Schelkunoff considers that the field is attenuated by a combination of absorption losses "A", depending on the number of skin depths through the wall, and reflection losses "R" depending on the ratio between the wave and barrier impedances (i.e. mismatch).

$$SE_{dB} = R_{dB} + A_{dB} + B_{dB} \tag{18}$$

The last term "B" accounts for the re-reflection effects and only plays role if $t < \delta$ where "t" is the thickness of the wall and δ is the skin depth. This simple model assumes that the coupling between the screen currents and the source of the incident field is negligible, which is not guaranteed in practice.

FUTURE RESEARCH DIRECTIONS

Although several measurement techniques are already included in the standard procedure of EMC testing, complementary methods that are presented in this chapter have to be evaluated and possibly adopted in future releases of the aforementioned standards since they cover frequency ranges that are not discussed until now. Electromagnetic cleanliness is a key parameter for the success of the majority of the space missions. For that reason future development towards this direction is essential and shall be the scope of the research of all space focused teams and engineers.

CONCLUSION

In this chapter, the main issues regarding ELF electric field measurement and modeling techniques for a spacecraft unit are briefly discussed. A modeling methodology is reviewed and analyzed focusing on the open issues that need further exploration and development. Space engineering and aircraft design procedures cope with stringent requirements regarding electromagnetic compatibility and interference since electromagnetic cleanliness is a critical aspect in the majority of the space missions. Searching and developing new methodologies in this field seems crucial. This work highlight the importance of this issue and urge the researchers and engineers to include in future development all the contributing additional phenomena, such as harness topologies and solar arrays that cannot be treated as electric dipoles.

ACKNOWLEDGMENT

Author thank Thales Alenia Space (TAS -I) and European Space Agency (ESA) for the permission to publish Table 1 and Figure 1.

REFERENCES

Airbus Defence & Space SAS. (2014). *Medium Power PCDU, POW2 datasheet.* Author.

Antonucci, F., Armano, M., Audley, H., Auger, G., Benedetti, M., Binetruy, P., & Zweifel, P. et al. (2011). LISA Pathfinder: Mission and status. *Classical and Quantum Gravity*, *28*(9), 9. doi:10.1088/0264-9381/28/9/094001

Baklezos, A. T., Nikolopoulos, C. N., & Capsalis, C. N. (2017). An equivalent dipole method with novel measurement positioning for modeling electric emissions in space missions. *Electromagnetics*, *37*(7). doi:10.1080/02726343.2017.1376902

Bayle, O., Lorenzoni, L., Blancquaert, T., Langlois, S., Walloschek, T., Portigliotti, S., & Capuano, M. (2011). *Exomars Entry Descent and Landing Demonstrator Mission and Design Overview*. NASA Solar System.

Benkhoff, J., Casteren, J., Hayakawa, H., Fujimoto, M., Laakso, H., Novara, M., & Ziethe, R. et al. (2010). BepiColombo—Comprehensive exploration of Mercury: Mission overview and science goals. *Planetary and Space Science*, *58*(1), 2–20. doi:10.1016/j.pss.2009.09.020

Bojarski, N. (1982). Low frequency inverse scattering. *IEEE Transactions on Antennas and Propagation*, *30*(4), 775–778. doi:10.1109/TAP.1982.1142846

Brownlee, J. (2011). *Clever algorithms: Nature-inspired programming recipes*. Jason Brownlee.

Bureau International des Poids et Mesures. (2008). Evaluation of measurement data - Guide to the expression of uncertainty in measurement. GUM 1995 with minor corrections. *JCGM*, *100*, 2008.

Carrubba, E., Junge, A., Marliani, F., & Monorchio, A. (2012, May). Particle swarm optimization to solve multiple dipole modelling problems in space applications. In *Aerospace EMC, 2012 Proceedings ESA Workshop on* (pp. 1-6). IEEE.

Danzmann, K. (2008). LISA mission overview. *Advances in Space Research*, *25*(6), 1129–1136. doi:10.1016/S0273-1177(99)00973-4

Dawar, D., & Ludwig, S. A. (2014, December). Differential evolution with dither and annealed scale factor. In *Differential Evolution (SDE), 2014 IEEE Symposium on* (pp. 1-8). IEEE. doi:10.1109/SDE.2014.7031528

Drăgoi, E. N., Curteanu, S., & Vlad, D. (2012, October). Differential evolution applications in electromagnetics. In *Electrical and Power Engineering (EPE), 2012 International Conference and Exposition on* (pp. 636-640). IEEE. doi:10.1109/ICEPE.2012.6463801

Drinkwater, M. R., Floberghagen, R., Haagmans, R., Muzi, D., & Popescu, A. (2003). GOCE: ESA's first Earth Explorer Core mission. In Earth gravity field from space—From sensors to earth sciences, (pp. 419-432). Springer Netherlands.

Duffy, A. P., Martin, A. J. M., Orlandi, A., Antonini, G., Benson, T. M., & Woolfson, M. S. (2006). Feature Selective Validation (FSV) for validation of computational electromagnetics (CEM). Part I – The FSV method. *IEEE Trans. on Electromagnetic. Compatibility*, *48*, 449–459.

Escoubet, C. P., Schmidt, R., & Goldstein, M. L. (1997). Cluster-Science and mission overview. In The Cluster and Phoenix Missions, (pp. 11-32). Springer Netherlands. doi:10.1007/978-94-011-5666-0_1

European Cooperation for Space Standardization (ECSS) Secretariat. (2008). *ECSS-E-ST-50-12C, Space Engineering – SpaceWire- Links, nodes, routers and networks, ESA-ESTEC*. Noordwijk, The Netherlands: Requirements & Standards Division.

European Cooperation for Space Standardization (ECSS) Secretariat. (2012). *ECSS-E-HB-20-07A. Space engineering – Electromagnetic compatibility Handbook*. ESA-ESTEC. Noordwijk, The Netherlands: Requirements & Standards Division.

European Cooperation for Space Standardization (ECSS) Secretariat. (2012). *ECSS-E-ST-20-07, C Rev. 1. Space engineering, Electromagnetic compatibility*. Author.

Friis-Christensen, E., Lühr, H., Knudsen, D., & Haagmans, R. (2008). Swarm–an Earth observation mission investigating geospace. *Advances in Space Research*, *41*(1), 210–216. doi:10.1016/j.asr.2006.10.008

IEEE Standard P1597. (2008). *Standard for Validation of Computational Electromagnetics Computer Modeling and Simulation – Parts 1&2*. IEEE.

Jackson, J. D. (1999). *Classical Electrodynamics*. New York: John Wiley and Sons.

Kennedy, J. (2011). Particle swarm optimization. In Encyclopedia of machine learning (pp. 760-766). Springer US.

Koutantos, G. I., Nikolopoulos, C. D., Baklezos, A. T., & Capsalis, C. N. (2016). On the Modeling of ELF Electric Fields for Space Mission Equipment. *IEEE Transactions on Electromagnetic Compatibility*, *58*(8), 1782–1786.

Li, P., Li, Y., Jiang, L. J., & Hu, J. (2013). A wide-band equivalent source reconstruction method exploiting the Stoer-Bulirsch algorithm with the adaptive frequency sampling. *IEEE Transactions on Antennas and Propagation*, *61*(10), 5338–5343. doi:10.1109/TAP.2013.2274032

Mikki, S. M., & Antar, Y. M. (2012). Near-field analysis of electromagnetic interactions in antenna arrays through equivalent dipole models. *IEEE Transactions on Antennas and Propagation*, *60*(3), 1381–1389. doi:10.1109/TAP.2011.2180318

Müller, D., Marsden, R. G., Cyr, O. S., & Gilbert, H. R. (2013). Solar orbiter. *Solar Physics*, *285*, 25–70. doi:10.1007/s11207-012-0085-7

Obiekezie, C. S., Thomas, D. W., Nothofer, A., Greedy, S., Arnaut, L. R., & Sewell, P. (2014). Complex locations of equivalent dipoles for improved characterization of radiated emissions. *IEEE Transactions on Electromagnetic Compatibility*, *56*(5), 1087–1094. doi:10.1109/TEMC.2014.2313406

Oliveri, G., Rocca, P., & Massa, A. (2012, March). Differential evolution as applied to electromagnetics: Advances, comparisons, and applications. In *Antennas and Propagation (EUCAP), 2012 6th European Conference on* (pp. 3058-3059). IEEE.

Orlandi, A., Duffy, A. P., Archambeault, B., Antonini, G., Coleby, D. E., & Connor, S. (2006). Feature selective validation (FSV) for validation of computational electromagnetics (CEM). part II-assessment of FSV performance. *IEEE Transactions on Electromagnetic Compatibility*, *48*(3), 460–467. doi:10.1109/TEMC.2006.879360

Paul, C. R. (1992). *Introduction to electromagnetic compatibility (EMC)* (2nd ed.). Introduction to Electromagnetic Compatibility.

Scheeres, D. J., Marzari, F., Tomasella, L., & Vanzani, V. (1998). ROSETTA mission: Satellite orbits around a cometary nucleus. *Planetary and Space Science*, *46*(6), 649–671. doi:10.1016/S0032-0633(97)00200-6

Storn, R., & Price, K. (1995). *Differential evolution–a simple and efficient adaptive scheme for global optimization over continuous spaces: technical report TR-95-012*. Berkeley, CA: International Computer Science.

Tsitsas, N. L. (2013). A Low-Frequency Electromagnetic Near-Field Inverse Problem for A Spherical Scatterer. *Journal of Computational Mathematics*, *31*(5), 439–448. doi:10.4208/jcm.1304-m4388

Wang, Y. M., & Chew, W. C. (1989). An iterative solution of the two-dimensional electromagnetic inverse scattering problem. *International Journal of Imaging Systems and Technology*, *1*(1), 100–108. doi:10.1002/ima.1850010111

Zaiping, N., & Yerong, Z. (1998). Hybrid Born iterative method in low-frequency inverse scattering problem. *IEEE Transactions on Geoscience and Remote Sensing*, *36*(3), 749–753. doi:10.1109/36.673668

KEY TERMS AND DEFINITIONS

Differential Evolution Algorithm (DEA): Is a method that optimizes a problem by iteratively trying to improve a candidate solution with regard to a given measure of quality.

Electric Cleanliness: The generated electric field from the spacecraft is required to not exceed specific thresholds at the location of the electric sensors and therefore, disturb their measurements.

Electric Dipole Model (EDM): A method that optimizes a problem by iteratively trying to improve a candidate solution with regard to a given measure of quality. Such methods are commonly known as metaheuristics as they make few or no assumptions about the problem being optimized and can search very large spaces of candidate solutions.

Electrical Ground Support Equipment (EGSE): Is an integrated suite of electrical satellite testing solutions to make sure that the satellites you are launching work perfectly.

Electromagnetic Compatibility (EMC): The branch of electrical engineering concerned with the unintentional generation, propagation and reception of electromagnetic energy which may cause unwanted effects such as electromagnetic interference (EMI) or even physical damage in operational equipment.

Electromagnetic Interference (EMI): Also called radio-frequency interference (RFI) when in the radio frequency spectrum, is a disturbance generated by an external source that affects an electrical circuit by electromagnetic induction, electrostatic coupling, or conduction.

Equipment Under Test (EUT): Any instrument or unit related to the space mission that produces an electric field. These units can be inside (e.g., power conditioning and distribution unit, RF switch, battery, etc.) or outside the main body of the spacecraft (e.g., solar arrays, boom, etc.) and need to be tested and electrically characterized.

European Space Agency (ESA): Is an intergovernmental organization of 22 member states dedicated to the exploration of space. Established in 1975 and headquartered in Paris.

Extremely Low Frequency (ELF): Is the ITU designation for electromagnetic radiation (radio waves) with frequencies from 3 to 30 Hz, and corresponding wavelengths of 100,000 to 10,000 kilometers, respectively.

Multi-Frequency Electric Dipole Model (MFEDM): A method using electric dipoles to predict the radiated emission of an EUT.

Turbulence Heating ObserveR (THOR): A plasma mission aiming at understanding the physical mechanisms underlying the energy dissipation of turbulent fluctuations in plasmas.

Chapter 2
SpaceWire:
The Standard and EMC/EMI Aspects

Anargyros T. Baklezos
National Technical University of Athens, Greece

Christos N. Capsalis
National Technical University of Athens, Greece

ABSTRACT

SpaceWire is a point-to-point bit shipping protocol for high-speed data communication links and networks providing equipment compatibility and seamless component reusability. It has found great application in many space missions reducing the development cost, offering architectural flexibility and improving reliability. This chapter delves into the standard describing the SpaceWire, focusing on the lower levels that play a key role in the electromagnetic behavior of the system and concern cable assemblies, shielding, bonding, and grounding. Findings regarding emissions affecting spacecraft components are presented as well as other EMC issues that have an impact on the system performance. Recent developments and upcoming updates to the standard are also presented and discussed.

INTRODUCTION

SpaceWire is a standard for high speed data communication links and networks for use on spacecraft. SpaceWire was defined in the European Cooperation for Space Standardization (ECSS) standard ECSS-E50-12A in 2003. The SpaceWire standard was authored by Steve Parkes, University of Dundee with contributions from many individuals within the SpaceWire Working Group from European Space Agency (ESA), European Space Industry, Academia and NASA. It was replaced by ECSS-E-ST-50-12C (ECSS Secretariat, 2008) in order to ensure the standard was in

DOI: 10.4018/978-1-5225-5415-8.ch002

accordance with the new ECSS numbering system. There are no technical content differences between ECSS-E-50-12A and ECSS-ST-E-50-12C and any information relating to the 'old' ECSS-E-50-12A is still applicable to newer ECSS-ST-E-50-12C ("ECSS-E50-12A vs ECSS-E-ST-50-12C, what's the difference? | STAR-Dundee", 2017; "SpaceWire - The Standard", 2017).

The Standard specifies the physical interconnection media and data communication protocols to enable the reliable sending of data at high-speeds, between 2 Mb/s and 400 Mb/s, from one unit to another. SpaceWire links are full-duplex, point-to-point, and serial data communication links. Since the SpaceWire standard was first published by the European Cooperation for Space Standardization, it has been adopted by ESA, National Aeronautics and Space Administration (NASA), Japan Aerospace Exploration Agency (JAXA) and Roscosmos State Corporation for Space Activities (ROSCOSMOS) for many missions and is being widely used on scientific, Earth observation, commercial and other spacecraft. High-profile missions using SpaceWire include: Gaia, ExoMars, Bepi-Colombo, GOES-R, Lunar Reconnaissance Orbiter, ASTRO-H and James Webb Space Telescope, (Parkes, 2012).

The needs of spacecraft on-board data handling applications has given rise to SpaceWire technology. The Standard provides a formal basis for the exploitation of SpaceWire in a wide range of on-board processing systems. For example the integration and testing of complex on-board subsystems with ground support equipment connecting directly into the on-board data-handling subsystem. SpaceWire aims to offer equipment compatibility and reusability for components and subsystems. This means directly connecting a component made for one subsystem to another and operating without issues and readily use systems in one mission that were developed for another mission, thus reducing the cost of development, offering architectural flexibility, improving reliability and most importantly increasing the amount of scientific work that can be achieved within a limited budget (ECSS Secretariat, 2008).

Since the publication of ECSS-E-ST-50-12C standard (ECSS Secretariat, 2008), the engineering and scientific community have applied the guidelines that it dictates in many missions. Practical applications highlighted best practices as well as problems. Different applications called for different designs, some of them were hard to implement while maintain conformity with the standard. This fact led the European Cooperation for Space Standardization in 2012 to start the long process of updating the Standard including the experience of the past and the needs of the future. Working Group ECSS-E-ST-50-12-C Rev. 1 WG. is preparing Revision 1 of the standard (ECSS-E-ST-50-12C Rev. 1 Working Group, 2015). While the draft of the Revision. 1, currently under public review, is not a formal standard and cannot be referenced as such, it is a clear indicator of the direction the Standard is headed, its evolution to overcome many of the earlier version's shortcomings. Additionally since the Standard's publication at 2008, the ESCC Detail Specification No. 3902/003 was updated and ESCC Detail Specification No. 3902/004 was published. Revision 1 of ECSS-E-ST-50-12C is expected to reflect these changes in the normative framework.

Taking into consideration the widespread use of the SpaceWire, personnel active in the space mission field is bound to come across some implementation of a system incorporating SpaceWire. The main objective of the chapter is to discuss the SpaceWire standard ECSS-E-ST-50-12C with regard to Electromagnetic Compatibility and Immunity issues. Authors aim to highlight the levels of the standard that define the electromagnetic compatibility (EMC) behavior of the system. This can be beneficial for the engineer or scientist either from a system modelling or a radiated emissions measurements standpoint. Students interested in space missions can also benefit from the concentrated information found in this chapter.

BACKGROUND

Before there was a standard for interconnecting units, spacecraft and equipment manufacturers used their own proprietary communication interfaces. As a result, several different communication links were used on a spacecraft, increasing the cost and extending the time required for spacecraft integration and test. There was a clear need for a standard on-board communication link that would simplify spacecraft development. (Parkes, 2012)

The work on what later became SpaceWire, started in 1992. There was considerable interest in high performance digital signal processing systems at that time. However, the single-chip devices could not satisfy the demands. This lead to the investigation of the use of parallel processing and in return gave place to research concerning the interconnecting network of the individual processing elements. The Inmos Transputer, a microprocessor designed for parallel processing was studied, and the serial communication links being developed for the T9000 Transputer were identified as being an attractive solution for spacecraft on-board networking. This serial link technology was subsequently published as IEEE 1355-1995. Several radiation tolerant devices were developed using it and the IEEE 1355-1995 standard was used on some space missions. However, there were many problems with this standard, which had to be resolved if this technology was to continue to be used for ESA spacecraft. University of Dundee received a contract from ESA to examine and solve these problems. The result of this contract was the SpaceWire standard, a brief description of which follows (Parkes, 2012).

SpaceWire links are point-to-point data links that connect a SpaceWire node to another node or to a switch. Information can be transferred in both directions of the link at the same time. Links are full-duplex, bidirectional able to operate at rates between 2 Mbps and 400 Mbps.

The SpaceWire Standard ECSS-E- ST-50-12C (ECSS Secretariat, 2008) uses Low Voltage Differential Signaling (LVDS), defined in ANSI/TIA/EIA-644, Electrical Characteristics of Low Voltage Differential Signaling Interface Circuits (Telecommunications Industry Association, 2001), as the physical level due to its capability of low power, low noise, and high-speed point-to-point communications.

Information is sent across a link as a serial bit stream with the use of two signals, Data (D) and Strobe (S). Standard TIA-644-A requires that both Data and Strobe signals are driven across the link using LVDS and two wires for each signal; so SpaceWire cable contains four screened twisted pairs, one for each signal (D or S) for each direction.

SpaceWire achieves bit synchronisation by sending a clock signal along with the serial data. To reduce the maximum clock to data skew requirements, the clock signal is encoded into a Strobe signal so that a XOR function on the Data and Strobe signal recovers the clock signal. Data is transmitted as a stream of elementary characters. Character synchronisation is only performed at the start of the link. The detection of parity errors means loss of character synchronisation, a restart of the link is the sole mean to recover the characters synchronization.

There is a simple mechanism to start a link, keep the link running, send data over the link, ensure that data is only sent if the other end of the link is ready, and to recover from any errors on the link. A link state-machine in the SpaceWire interface that is transparent to the user application handles all these functions.

The SpaceWire (SpW) protocol is comprised of six levels as depicted in Figure 1:

- **Physical Level:** Defines connectors, cables, cable assemblies, etc.
- **Signal Level:** Defines encoding, voltage levels, noise margins, and signaling rates.
- **Character Level:** Defines the data and control characters used to manage data across a link.
- **Exchange Level:** Defines the protocol for link initialization, flow control, link error detection and link error recovery.
- **Packet Level:** Manages data packetization and transmission over a SpW link.
- **Network Level:** Defines the structure of a SpaceWire network and source to destination node data transfers. Also defines link errors and network level error management.

The SpaceWire physical level uses Data-Strobe (DS) Encoded LVDS to communicate serial, full-duplex, bidirectional data. ANSI/TIA/EIA-644 defines the electrical signal levels only; it does not define a protocol. Instead, the protocol is defined in the SpaceWire Standard specification ECSS-S-ST-50-12C (ECSS Secretariat, 2008). LVDS is a method to transmit bidirectionally over differential

Figure 1. SpaceWire ECSS-E-ST-50-12C Levels

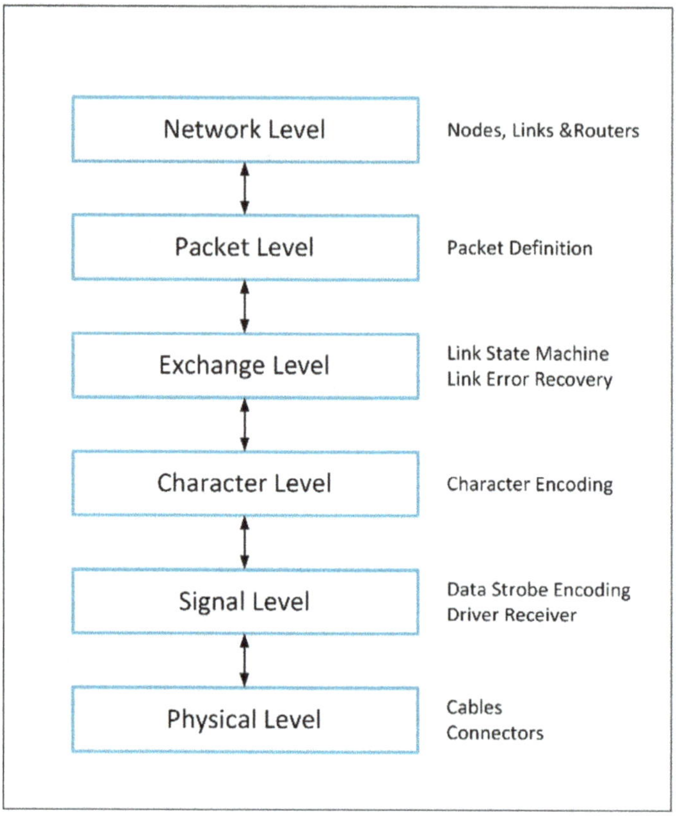

media using a low voltage signal swing using a driver and a receiver. SpaceWire uses Data Strobe (DS) encoding to send information over the LVDS links. The data value is transmitted directly and the Strobe signal changes state whenever the Data remains unaltered from one data bit interval to the next. A XOR function on the Data and Strobe signals result in the clock signal. There is a slight delay between edges of D and S and recovered clock. Data is Non Return to Zero, (NRZ) so a 100MHz clock equates to a data rate of 200Mbps.

Two categories of characters are defined in ECSS-E-ST-50-12C standard, data and control characters. These characters are further defined as either link characters or normal characters. The Exchange level does not pass a link character to the Packet Level, meaning that a link character does not get entered into the memory space of a SpaceWire node. Flow control token (FCT), escape characters (ESC) as well as NULL control code (ESC followed by FCT) and the Time code characters (ESC followed by data character) are link characters. A Normal Character ends with an EOP or EEP and passes through a SpW network at a packet level. Data characters contain a parity bit, data-control flag, and an eight-bit data value, transmitted LSB to MSB. The even or odd parity bit is calculated by adding the number of ones that are contained in the previous 8-bits data. Meaning if the number of 1's in bits added together is even, the data character is said to have even parity. There are two Control Characters that are formed using control characters and data characters, these are the NULL and Time Code Characters. NULL characters are used during the initialization sequence and are also used to keep the LVDS SpW lines active when no actual data is sent. Distributing system time is accomplished using Time Codes. The use of time codes is optional.

The exchange level defines the mechanisms for link initialization, link flow control, link error detection, and link error recovery. The state machine causes a SpW link to continuously attempt initialization as long as the link is active. Disconnection errors are detected after 850ns without Data and Strobe transitions. A parity error is detected when the first bit of the next character is received and link recovery starts. Initialization is executed upon the initial start of the link or reset when an error is detected (Parity Error, Escape Error, etc), or a link disconnection occurs. Under normal operation, a SpaceWire host node manages data by sending tokens to a destination node on the other end of the link. Each token indicates to the destination node that the host has 8 bytes of available buffer space. The destination node has a corresponding counter that records each token that is received. The destination node can send data to the host, according to the tokens of available buffer space at the host side it has received.

SpaceWire packets have a very flexible format. Each packet simply needs a Destination Address, a Cargo and an End of Packet Marker (EOP) that signals the end of Cargo transmission. Destination Address consists of a number of bytes of data characters representing the destination path of the packet and the Cargo consists of the transmission data without a limit to the size or format of the data. SpaceWire supports data rates from 2Mbps up to 400 Mbps. Packets travel through a Network composed of Nodes, Links, and Routers. SpaceWire nodes can be routers, sensors, memory units, processors, and telemetry sub-systems. Nodes represent the source and destination of SpW packets and provide the interface to the application systems. The SpaceWire Standard supports Wormhole Routing, Path Addressing, Logical Addressing, and Group Adaptive Routing for networking. Topologies like Distributed and Centralized network offer the designer a great deal of freedom. Distributed networks require look-up tables to be configured and provide lower harness mass, good throughput performance and increased reliability through redundant paths. A redundant network design ensures that the system remains functional if one or more of the systems components fail. Centralized networks require higher harness mass and offer high throughput performance and simple configuration. With centralized networks, there is heavy reliance on the central router decreasing the system's fault tolerance. Other topologies are also possible each with each own considerations.

THE SPACEWIRE STANDARD

This chapter discusses the SpaceWire standard from an EMC perspective. While a short overview of the whole standard is already presented, authors aim to provide insight into the levels of the standard that mainly affect the Electromagnetic Emissions of the corresponding system, in particular the physical, signal and character levels. This is necessary in order to enable for example emissions modelling or interference minimization. In order to accomplish that, a more thorough presentation of the corresponding levels is mandatory.

Figure 2. Physical layer components, Z is the 100Ω termination resistor

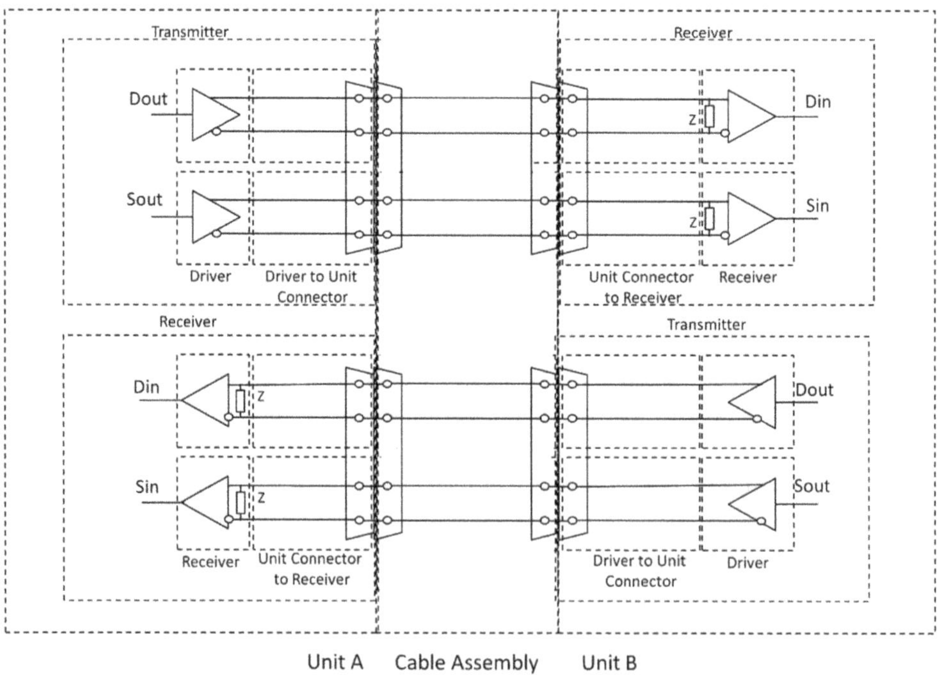

Physical Level

The physical level of the SpaceWire standard covers cables, connectors, cable assemblies and printed circuit board tracks (Figure 2). SpaceWire was developed to meet the EMC specifications of typical spacecraft. Some details concerning the construction of the cable and the connectors is useful to be provided at this point in order to more clearly present the physical level structure

Cables

Typical SpaceWire cable is constructed according to ESCC 3902/003. The SpaceWire cable comprises four twisted pair wires with a separate shield around each twisted pair and an overall shield as depicted in Figure 3 (ECSS Secretariat, 2008).

Figure 3. Transverse cut of SpaceWire cable construction (ECSS Secretariat, 2008)

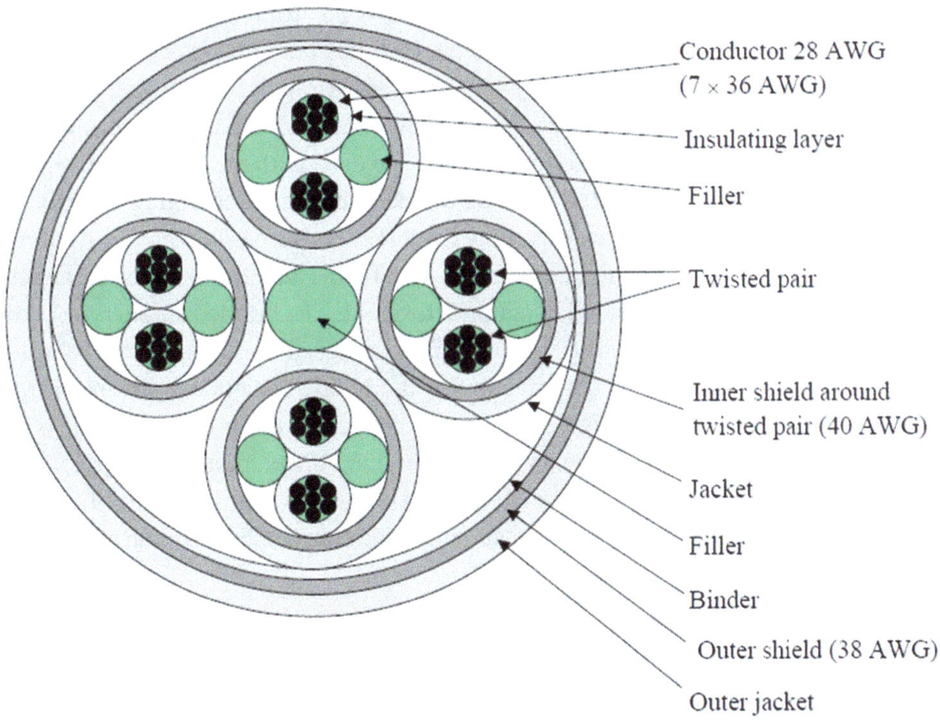

Inner Conductors

Each signal wire is 28 AWG, constructed from seven strands of 36 AWG silver-coated, high-strength copper alloy. The thickness of the silver coating is minimum 2,0 μm. The maximum acceptable DC resistance of the inner conductor is 256 Ω/km. Each signal is insulated using white, expanded, microporous PTFE without any except those for processing and pigmentation. (ECSS Secretariat, 2008; ECSS Secretariat, 2003 and ECSS Secretariat, 2014a). In ECSS Secretariat, (2014a) two identically structured variant exist with the exception of the diameter of the signal wire for the inner conductor. Variant 2 is 26AWG, hence all dimensions are little larger. Since ECSS-E-ST-50-12C precedes ECSS Secretariat, (2014a) there is no mention of Variant 2 in the Standard, so in this chapter Variant 1 (28 AWG) is always considered. It should be clarified that Variant 1 of Secretariat, (2003) is identical to Secretariat, (2014a).

Twisted Pair

The characteristic impedance of each differential signal pair is $(100 \pm 6)\,\Omega$ differential impedance with a skew value between each signal in each differential signal pair less than 0,1 ns/m. Regarding the lay-length, it is stated in ECSS-E-ST-50-12C the lay-length of the two insulated conductors comprising a differential signal pair must not be less than 12 times and not more than 16 times the outside diameter of the unshielded twisted pair, yielding a lay-length of 24-32mm corresponding to 32-41 twists/m (ECSS Secretariat, 2014a). Also Fillers, made of expanded microporous PTFE with only those additives for processing, are used with the differential signal pairs so as to ensure a smooth and uniform diameter under the shielding in order to contribute to a uniform impedance over the cable. Fillers have a diameter of 1,0 mm. Also in the matter of inner shield, the ECSS-E-ST-50-12C (2008) states that each differential signal pair is shielded by a braided shield. The braided shield type is of push-back type and provide not less than 90% coverage. The shield wire size is 40 AWG. All strands used in the manufacture of the braided shield are silver-coated, soft or annealed oxygen-free high conductivity copper with minimum thickness of silver 2,5 μm. The protective sheath for the shielded differential signal pairs is a white layer of extruded fluorpolymer PFA with a nominal wall thickness of 0,15 mm and only those additives for processing and pigmentation (ECSS Secretariat, 2008).

Complete Cable

The complete cable has a diameter of 7 mm maximum and a maximum weight of 80 g/m. It is constructed by four sets of differential signal pairs, twisted together not less than 12 times and not more than 16 times the outside diameter of two shielded and jacketed differential signal pairs, yielding a 55-77mm lay-length (ECSS Secretariat, 2014a). A filler of diameter of 1,0 mm, is used in the center of the four differential signal pairs so as to ensure a smooth and uniform diameter under the shielding in order to contribute to a uniform impedance over the cable. The filler material as used for the complete cable is microporous PTFE with only those additives for processing. A binder, of the same material as the filler, is applied over the four differential signal pairs and central filler to keep the signal pairs and filler together in a fixed position, wrapped with 50% maximum overlap (ECSS Secretariat, 2008).

The set of four jacketed and screened differential signal pairs are shielded by an outer braided shield of push-back type that provides not less than 90% coverage. The shield wire size is 38 AWG. All strands used in the manufacture of the braided shield are silver-coated, soft or annealed oxygen-free high conductivity copper with a minimum thickness of silver of 2,5 μm. It should be noted that the twisted pair shields do not make contact with one another nor with the outer shield. The outermost jacket over the four twisted screened and jacketed differential signal pairs is a white layer of extruded Fluoropolymer PFA with only those additives for processing and pigmentation and no identifying marking since pressure applied to the cable during the marking process could adversely affect the electrical properties of the cable The nominal wall thickness of the jacket for the shielded differential signal pair is 0,25 mm (ECSS Secretariat, 2008).

The skew between the parts of the differential signal in one differential signal pair is allowed to be maximum 0,1 ns/m. The skew between one differential signal pair and each other differential signal pair within the cable must be less than the maximum of 0,15 ns/m (ECSS Secretariat, 2008).

The maximum length of the cable assembly is set to 10 m in order to ensure that the end to end skew and jitter introduced by the cable assembly does not exceed the maximum budget for the cable. Longer length cables can be used at slow data signaling rates provided that the signal attenuation and system jitter and skew limits are not violated at the operating data signaling rate (ECSS Secretariat, 2008). Detailed information can be found in clause 6 of the ECSS-E-ST-50-12C standard.

Shield Details

The coverage factor $K\left(\%\right)$ of the braided shield is calculated with regard to number of strands N, the serving pitch P, the shield stand diameter d in mm, the angle a of shield with the cable axis in degrees, the effective diameter D of core under the shield in mm and the number of carriers C as follows (ECSS Secretariat, 2014a):

$$K = (2F - F^2)$$

$$F = \frac{N \cdot P \cdot d}{\sin a}$$

$$\tan a = 2\pi \frac{(D \cdot 2d) \cdot P}{C}$$

The shield effectiveness must abide to the limits depicted in Figure 5 for Variant 01 (ECSS Secretariat, 2014a)

Connectors

The SpaceWire connector is a nine contact micro-miniature D-type with solder contacts, as defined in ESCC 3401/071, or crimp contacts. Connector contact identification as viewed from rear of receptacle or front of plug in depicted in Figure 4 (ECSS Secretariat, 2008).

The inner shield connection is connected to the inner shield of the SpaceWire cable. This inner shield of the SpaceWire cable is to be connected to signal ground according to the EMC requirements of the mission via a parallel resistor and capacitor (ECSS Secretariat, 2008).

Figure 4. SpaceWire connector contact identification (ECSS Secretariat, 2008)

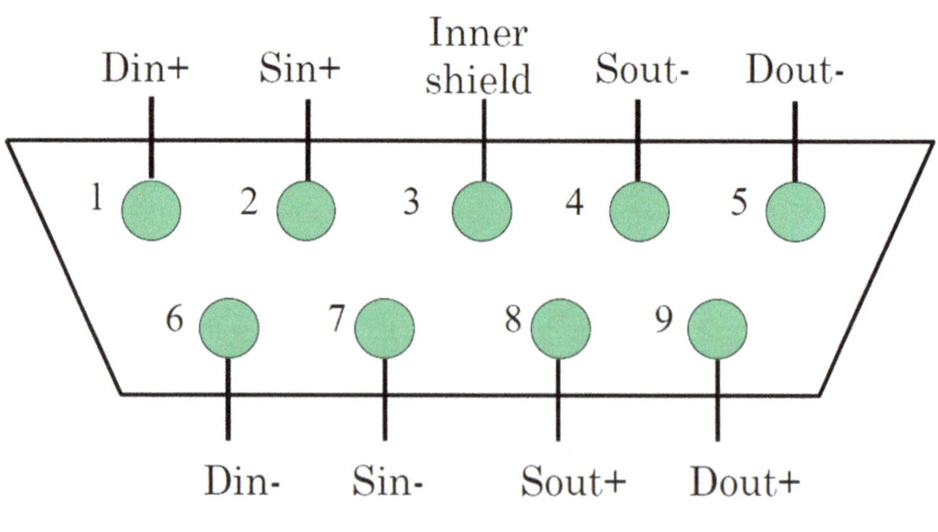

Figure 5. Minimum Shielding Effectiveness (ECSS Secretariat, 2014a)

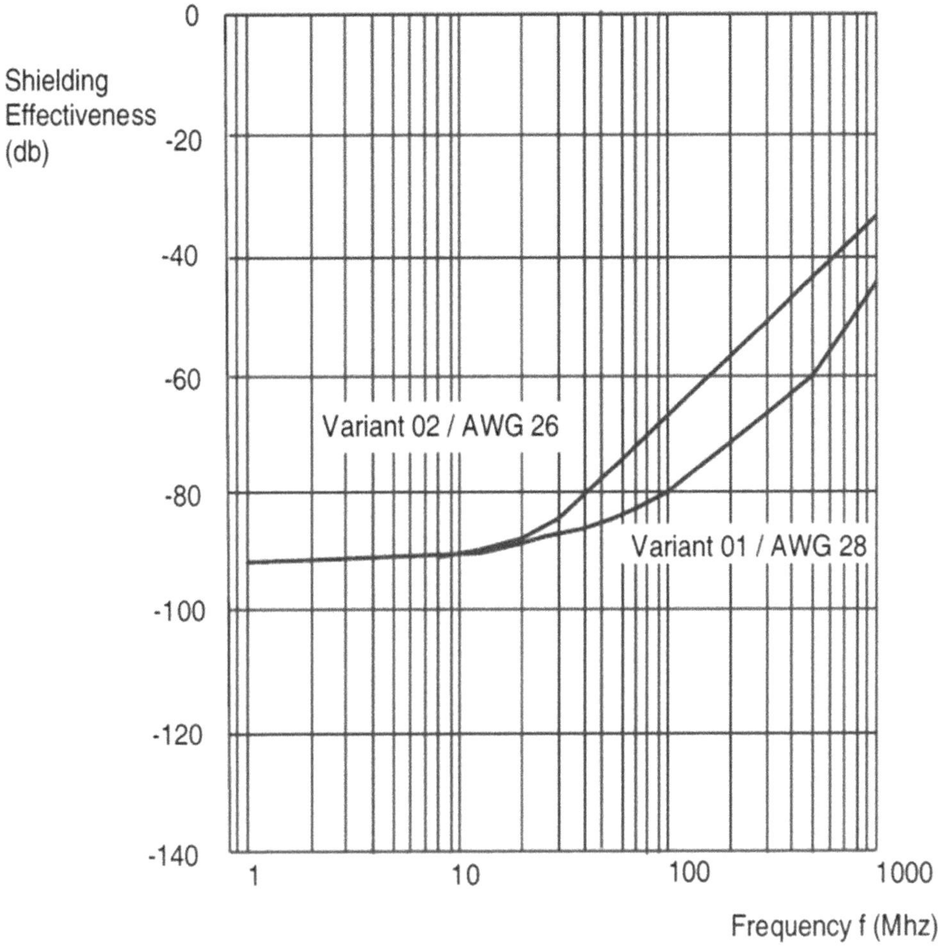

In case of connections to a PCB, flying lead connectors are used. These connectors must have all the leads cropped to the same short length (less than 25 mm) and the wires comprising the differential signal pairs must be twisted together in order to minimize the discontinuity in impedance caused by the connector (ECSS Secretariat, 2008).

PCB mounting right-angled connectors use is generally discouraged. In case they are used, signal path length compensation is mandatory to be performed by adjusting the length of tracks on the PCB. The signals connected to the top row must be given correspondingly shorter PCB track lengths than tracks going to the bottom row since the topmost row of pins on the right-angled connect have longer leads than the bottom row. Also in this case, track length compensation is mandatory to be performed at the connector end of the PCB tracks to maintain the differential signal across the PCB (ECSS Secretariat, 2008).

Cable Assemblies

SpaceWire cable assemblies are made from a length of SpaceWire cable of up to 10 m length terminated at each end by nine-pin micro-miniature D-type connectors. Longer length cables are allowed at slow data signaling rates provided that the signal attenuation, system jitter and skew limits are not violated at the operating data signaling rate. The cable signal wires cross over to achieve a transmit-to-receive interconnection, e.g. Dout+ is connected to Din+ (ECSS Secretariat, 2008).

The SpaceWire cable assembly includes an outer-shield which is 360° terminated to the connector backshell at each end of the cable. The individual shields of the differential signal pairs carrying the output signals Dout+, Dout- and Sout+ and

Figure 6. SpaceWire cable assembly (ECSS Secretariat, 2008)

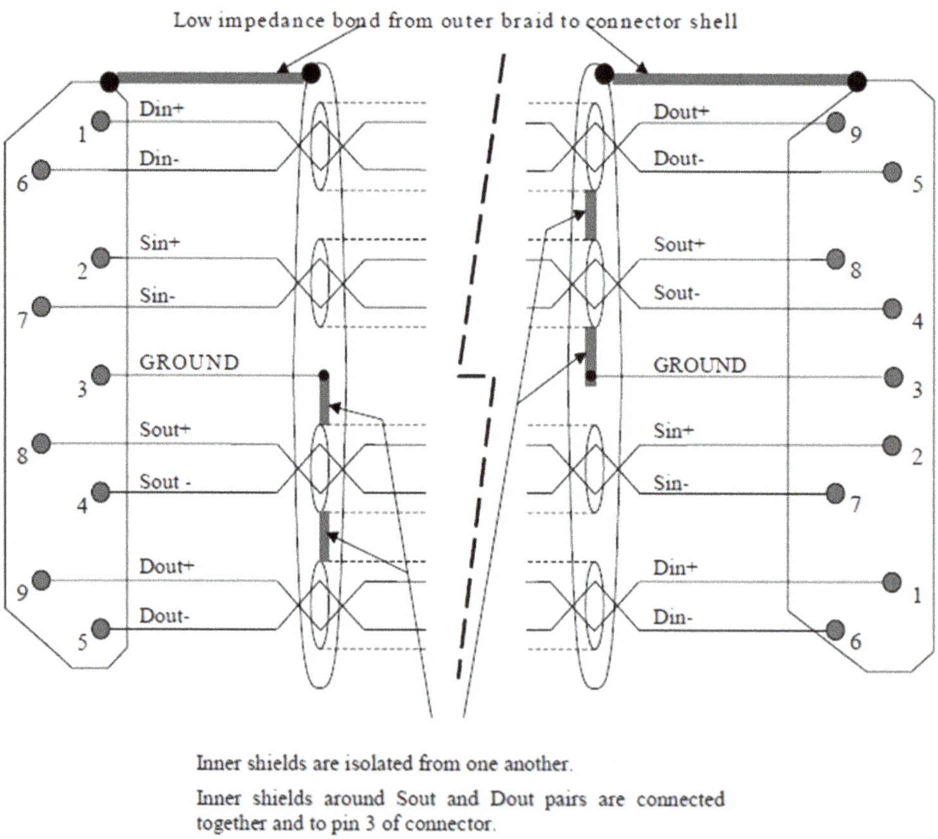

Table 1. Cable assembly signal wire connections (ECSS Secretariat, 2008)

Signal at A end	Pin at A end		Pin at B end	Signal at B end
A-Din+	1	Connection	9	B-Dout+
A-Din-	6	Connection	5	B-Dout
A-Sin+	2	Connection	8	B-Sout+
A-Din-	7	Connection	4	B-Sout-
A-(Drain of pairs 5,9 and 4,8)	3	No Connection	3	B-(Drain of pairs 5,9 and 4,8)
A-Sout+	8	Connection	2	B-Sin+
A-Sout-	4	Connection	7	B-Sin-
A-Dout+	9	Connection	1	B-Din+
A-Dout-	5	Connection	6	B-Din-
A-Shield	Shell	Connection	Shell	B-Shield

Sout- are connected together and to the pin 3 of the connector as depicted in Figure 6 and tabulated in Table 1. This way two of the differential pairs are connected at one end of the cable and the remaining two at the other end (ECSS Secretariat, 2008).

A metal shell is bonded to the main body of the connector via a low impedance connection (less than 1 Ω) in order to provide the necessary shielding to the connector. The connector shell is bonded to the outer shield of the cable via a low impedance connection (less than 1 Ω) (ECSS Secretariat, 2008).

Printed Circuit Board Tracks

SpaceWire can also be run over printed circuit boards (PCBs) or backplanes. Only point to point connections are supported on a PCB or backplane, not multi-drop bus structures. Bus type structures are built from point to point connections between nodes on the backplane. High-speed SpaceWire signals have a bandwidth of over 1 GHz (frequency of the signal edges); the PCB layout needs to be designed with some precautions. Among them, the PCB tracks must have $(100 \pm 6)\ \Omega$ differential impedance, track pairs need to be well separated from other tracks, elimination of right-angle turns, minimal use of vias, each signal of the differential pair to be tracked identically and kept to the same length (difference in tracks less than 5 mm and less than 5% of the track length) to minimize skew of the differential signal, Data and Strobe signal pairs to be of the same length (less than 5 mm and less than 5% of the track length) to minimize the skew introduced between the Data and Strobe (ECSS Secretariat, 2008).

Signal Level

The signal level part of the SpaceWire standard covers signal voltage levels, noise margins and signal encoding.

Low Voltage Differential Signaling (LVDS) as defined in ANSI/TIA/EIA-644 is specified as the signaling technique for SpaceWire. LVDS uses balanced signals to provide very high-speed interconnection using a low voltage swing (350 mV typical). The balanced or differential signaling provides adequate noise margin to enable the use of low voltages in practical systems. The low voltage swing results in relatively low power consumption at high speed. LVDS is appropriate for connections between chips on a board, boards in a unit, and unit to unit interconnections over distances of 10 m or more (ECSS Secretariat, 2008). The LVDS signaling levels are illustrated in Figures 7 and 8.

A typical LVDS driver and a receiver are shown in Figure 9, connected by a medium (cable or PCB traces) of 100 ohm differential impedance. The LVDS driver uses current mode logic. A constant current source of around 3.5 mA provides the current that flows out of the driver, along the transmission medium, through the 100 ohm termination resistance and back to the driver via the transmission medium. Two pairs of transistor switches in the driver control the direction of the current flow through the termination resistor. When the driver transistors marked "+" are turned on and those marked "-" are turned off, current flows as indicated by the arrows on the diagram creating a positive voltage across the termination resistor. When the two driver transistors, marked "-", are turned on and those marked "+"

Figure 7. LVDS Voltage Swing across 100Ω termination resistor (ECSS Secretariat, 2008)

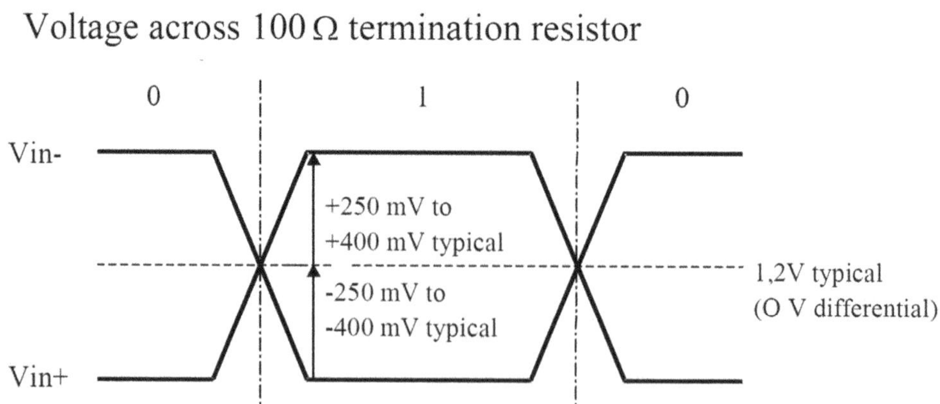

Figure 8. LVDS Receiver Input Thresholds (ECSS Secretariat, 2008)

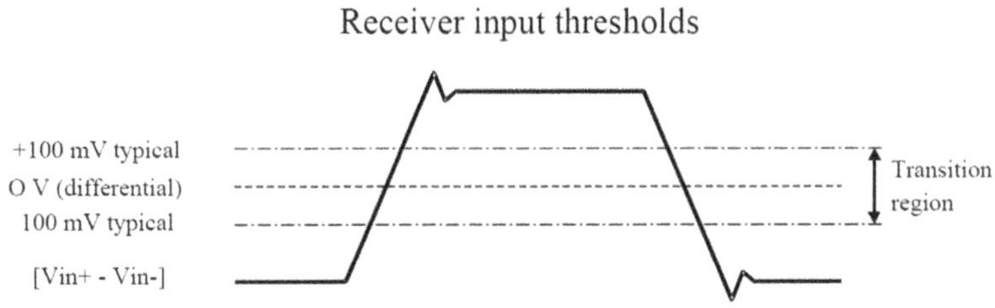

are turned off, current flows in the opposite direction producing a negative voltage across the termination resistor (ECSS Secretariat, 2008).

LVDS has several features that make it very attractive for data signaling (ECSS Secretariat, 2008):

- Near constant total drive current (+3.5 mA for logic 1 and -3.5 mA for logic 0) which decreases switching noise on power supplies.
- High immunity to ground potential difference between driver and receiver, LVDS can tolerate at least ±1 V ground difference.
- High immunity to induced noise because of differential signaling, normally using twisted-pair cable.

Figure 9. LVDS operation (ECSS Secretariat, 2008)

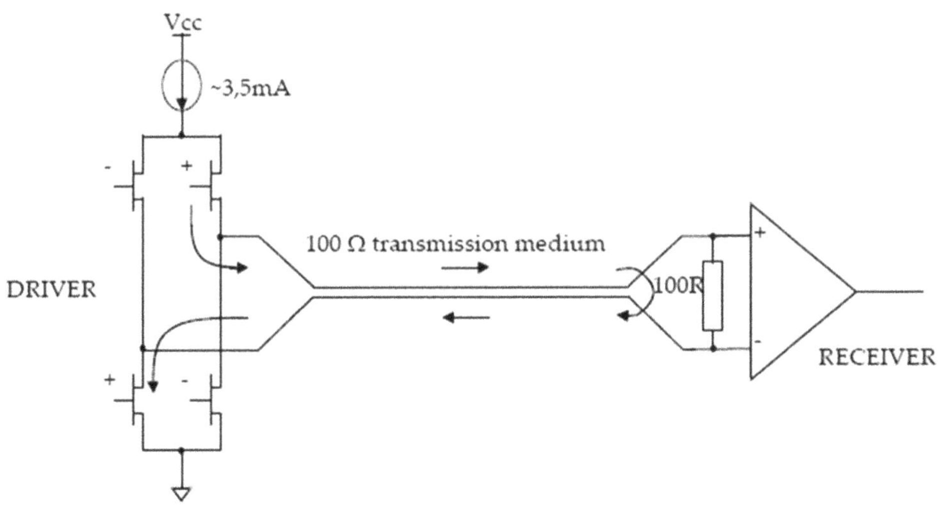

- Low EM emission because small equal and opposite currents create small electromagnetic fields which tend to cancel one another out.
- Failsafe operation
- Not dependent upon particular device supply voltage(s).
- Simple 100 ohm termination at receiver.
- Power consumption is typically 50 mW per driver/receiver pair for LVDS compared to 120 mW for ECL or PECL signals used in IEEE Standard 1355-1995.

Short Description of the Failsafe Operation of LVDS

In order to prevent the system operating under erroneous conditions and prevent undesired behavior the following must be incorporated into the design of the driver and the receiver (ECSS Secretariat, 2008):

- When the driver is not powered its output should be high impedance (greater than 100 kΩ).
- When the receiver is not powered its input should be high impedance (i.e. greater than 100 kΩ).
- Assuming a noise threshold of 10 mV is not exceeded at the receiver input, the receiver outputs must be locked to high state (inactive) and not oscillate when:
    - Driver is not powered, disabled or not connected to the receiver
    - Receiver inputs are shorted or open circuit, in case of cable or a wire in cable is disconnected

Figure 10. Data-Strobe (DS) encoding (ECSS Secretariat, 2008)

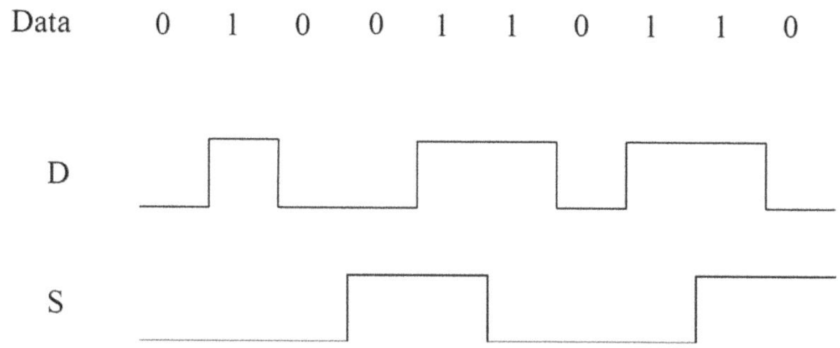

Signal Coding

SpaceWire uses Data-Strobe (DS) encoding. This is a coding scheme which encodes the transmission clock with the data into Data and Strobe so that the clock is recovered (bit synchronization) by performing a XOR function on the Data and Strobe lines. The Data signal follows the data bit stream, i.e. is high when the data bit is 1 and low when the data bit is 0, and the Strobe signal changes state whenever the Data does not change state from one bit to the next (ECSS Secretariat, 2008). This coding scheme is illustrated in Figure 10.

DS encoding improves the skew tolerance to almost 1-bit time, as opposed to 0.5 bit time for simple data and clock encoding. A link utilizes two pairs of differential signals, one pair transmitting Data and Strobe signals in one direction and the other pair transmitting Data and Strobe in the opposite direction, totaling eight wires for each bi-directional link. (ECSS Secretariat, 2008).

Data Signaling Rate

The minimum data signaling rate at which a SpaceWire link is allowed to operate is 2 Mb/s. On the contrary the maximum data signal rate is the highest data signaling rate at which a SpaceWire link can operate and is defined by consideration of signal skew and jitter. The link in one direction can operate at a different data signaling rate to the same link in the opposite direction. Links within a system can operate at different data signaling rates. A SpaceWire link can operate at any data signaling rate between the minimum data signaling rate and the maximum possible data signaling rate. After a reset or disconnect the SpaceWire link transmitter initially commences operating at a data signaling rate of (10±1) Mb/s and continues at that rate until commanded otherwise (ECSS Secretariat, 2008).

Skew and Jitter Overview

The maximum data signaling rate that can be achieved is different from one system to another, depending on several factors such as cable length, driver-receiver technology, and encoder-decoder design, and is limited by skew and jitter. Usual contributors to skew and jitter are depicted in Figure. 11. A data rate is achievable when the total system skew and jitter added to the minimum edge separation of the receiver is lower than the bit period for this data rate (ECSS Secretariat, 2008).

Figure 11. Contributors to skew and jitter (ECSS Secretariat, 2008)

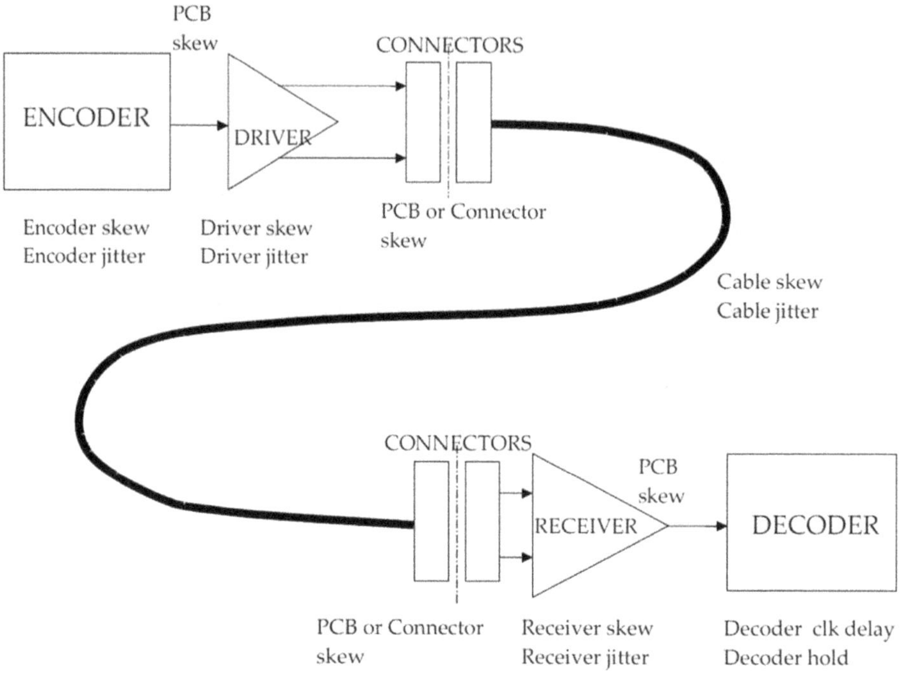

Figure 12. Parity Coverage (ECSS Secretariat, 2008)

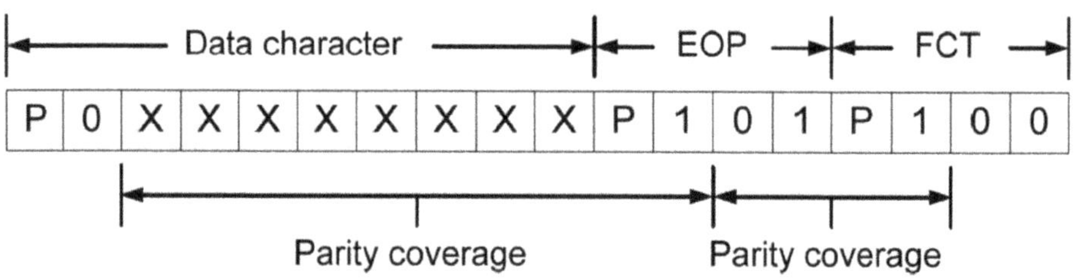

Character Level

SpaceWire has two types of characters, data and control characters which are illustrated in Figure 13.

Data Characters

Data characters contain a parity-bit, a data-control flag and the eight-bits of data transmitted least-significant bit first. The parity-bit covers the previous eight-bits

Figure 13. Control characters &codes and Data characters (ECSS Secretariat, 2008)

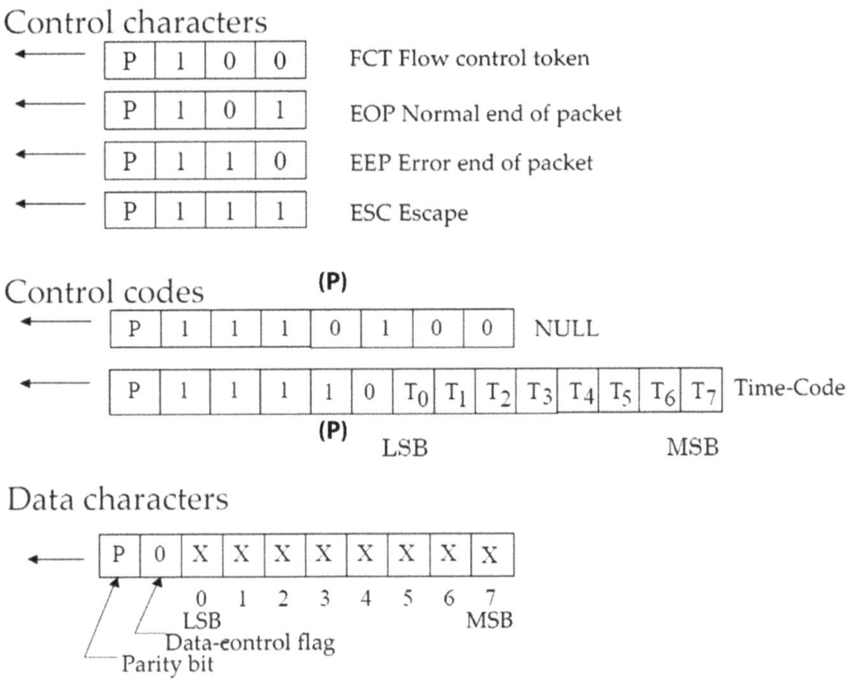

of a data character or the two bits of a control character, the current parity-bit and the current data-control flag and is depicted in Figure 12. The parity bit is set to produce odd parity so that the total number of 1's in the field covered is an odd number. The data-control flag is set to zero to indicate that the current character is a data character (ECSS Secretariat, 2008).

Control Characters and Codes

Control characters contain a two-bit control code. Each control character is formed from a parity-bit, a data-control flag and the two-bit control code. The data-control flag is set to one to indicate that the current character is a control character. Parity coverage is similar to that for a data character. One of the four possible control characters is the escape code (ESC). This can be used to form longer control codes. Two of these codes are the NULL code and the Time-Code, as depicted in Figure 13(ECSS Secretariat, 2008).

- NULL control code is formed from ESC followed by the flow control token (FCT). NULL is transmitted, whenever a link is not sending data or control

Figure 14. SpaceWire Link Start-Up Bit Sequence (ECSS Secretariat, 2008)

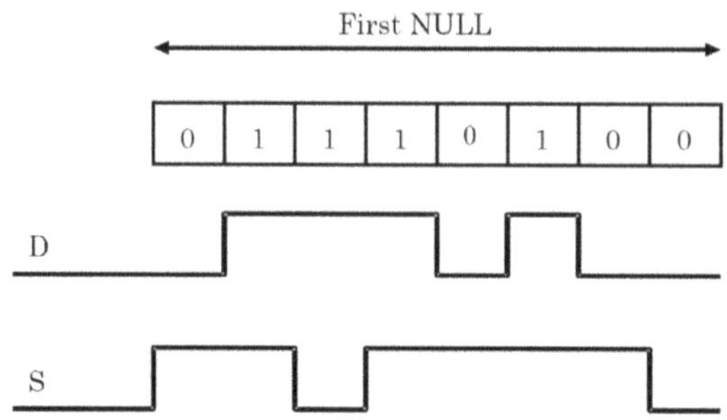

tokens, to keep the link active and to support link disconnect detection. The parity bit (P) in the middle of the control code is zero.

- The Time-Code is used to support the distribution of system time across a network. A Time-Code is formed by ESC followed by a single data-character.

Character Synchronization

Character synchronization is performed only once when a link is started or re-started following a link disconnection. After reset or link error the Data and Strobe signals are set to zero. When the transmitter is enabled after reset the first bit that is sent is a parity bit set to zero so that the first transition is on the Strobe line as depicted in Figure 14 (ECSS Secretariat, 2008).

With all the provided information regarding SpaceWire Standard as well as the systems built conforming to it, it is the authors' belief that readers have all the necessary data to start exploring the SpaceWire ecosystem, at least from an EMC point of view. Exchange level and higher, according to Figure 1, are out of the main scope of this chapter and they will not be analyzed further. Instead, some grounding strategies are presented and discussed. In addition to this, SpaceWire emissions, as well as some considerations to minimize them are discussed later in this chapter.

Bonding & Grounding

The grounding strategy is a very important issue affecting the design of the spacecraft. A low-inductance and low-resistance ground path is necessary providing a return path for fault currents and for currents induced by electromagnetic interference (EMI). This is usually made by Aluminum, but as Space Agencies ask for lightweight structures the standard aluminum panels are replaced by lighter materials, like Carbon-Fiber-Reinforced polymer (CFRP). Composite materials are lighter than aluminum and are commonly used as an alternative for the spacecraft's haul.

The bonding methods can be classified in direct and indirect. In direct bonding, the junctions are permanently fixed together without additional joints. A lower impedance path is obtained, increasing the electromagnetic compatibility (EMC) performance. Therefore, direct bonding is preferable, but indirect bonding cannot be avoided when units are installed on partially/non- conductive surfaces like CFRP. The grounding network - usually composed of strips or foils of aluminum composing a “rail” structure as a mass saving measure, in order to reduce the inductance of the ground connection and the common mode current - is an additional item of the spacecraft composed of grounding rails routed under the cable bundles, or ground planes which are mounted on the facesheet. This solution, however, is not very flexible when the routing of cable bundles has to be adapted for late design modifications (Nicoletto, Boschetti, &Savi, 2014). Moreover, their successful exploitation must also take into account possible impacts on the electromagnetic compatibility performance of the on-board electronic systems. Consequently, electromagnetic interference issues have to be analyzed when such materials are used, using the radiated Electric Field (E-field) from high speed digital lines as the parameter for this assessment.

Authors of Leininger, Thurecht, Pfeiffer, and Ruddle (2012) discuss the use of network of round wires over a CFRP sheet as a replacement of the classical grounding rail system. They demonstrate that the electrical conductivity of carbon fibers can be used to simplify the grounding network. A low-resistance, low inductance network of metallic conductors is employed in order to divert fault currents from flowing through the CFRP structure. As an optimal alternative the metallization of the CFRP facesheet is proposed, otherwise ground conductors can be used. The use of rails, unless they are wide and track the entire harness, has no better results than the use of round wires, which can replace them and are easier to implement. In addition in order to really benefit of the ground plane properties of CFRP facesheet panels, low inductance connections of the electronic units to the panels need to be implemented (Leininger, Thurecht, Pfeiffer, &Ruddle, 2012).

The authors of Nicoletto et al. (2014) compared the simulated radiated E-field for the three ground planes; an aluminum plane with a 2.5:1 ratio between length and width. The second is a plane made of CFRP. The third is a long strip (2.5x0.03x0.001 m) routed on top of a CFRP panel. This last case corresponds to a typical implementation of a grounding rail network. They concluded that the grounding plane made of aluminum, due to its intrinsic lower impedance, represents the best solution in terms of minimization of the radiated E-Field, in the whole frequency range analyzed. The pure CFRP plane, compared with the one implementing a grounding rail network, induces higher levels of radiated E-field, mainly observable at lower frequencies (below 200 MHz). Nevertheless, the radiated electric field levels are well inside the typical requirements applicable to Space equipment. Based on these results, CFRP panels may be used for common mode current return, avoiding the mass penalties induced by the implementation of a grounding network which tracks the entire harness (Nicoletto et al., 2014).

Emissions & Considerations

Authors of Cook & Walker (2007) approached emissions of the SpaceWire from the signal spectrum on the cable perspective. They discussed the lowest frequency generated in any communications system and that is inverse of the longest periodicity found in the signal. This is several bit times – since the usual unit of transmission is a group of bits such as an 8-bit byte. This is a sub-multiple of the bit-rate. Random data can further lower the minimum frequency by extending the periodicity, however basic bit grouping of the system can still produce strong emissions. Harmonics of actual digital signals are limited by the rise and fall times. In general, the level drops by some 20dB/decade (6dB/octave) up to some 'knee' frequency where the rate of fall increases. The knee frequency is determined by the rise and fall times, being 0.5 / the rise (fall) time. For a rise/fall time of 1ns the knee frequency is 500MHz. Real data is not totally random and the spectrum loses its smoothness by gaining 'spikes' at significant frequencies as determined by the coding scheme being used. Authors argue that a reason for increased emissions are the fast LVDS drivers, which are faster than actually needed. This great speed provides increased working margin at the expense of a greatly extended emission spectrum. If the pulse rise and fall times are matched to the data rate the highest frequencies produced can be limited. One possible way to do this is to control edge rates by capacitive loading. Adding a capacitor across each of the transmit pairs has a very significant effect in limiting the spectrum of the emissions. Another technique they propose is alter the edge to edge spacing, the bit period, in a random sequence also known as spread spectrum clocking. This spreads the energy in single-frequency spikes over a range of nearby frequencies and reduces the peak levels significantly. It can easily be applied to SpaceWire since it has no restriction on the rate of change of data rate and already has the capacity to do so since a link must start at 10Mb/s and then change – mid-bit-stream – to the working rate. The effect of spreading – rate randomization – is to smear the energy of single frequency peaks over a band of frequencies. This reduces the average level. The reduction depends on the amount of spread and on the spikiness of the signal. This technique also provides power conservation by reducing the transmit speed. The two above techniques combined result is a very significant reduction in the generated spectrum and thus reduced emissions within the system. Authors provide measurements to back up their arguments. They measured a system including a SpaceWire module feeding 2.5m cable with a loopback connector at the end. A passive network was used to attenuate and combine D and S with a transformer performing balanced to unbalanced conversion. (Cook & Walker, 2007).

A similar concept is presented in the work of Baklezos, Nikolopoulos, Capsalis, &Tsatalas, (2017), with actual radiated emissions. Authors aimed to measure the radiated emissions of a half-duplex system. In order to simulate the operation of the SpW/LVDS link, a small PCB, acting as a complete LVDS interface, was implemented to drive the signal from the 50 Ω output of the generator to the 100 Ω SpW cable. This LVDS driver device was fed by the Agilent 81130A outputs with the Data and Strobe signals and provided LVDS output. Additionally, this driver had termination resistances to terminate signals fed from the other side of the SpW/LVDS link. The LVDS interface PCB as well as the termination resistors were both encased in a shielded box grounded to the aluminum plane. The cable was terminated with 100 Ω resistors simulating the half-duplex operation of the SpW/LVDS link. For all the measurements in the frequency range from 30 MHz – 1 GHz, a 6 dB bandwidth of 100 KHz was selected as well as a measurement time of 0.015 s and the minimum measurement time is 0.15 s / MHz as specified in ECSS-E-ST-20-07C Rev. 1: Space Engineering – Electromagnetic Compatibility (ECSS Secretariat, 2012). Two different code words were used, one of 8 bit and one of 16

Figure 15. Comparative EMI emissions at pulse rate of 50MHz, half duplex operation. (Baklezos et al., 2017)

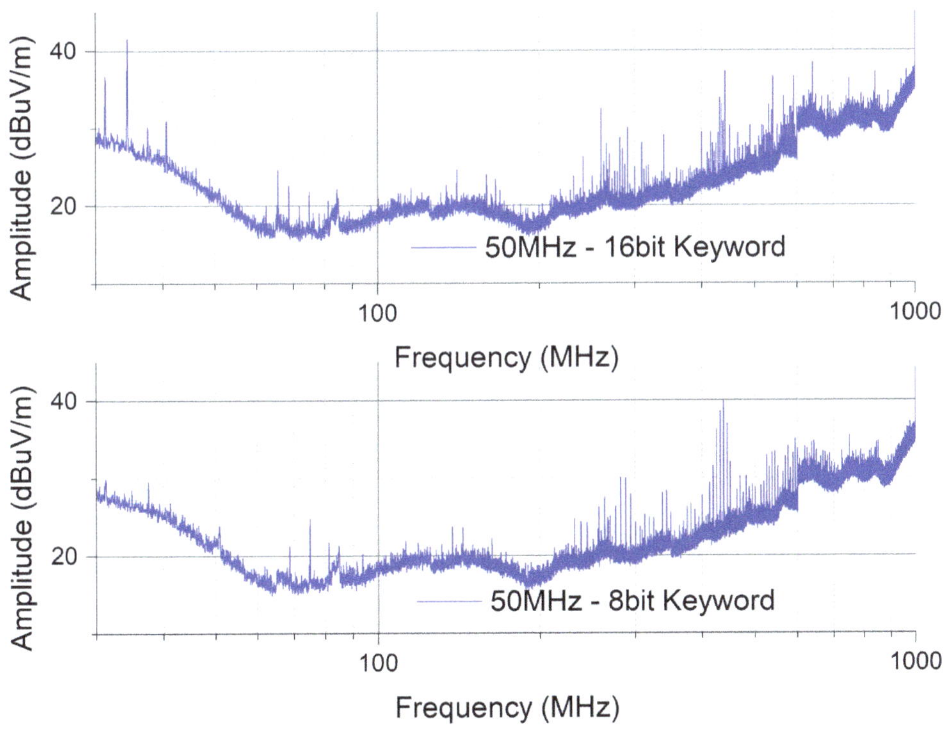

Figure 16. Comparative EMI emissions at pulse rate of 100MHz, half duplex operation (Baklezos et al., 2017)

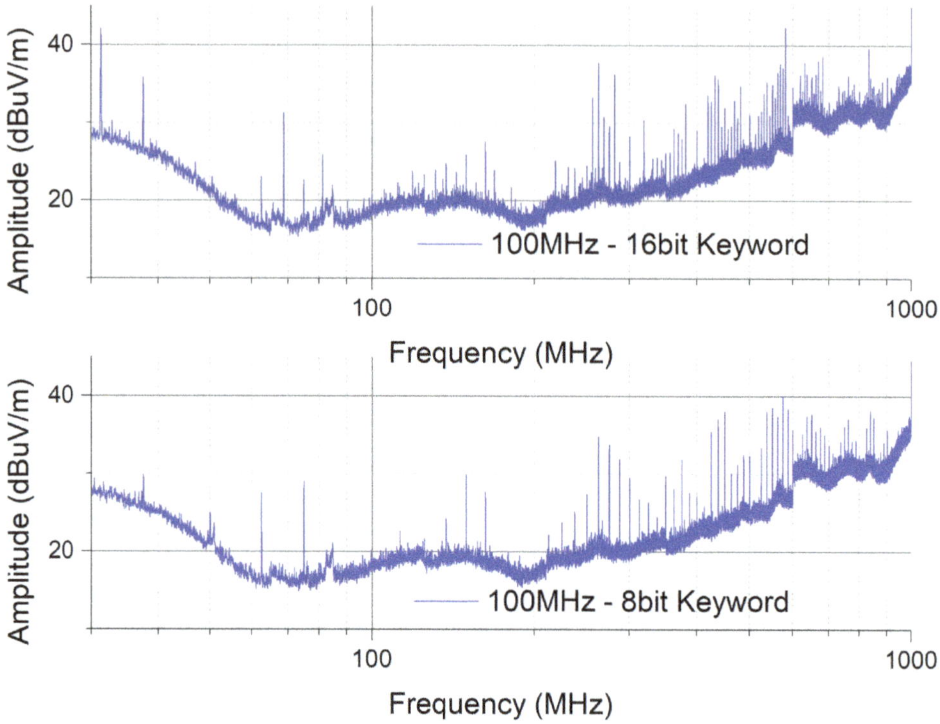

Bit. The emissions of the two pattern lengths scenarios were measured for the pulse frequencies of 50, 100 and 150 MHz. Comparative results of the radiated emissions measurements in the case of 50MHz pulse rate, half duplex operation for 8 bit and 16 bit pattern length, are presented in Figure. 15. It is clear that the longer pattern (16 bit) exhibits reduced magnitude and spacing compared to the 8 bit pattern. This behavior is repeated regardless of the link speed (pulse rate) as depicted in Figure 16 and Figure. 17.

Comparative results of the radiated emissions measurements in the case of 100MHz pulse rate, half duplex operation for 8 bit and 16 bit pattern length, are presented in Figure 12. It easily observed in Figures. 15, 16 and 17 that as link speed increases the radiated emission levels are increased throughout the spectrum. This well-defined effect of the link speed variation as well as the effect of the pattern length, can be used in order to reduce EMI at a frequency region of interest. An undesired spectral line can be shifted away from a particularly sensitive frequency range or have its magnitude reduced with the variation of the length of the code word *(Baklezos et al., 2017)*.

Figure 17. Comparative EMI emissions at pulse rate of 150MHz, half duplex operation (Baklezos et al., 2017)

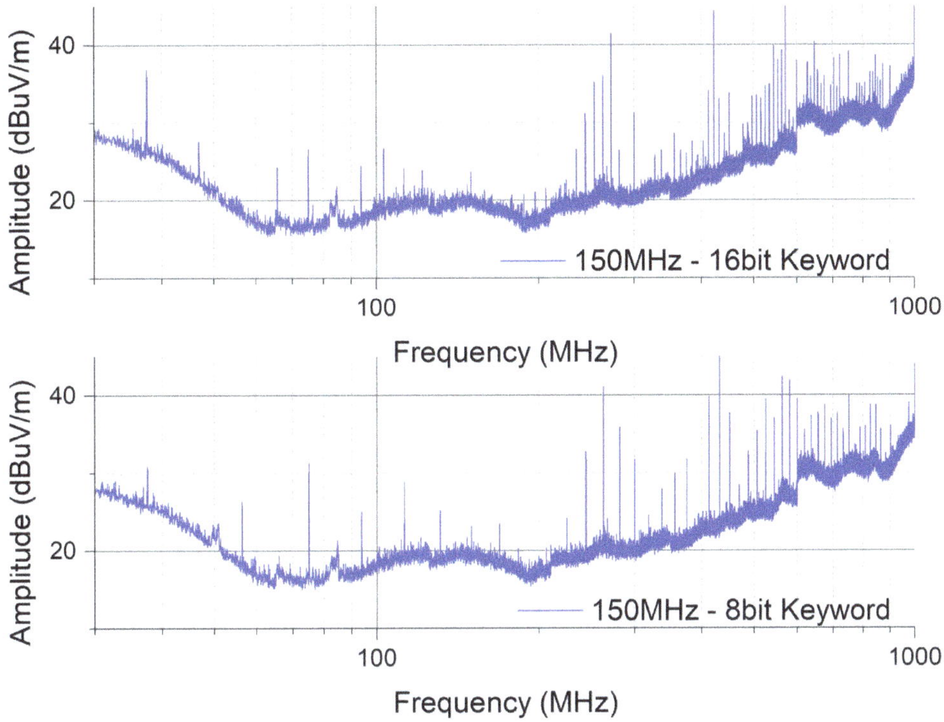

Measurements in Baklezos et al. (2017) were conducted in in accordance to ECSS-E-ST-20-07C Rev. 1: Space Engineering – Electromagnetic Compatibility (ECSS Secretariat, 2012) in order to maximize measurement repeatability and reliability. However some discrepancies were observed during the measurement period. A possible source of the deviations is that during the measurements the SpaceWire inner shields were terminated by pairs at pin 3 of the connector according to ECSS-ST-50-12. However that is now recognized as a not optimal setup and it is recommended that all shields are terminated to the backshell and pin 3 at both ends of the connector. Another possible contributor to these deviations at emissions, beyond the normal EMC margin, can be traced to the 9-pin micro miniature D-type connector specified in the SpaceWire standard that is not designed for balanced signals and causes emissions especially on higher frequencies.

The performance of the micro D-type connector is a known issue. To overcome that, Axon' Cable was selected to carry out the development of a new, compact impedance-matched SpaceWire connector under an ESA Technology Research Project. The findings of this project are presented in Enouf &Hemant, (2016).

Authors presented a compact, impedance-matched connector with a desirable size coupled with significantly improved electrical and EMC performances, namely "MicroMach". A trade-off was made with regard to the cable shield termination to the connector backshell (not full 360° screening) to minimize the size. Preliminary measurements are reported for bonding resistance between male & female contacts less than 5 mΩ, Near End and Far End Crosstalk (NEXT/FEXT) less than -50 dB up to 1 GHz, and Return loss less than -20 dB up to 1 GHz. The new MicroMach SpaceWire connector offers, at a size marginally larger than the current micro-D, significantly improved performance in data rate, EMC and crosstalk compared to any of the current market solutions (Enouf &Hemant, 2016).

SpaceWire Revision 1

The SpaceWire Standard remains unaltered since its first publication in 2003. The European Cooperation for Space Standardization in 2012 decided it was time to update it and to include the experience of almost ten years of its application. Working Group ECSS-E-ST-50-12C Rev.1 WG is preparing Revision 1 of the standard. The draft of the Revision. 1 (ECSS-E-ST-50-12C Rev. 1 Working Group, 2015) is currently under public review and expected to be published in its final form by the end of 2017 or early 2018. It will include a number of changes as well as new additions, many of which are out of the scope of this chapter. In this section authors will try to include the most important updates with a main focus to the cables and assemblies.

Values for maximum allowed signal loss through each differential pair of the cable are provided, according to Table 2, with regard to for the cable only without considering the connectors of the assembly (ECSS-E-ST-50-12C Rev. 1 Working Group, 2015).

Specifications for the maximum levels of Power Sum Near End Crosstalk (PSNEXT) and Power Sum Equal Level Far End Crosstalk (PSELFEXT) are also

Table 2. Maximum allowed Insertion loss values for a cable

Data rate (Mbps)	Third Harmonic (MHz)	Maximum Insertion Loss (dB/m)
10	15	0,27
100	150	0,45
200	300	0,65
400	600	0,94

Table 3. Cable PSNEXT specification

Frequency (MHz)	PSNEXT (dB)
100	73,0
500	63,5
1000	39,0

Table 4. Cable PSELFEXT specification

Frequency (MHz)	PSELFEXT (dB)
100	65,0
500	50,5
1000	41,5

included in the future revision 1 as tabulated in Tables 3 and 4 (ECSS-E-ST-50-12C Rev. 1 Working Group, 2015).

Connectors

In the future standard the connectors for the SpaceWire cable assembly are either type A or type B. The type A connector is a microminiature Dtype connector with nine crimp or solder contacts, as defined in ESCC 3401/029 and ESCC 3401/071. In addition to type A connectors, other connectors (type B) are allowed, provided that they meet the requirements. Type B connectors must have the following characteristics (ECSS-E-ST-50-12C Rev. 1 Working Group, 2015):

- Contact pairs with differential impedance of (100 ±6) Ω,
- Compatible with ESCC 3902/003 cables or balanced cables in ESCC 3902/002,
- Compatible with the space environment as defined in ESCC 3401.

A connector with female contacts must include a conductive gasket for EMI improvement. Connectors with male contacts will be used on cable assemblies. The outer shield of the SpaceWire cable must be circularly terminated to the connector body via an EMI backshell. The electrical resistance of the connection between the outer shield of the cable and the body of the connector with male contacts must be less than 10 mΩ at DC. The inner shields of the SpaceWire cable must be connected to the connector body by an impedance of less than 10 mΩ at DC. Any pigtails used to connect the inner shields to the backshell must be less than 10 mm in length. When type A connectors with female contacts are used to connect two cables, pin 3 must be connected to the body of the connector. Pin 3 of type A connectors with female contacts is connected to Circuit Ground also when used with type A cables in order to prevent charge accumulation (ECSS-E-ST-50-12C Rev. 1 Working Group, 2015).

In case of Type B connector the assignment of signals to connector contacts must be defined by the supplier but with the limitation that the two signals in each pair of conductors (Din+ and Din-, Sin+ and Sin-, Dout+ and Dout-, Sout+ and Sout-) must be assigned to adjacent pins of the connector. No connector pin will be connected to the cable shield. In the case where an outer shield is present, it must be circularly connected to the connector body. The inner shields must be circularly connected to the connector body when possible (ECSS-E-ST-50-12C Rev. 1 Working Group, 2015).

Cable Assemblies

Cable assembly type AL (type A Legacy) is the cable assembly specified in the current issue of the SpaceWire standard (ECSS Secretariat, 2008), prior to the revision 1. Cable assembly type AL, after revision 1 is issued, will not be recommended for use in new designs (ECSS-E-ST-50-12C Rev. 1 Working Group, 2015).

Acceptable cable assemblies in the future standard revision will consist of

- Two identical connectors of type A with male contacts joined by a length of cable,
- Two connector of Type B joined by a length of cable, or
- A length of cable in between one Type A connector with male contacts at one end, and one Type B connector at the other end.

The insertion loss through each differential pair in a cable assembly must be less than 7 dB at frequencies up to 1,5 times the data signalling rate.(i.e. for a 200 Mbps signal, the frequency at which the insertion loss is measured is 300 MHz) . (ECSS-E-ST-50-12C Rev. 1 Working Group, 2015).

Cable assembly type A will use type A connectors and cable with an outer shield. Pin 3 of the connector must be left unconnected. The individual shields of each of the four differential signal pairs will be bonded to the connector shell via a low impedance connection of less than 10 mΩ. A metal backshell will be used for each connector to create, together with the outer shield, a continuous conductive barrier and to facilitate the termination of the inner shields. The outer shield of the cable must be 360° terminated to the connector backshell via a low impedance connection.

The connection impedance between the main body of the connector and the backshell must be less than 10 mΩ. The differential far end and near end crosstalk (FEXT and NEXT) between any two pairs in the SpaceWire cable assembly must be less than 20 dB up to 1 GHz. (ECSS-E-ST-50-12C Rev. 1 Working Group, 2015).

Cable assembly type B will be a cable assembly with an outer shield and matched impedance type B connectors as defined by the supplier. The cable in Cable assembly type B may be another form of cable with the following properties:

- Differential impedance of (100 ± 6) Ω,
- Compatibility with ESCC 3902/003 cables or balanced cables in ESCC 3902/002,
- Compatibility with the space environment as defined in ESCC 3401.

The individual shields of each of the four differential signal pairs must be bonded to the connector shell via a low impedance connection (less than 10 mΩ). A metal backshell must be used for each connector to create, together with the outer shield, a continuous conductive barrier and to facilitate the termination of the inner shields. The outer shield of the cable must be 360° terminated to the connector backshell via a low impedance connection. The main body of the connector must be connected to the backshell via a low impedance connection (less than 10 mΩ). The differential far end and near end crosstalk (FEXT and NEXT) between any two pairs in the SpaceWire cable assembly must be less than 50 dB up to 1 GHz including the contribution of the cables and of the connectors. (ECSS-E-ST-50-12C Rev. 1 Working Group, 2015).

FUTURE RESEARCH DIRECTIONS

Space missions is an active field in engineering. The SpaceWire ecosystem is full with new challenges. An issue that is worth the focus is the research on the applicability and effectiveness of Galvanic Isolation of SpaceWire as it promises reduced sensitivity to common mode noise compared to LVDS. Another relative issue worth exploring is DC balanced SpaceWire. A Direct-Current line-balanced interface provides the ability to isolate the physical layer to achieve common mode voltage rejection or even the complete galvanic isolation. Moreover, it could minimize the number of conductors and transceivers by completely eliminating the need for a Strobe line. Moreover, in the SpaceWire ecosystem there is a new protocol, SpaceFibre, which supports both electrical and fiber-optic cables. It was developed to cover the need for higher and higher data rates offering multi-Gigabit/s speeds. SpaceFibre is in the stage of finalizing standardization. Another important aspect to consider is the use of SpaceWire in robotics systems in space. Current SpaceWire standard is not tailored to robotic use, for instance cables and connectors. Research that contributes on any of those areas will greatly affect the future the field of space missions.

CONCLUSION

SpaceWire is a point-to-point bit shipping protocol using LVDS as the physical layer. Given the wide use of the SpaceWire, due to the standardization, it is near impossible to work or study in the space mission field without some level of familiarity with SpaceWire. The main objective of the chapter is to discuss the SpaceWire standard ECSS-E-ST-50-12C from an EMC engineer's point of view, especially with regard to Electromagnetic Compatibility and Immunity issues. This chapter describes the SpaceWire standard, focusing on the levels with a major role to electromagnetic behavior of the system and concern cable assemblies, shielding, bonding and grounding. Findings regarding emissions affecting spacecraft components are presented as well as other EMC/EMI issues that have an impact on the system performance. The levels of the standard that define the EMC behavior of the system were highlighted. Engineers or scientists regardless of their approach to the standard, theoretical or implementation, can only gain from the information herein. Students interested in space missions can also benefit from the concentrated information found in this chapter. Finally, by including a section in this chapter that presents changes regarding upcoming Revision 1 of the standard authors hope to better prepare the readers for the future.

REFERENCES

Baklezos, A. T., Nikolopoulos, C. D., Capsalis, C. N., & Tsatalas, S. (2017, March). Effect of LVDS link speed and pattern length on spectrum measurements of a Spacewire harness. In *Antenna Technology: Small Antennas, Innovative Structures, and Applications (iWAT), 2017 International Workshop on* (pp. 38-41). IEEE. doi:10.1109/IWAT.2017.7915291

Cook, B., & Walker, P. (2007). Reducing Electromagnetic Emissions From SpaceWire. In *DAta Systems in Aerospace*. Naples: DASIA.

ECSS-E50-12A vs ECSS-E-ST-50-12C, what's the difference? | STAR-Dundee. (2017). Retrieved 15 June 2017, from https://www.star-dundee.com/knowledge-base/ecss-e50-12a-vs-ecss-e-st-50-12c-whats-difference

ECSS-E-ST-50-12C Rev. 1 Working Group. (2015). *ECSS-E-ST-50-12C Rev. 1 Draft (DIR3), Space Engineering – SpaceWire- Links, nodes, routers and networks, ESA-ESTEC*. Noordwijk, The Netherlands: Requirements & Standards Division.

Enouf, K., & Hermant, S. (2016, October). Compact, impedance-matched SpaceWire connector development—"MicroMach SpaceWire": SpaceWire components, long paper. In SpaceWire Conference (SpaceWire), 2016 International (pp. 1-7). IEEE.

European Cooperation for Space Standardization (ECSS) Secretariat. (2003). *ESCC Detail Specification No. 3902/003, Cable, "Spacewire", Round, Quad Using Symmetric Cables, FlexiblE, -200 TO +180 °C, ESA-ESTEC*. Noordwijk, The Netherlands: Requirements & Standards Division.

European Cooperation for Space Standardization (ECSS) Secretariat. (2008). *ECSS-E-ST-50-12C, Space Engineering – SpaceWire- Links, nodes, routers and networks, ESA-ESTEC*. Requirements & Standards Division.

European Cooperation for Space Standardization (ECSS) Secretariat. (2012). *ECSS-E-ST-20-07C Rev. 1, Space Engineering –Electromagnetic compatibility, ESA-ESTEC*. Requirements & Standards Division.

European Cooperation for Space Standardization (ECSS) Secretariat. (2014). *ESCC Detail Specification No. 3902/003, Cable, "Spacewire", Round, Quad Using Symmetric Cables, Flexible, -200 To +180 °c, Esa-Estec*. Noordwijk, The Netherlands: Requirements & Standards Division.

Leininger, M. M., Thurecht, F., Pfeiffer, E., & Ruddle, A. (2012, May). Advanced grounding methods in the presence of carbon fibre reinforced plastic structures. In *Aerospace EMC, 2012 Proceedings ESA Workshop on* (pp. 1-6). IEEE.

Nicoletto, M., Boschetti, D., & Savi, P. (2014, September). High Speed Digital Lines routed on non-metallic Spacecraft structures. In *Electromagnetic Compatibility (EMC Europe), 2014 International Symposium on* (pp. 785-789). IEEE. doi:10.1109/EMCEurope.2014.6931011

Parkes, S. (2012). SpaceWire User's Guide. STAR-Dundee Limited.

SpaceWire - The Standard. (2017). Retrieved 15 June 2017, from http://spacewire.esa.int/content/Standard/Standard.php

Telecommunications Industry Association. (2001). *ANSI/TIA/EIA-644, Electrical Characteristics of Low Voltage Differential Signaling (LVDS) Interface Circuits*. Arlington, VA: Standards and Technology Department.

KEY TERMS AND DEFINITIONS

AWG: American wire gauge.

Data Strobe Encoding (D-S): Encoding scheme in which a sequence of data bits and clock is encoded as the original data bit sequence, together with another bit sequence (strobe) which changes state whenever the data bit sequence does not. Clock can be recovered by performing XOR function on the data and strobe.

Electromagnetic Compatibility (EMC): The ability of a system to function satisfactorily in its electromagnetic environment without introducing intolerable electromagnetic disturbances to anything in that environment.

Electromagnetic Immunity: The ability of a system to maintain its proper functionality under exposure to various sources of electromagnetic disturbance.

Jitter: Random errors in the timing of a signal.

Lay-Length: Number of twists per length unit expressed as the length between one complete turn of a single end in the cable.

Shielding: A method for the mitigation of electromagnetic disturbances that is based on the concept of creating an electrically continuous enclosure around the under protection structure, mainly by the use of conductive materials.

Skew: Difference in time between the edges of two signals which should ideally be concurrent.

SpaceWire: A Standard that specifies the physical interconnection media and data communication protocols to enable the reliable sending of data at high-speed (between 2 Mb/s and 400 Mb/s) from one unit to another. SpaceWire links are full-duplex, point-to-point, and serial data communication links.

Chapter 3
Magnetic Dipole Modeling for DC and Low Frequency AC Magnetic Fields in Space Missions

Sotirios Spantideas
National Technical University of Athens, Greece

Nikolaos Kapsalis
National Technical University of Athens, Greece

ABSTRACT

In this chapter, modeling methods of static and slowly varying magnetic field emissions that are generated by spacecraft equipment are analyzed and discussed. In particular, specific issues on the established methods for multiple magnetic dipole modeling (MDM) are investigated and validated via near field measurements of well-behaved magnetic sources. Moreover, a software-based calibration technique for measuring facilities, dedicated to magnetic characterization of spacecraft units, is described and implemented on a configuration consisting of 12 sensors. Due to increasingly strict magnetic cleanliness demands, the modeling of units' induced DC magnetic behavior has become a necessary requirement for various space missions. Therefore, a baseline methodology regarding the measurements and modeling of induced magnetic fields is presented. Finally, DC methods are complemented to cover AC magnetic cleanliness requirements for on-ground verification of low-frequency magnetic fields, including AC induced magnetization effects.

DOI: 10.4018/978-1-5225-5415-8.ch003

INTRODUCTION

In most space missions, dedicated sensitive instruments are mounted on the spacecraft in order to perform in situ measurements of interplanetary magnetic fields (Acuna, 2002). Fluxgate and Search Coil Magnetometers are commonly placed on a long boom that extends beyond the main structure of the spacecraft. The sensitivity (resolution) of these magnetometers is very high for purposes of measuring very weak variations of magnetic flux density (Musmann, 2010). Due to their high sensitivity, however, these sensors are susceptible to magnetic field emissions generated by the spacecraft. Units and equipment on-board, as well as components of the spacecraft itself produce a combined magnetic field at the location of the magnetometers, thus interfering with the output of the instruments. Unquestionably, one of the critical requirements for these instruments is to operate in a magnetically "clean" environment; typical requirements on the total DC magnetic field generated by the spacecraft at the location of the magnetometers are in the order of 0.1-1 nT (Primdahl, 1989; Mehlem, 1987). Moreover, current and upcoming space missions (Solar Orbiter, PLATO, JUICE), have even more stringent requirements on magnetic cleanliness of slowly time varying AC magnetic fields.

The "magnetic cleanliness" of the spacecraft, which is targeted during the development phase, is realized by application of "best engineering practices", i.e. careful design and construction of the spacecraft by using non - magnetic materials, magnetic shielding, etc. Therefore, the spacecraft residual magnetism is usually kept below the requirements determined by the scientific application and commensurate with the sensitivity of the on-board magnetometers. In most cases, however, magnetic fields emitted by subsystems and components of the spacecraft are still present at the location of the magnetometers. To overcome the magnetic fields' measurements contamination, it has been a standard practice to perform on-ground electromagnetic compatibility (EMC) test campaigns at unit and system level (Junge & Marliani, 2011). At early stages of the space missions, the main target is to capture the magnetic signature of individual units and/or subsystems that will actually be mounted on the spacecraft and characterize them. The magnetic cleanliness test campaign is finalized at system level, with magnetic measurements and characterization of the entire spacecraft.

The unit level measurements are performed on-ground in dedicated facilities, namely Magnetic Coil Facilities (MCF). These facilities usually employ a pair of Helmholtz coils in order to compensate Earth's magnetic field (ECSS, 2012). The Equipment under Test (EUT), is placed on a turntable that makes a full rotation of 360° while several magnetometers measure its (remanent) magnetic field at various angles (most commonly at 0°, 10°, …), ensuring that the unit's magnetic signature is well covered. Although the location of the sensors is in close proximity to the EUT in order to obtain an adequate Signal to Noise Ratio (SNR), practical experience has shown that measurement distances below 3-5 times the maximum dimension of the unit should in principle be avoided, since the magnetic material inside the unit in most cases is not uniformly distributed and composite field behavior that deviates from the dipole approximation may emerge at smaller observation distances.

Various approaches have been proposed for identification of on-board "magnetically clean" specification points, where the magnetometers can be placed. The predominant method indicates that each EUT can be represented by a small number of equivalent magnetic dipole sources and involves the generation of models that can be used to accurately reproduce the magnetic field of the original equipment. The captured magnetic signature of each individual EUT is therefore modeled with a multiple Magnetic Dipole Model (MDM); the parameters of the model (position and magnetic moment vectors of the dipole sources) can be estimated applying either deterministic or stochastic techniques (Weikert, Mehlem, & Wiegand, 2012; Carrubba, Junge, Marliani, & Monorchio, 2012; Dumond, & Bergé, 2012; Boschetti, Gervasio, & Marziali, 2012). The magnetic cleanliness of past and recent space missions for spacecrafts with instruments susceptible to magnetic fields has been largely dependent on MDM modeling at unit level, by capturing near field magnetic measurements of real spacecraft equipment.

Furthermore, the response of various units to an applied external field may additionally cause magnetic field effects. During most on-ground measurement tests, the Earth's magnetic field is present with a level of about 50 μT (in case that is not compensated), as well as fields from the electric power lines or from nearby equipment. In the presence of external field(s), several EUTs exhibiting conducting and/or magnetic properties generate induced magnetic fields that have also a significant impact on the measurements' contamination. For this reason, the MCFs typically employ another pair of Helmholtz coils that generates a (spatially) uniform external magnetic field at the volume of the EUT in order to measure and characterize the unit's induced magnetic behavior. The modeling involves the generation of equivalent magnetic dipoles that simulate the induced magnetization of the unit. In principle, the magnetic sources of an EUT can be categorized into remanent dipoles, which remain in absence of any external fields and induced dipoles, which are only present with non-vanishing external fields. The total magnetic field generated by the EUT is the superposition of the individual contributions of all its magnetic sources.

In the first part of this chapter, specific issues in the MDM method for DC magnetic field measurements are identified and investigated, including special concerns about the induced DC magnetic field measurements and modeling. Although the MDM method is well-established for unit level magnetic cleanliness issues in space missions, the measured magnetic field data from existing MCF facilities typically exhibit environmental noise, since the rotation of the unit requires a significant amount of time, fact that virtually reduces the accuracy of the predicted dipole models. The present analysis aims to illustrate stochastic techniques that are employed for dipole modeling in order to increase the reproducibility of the test results and eliminate uncertainty and ambient environmental variations during the measurement procedure. For these purposes, the presented test methodology and modeling procedure involve simultaneous measurements from multiple magnetometers that capture the magnetic signature of a unit, thus significantly enhancing the quality of the magnetic measurements, as well as the accuracy of the estimated MDM. The presented methodology is validated with both simulated and real measurement data. Apart from environmental conditions, one of the main reasons that cause uncertainty in the MDM prediction is the relative distance between the magnetometers and the EUT. Consequently, it is essential that the positions of the sensors in a measuring facility are accurately defined prior to performing the measurements. In the second section of the chapter, a calibration method that employs stochastic techniques, targeting to determine the position of the magnetometers in a measuring facility, is presented. It is worth mentioning that current calibration methods rely on the resolution of the facility's mechanical support system, typically obtaining accuracy levels in the range of 1 mm. The purpose of the proposed software based calibration technique is to reduce the uncertainty in the modeling procedure by significantly enhancing the precision of the sensors' position. The calibration method is evaluated with simulated results and verified with real magnetic field measurement.

In the last part of this chapter, existing DC magnetic cleanliness methods are complemented to cover AC cleanliness requirements for on-ground verification of low-frequency magnetic fields. To the best of the authors' knowledge, the test methodology for AC magnetic measurements involves the use of dual-magnetometer configuration (Güttler, Hillenmaier, & Auster 2009) and the modeling entail only worst case estimations of the radiated magnetic field. The illustrated test methodology in the present chapter involves simultaneous measurements from multiple magnetometers and employs stochastic techniques in order to estimate the parameters of virtual magnetic dipole sources, which in turn, can be used to accurately recreate the measured magnetic field spectrum and effectively emulate the signature of the original unit. Additionally, specific considerations about the AC induced magnetic dipole modeling are described and discussed.

BACKGROUND

The magnetic sources inside an EUT are assumed to be represented by a small number of dipoles (Paul, 2006). However, it should be noted that not all magnetic field sources inside a spacecraft are magnetic dipoles. For instance, electric currents flowing in the equipment and solar arrays need to be treated analytically with Biot - Savart law. For simplicity's sake, the EUT is assumed to be located at the origin in Cartesian coordinate system and its magnetic behavior may be modeled as a set of M magnetic dipoles positioned at $\mathbf{r'}_i = (x_i, y_i, z_i)$, $i = 1,2,\ldots M$, with magnetic moments:

$$\mathbf{m}_i = m_{i_x}\hat{\mathbf{x}} + m_{i_y}\hat{\mathbf{y}} + m_{i_z}\hat{\mathbf{z}} \tag{1}$$

Furthermore, the N_{MAG} observation points are located at $\mathbf{r}_{0j} = (x_{0j}, y_{0j}, z_{0j})$, $j = 1,2,\ldots N_{MAG}$, as depicted in Figure 1. The magnetic field generated by the magnetic dipole i at the observation point j can be expressed in components (Jackson, 1999):

Figure 1. Formulation of the MDM problem

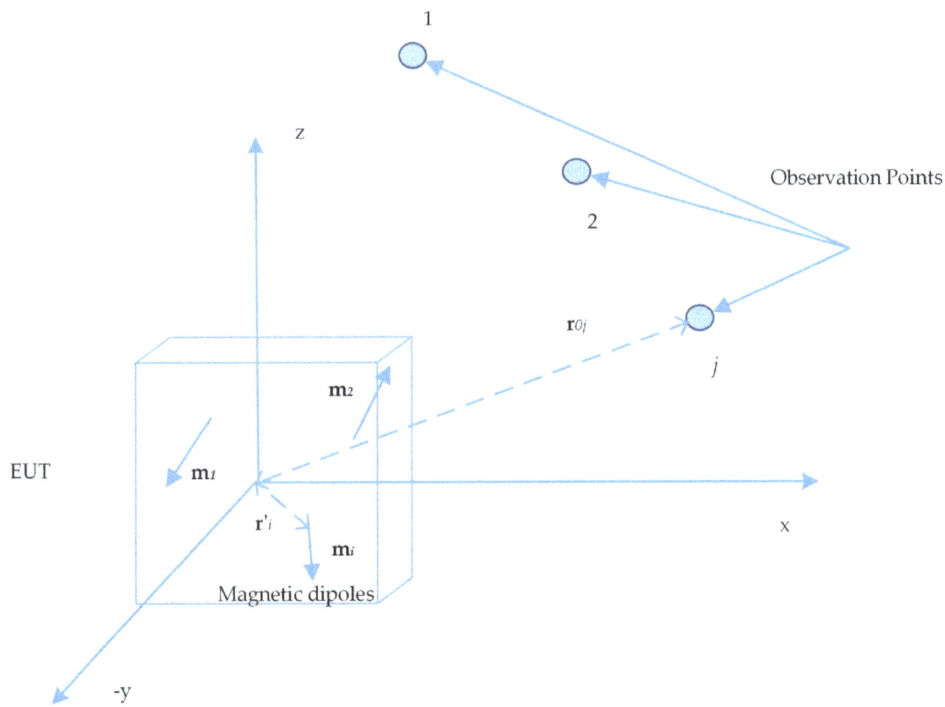

$$\begin{aligned} Bx_{ij} &= \frac{\mu_0}{4\pi}\left[\frac{3\left(x_{0j}-x_i\right)}{\left|\mathbf{r}_{0j}-\mathbf{r'}_i\right|^5}L_{ij}-\frac{m_{i_x}}{\left|\mathbf{r}_{0j}-\mathbf{r'}_i\right|^3}\right] \\ By_{ij} &= \frac{\mu_0}{4\pi}\left[\frac{3\left(y_{0j}-y_i\right)}{\left|\mathbf{r}_{0j}-\mathbf{r'}_i\right|^5}L_{ij}-\frac{m_{i_y}}{\left|\mathbf{r}_{0j}-\mathbf{r'}_i\right|^3}\right] \\ Bz_{ij} &= \frac{\mu_0}{4\pi}\left[\frac{3\left(z_{0j}-z_i\right)}{\left|\mathbf{r}_{0j}-\mathbf{r'}_i\right|^5}L_{ij}-\frac{m_{i_z}}{\left|\mathbf{r}_{0j}-\mathbf{r'}_i\right|^3}\right] \end{aligned} \tag{2}$$

where the dot product L_{ij} is:

$$L_{ij} = m_{i_x}\left(x_{0j}-x_i\right)+m_{i_y}\left(y_{0j}-y_i\right)+m_{i_z}\left(z_{0j}-z_i\right) \tag{3}$$

and $\left|\mathbf{r}_{0j}-\mathbf{r'}_i\right|$ is the distance between the observation point and the source:

$$\left|\mathbf{r}_{0j}-\mathbf{r'}_i\right| = \sqrt{\left(x_{0j}-x_i\right)^2+\left(y_{0j}-y_i\right)^2+\left(z_{0j}-z_i\right)^2} \tag{4}$$

The total magnetic field from the M magnetic dipoles at the j observation point will be the superposition of each individual contribution:

$$\mathbf{B}_j\left(t\right) = \sum_{i=1}^{M} Bx_{ij}\,\hat{\mathbf{x}} + \sum_{i=1}^{M} By_{ij}\,\hat{\mathbf{y}} + \sum_{i=1}^{M} Bz_{ij}\,\hat{\mathbf{z}}. \tag{5}$$

The magnetic field vector is expressed in components, since the measuring fluxgate sensors typically provide precision measurements of static and alternating magnetic fields in three axes $\left(B_x, B_y, B_z\right)$. Several types of sensors, however, measure the amplitude of the magnetic field:

$$|\,\mathbf{B}_j\,| = \sqrt{Bx_j^{\,2}+By_j^{\,2}+Bz_j^{\,2}}. \tag{6}$$

By employing the above equations (2) – (5), the magnetic signature of typical magnetic sources can be obtained. Specifically, the theoretical (without noise) magnetic field of a dipole and a quadrupole is calculated at a circle with radius 20 cm located at $z = 0$ cm. The observation points are chosen in order to simulate the existing mobile coil facilities (MCF), where rotational measurements with 3-axes fluxgate magnetometers are employed. Assuming that one dipole is located at (1, 1, 1) cm, having a magnetic moment of (0, 0, 30) mAm^2, the magnetic field at a circle of radius 20 cm can be calculated in components (B_x, B_y, B_z) at angles with 10° step. The resulting magnetic behavior of the virtual dipole source is depicted in Figure 2.

Similarly, the magnetic signature of a quadrupole magnetic source located at (1.35, 1, 1) cm can be calculated. The quadrupole is composed of two dipoles in close proximity (distance between them is 7 mm on the x-axis), having anti-parallel

Figure 2. Magnetic signature of a dipole source

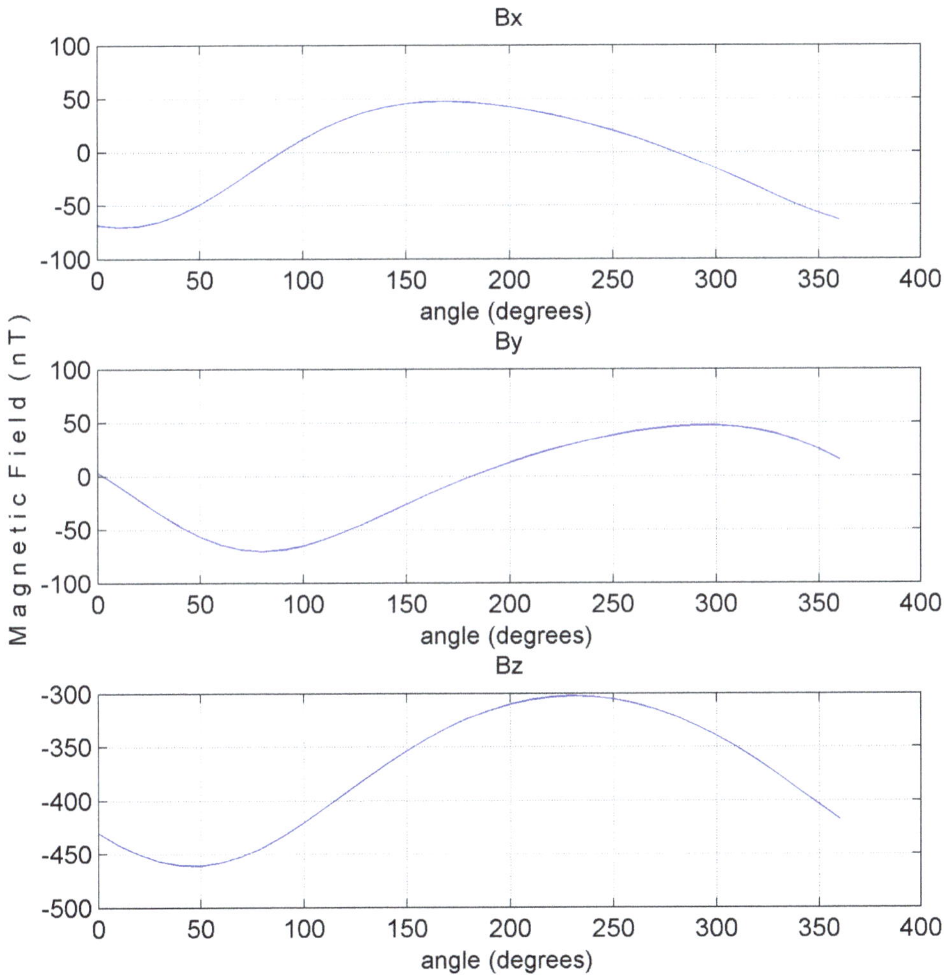

magnetic moments of (0, 0, 30) mAm^2 and (0, 0, -30) mAm^2 respectively. This configuration of the two dipoles gives a magnetic quadrupole moment of about 0.15 mAm^3 (Spantideas, Kapsalis, Kakarakis, & Capsalis, 2014). The magnetic quadrupole field at a circle of radius 20 cm and $z = 0$ cm can be calculated in components (B_x, B_y, B_z), as shown in Figure 3.

Evidently, the magnetic signatures of magnetic dipole and magnetic quadrupole differ significantly. Moreover, it is worth mentioning that the magnetic field fall-off of the dipole follows the r^{-3} distance power law, whereas the power law due to the quadrupole is r^{-4}. These simulation results show that the quadrupole's contribution

Figure 3. Magnetic signature of a quadrupole source

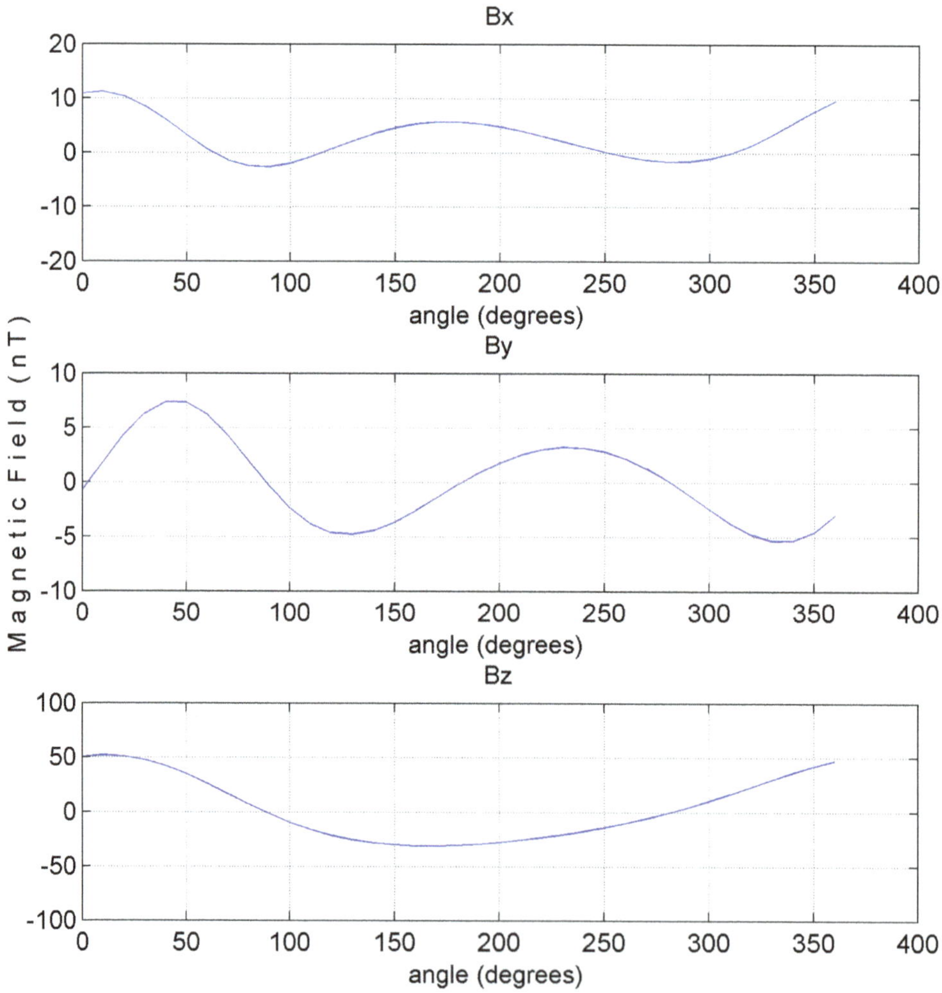

to the magnetic field cannot be neglected when demanding levels of magnetic cleanliness are required.

From a wider perspective, the MDM problem belongs to the category of inverse non-linear problems. The measured magnetic field data $(B_{j_x}, B_{j_y}, B_{j_z})$ at the location of the N_{MAG} measurement points $\mathbf{r}_{0j} = \left(x_{0j}, y_{0j}, z_{0j}\right)$, $j = 1,2,\ldots N_{MAG}$, are fitted to a small number of equivalent magnetic dipole sources. Therefore, the parameters of the dipole source(s) that generate an identical to the measured magnetic field need to be identified. The parameters of each dipole source include its position (x, y, z) and magnetic moment vectors (m_x, m_y, m_z).The fitting process can be performed with nonlinear programming methods(Mehlem, Wiegand, & Weikert, 2012)or stochastic techniques, e.g. (Kakarakis, Kapsalis, & Capsalis, 2015). A brief description of the Particle Swarm Optimization (PSO) entities and parameters is presented in the following section, along with the detailed discussion on the implementation of the stochastic method on the MDM problem and the evaluation criterion that is utilized to assess the resulting models (goodness of fit).

PARTICLE SWARM OPTIMIZATION (PSO) TECHNIQUE

PSO: A Short Overview

PSO is a stochastic optimization method based on the movement of a swarm of bees (particles) in a field (Kennedy, 1995). The particles (or agents) begin at random positions in a given search space and exhibit random velocities. Their velocities and positions are updated in each iteration of the algorithm, taking into consideration not only the personal optimum detected by each particle, but also the global optimum encountered by the entire swarm. Eventually, the particles converge towards the global optimum solution. The PSO related terms are briefly outlined below:

1. **Particle** or **Agent:** each individual in the swarm. All particles start from random locations and move towards a direction that is calculated taking into account both the (i) personal and (ii) global optimum.
2. **Position:** position refers to a particle's place in the search space, represented by a (M×6) vector **D**. The elements of vector **D** include the variables to be optimized. Specifically, in the case of MDM problem, these parameters are the position and magnetic moment vector of the magnetic dipole sources. Therefore, the vector $\mathbf{D}_i$ has elements $(x_i, y_i, z_i, m_{x_i}, m_{y_i}, m_{z_i})$, $i = 1,2,\ldots M$.

This dimensional space represents the solution space for the MDM optimization problem, where any set of M dipoles' parameters represents a solution.

3. **Goodness of Fit**: a function must be determined in order to assess the current position of each particle. This fitness function takes into consideration the positions of all agents in the solution space and returns a unique number evaluating the current solution. In the present work, the fitness function represents the relative deviation between the measured and modeled magnetic field values. The particles' goal is to minimize this function, thus converging to an adequate modeling solution.
4. ***pbest*:** the personal best indicates the position $(\mathbf{D}_i)$ with the lowest fitness encountered by each particle. As each agent explores the search space, it compares the fitness value of its current location to that of its personal best. If the current location has a lower fitness value, *pbest* is replaced with its current position. In every iteration of the algorithm, each agent exhibits only one *pbest,* i.e. the position with the lowest fitness that the individual has encountered.
5. ***gbest*:** the position $(\mathbf{D}_i)$ with the lowest fitness value encountered by all particles is known as the global best or *gbest*. As the particles move through the search space, they compare the fitness value of their location to that of the global optimum. If they find a position with lower fitness, *gbest* is then replaced with the current location. In each iteration of the PSO algorithm, the swarm has encountered only one global optimum solution with the lowest fitness value, i.e. *gbest*.

Implementation of PSO on the MDM Problem

After the evaluation of *pbest* (the fittest vector detected by each particle) and *gbest* (the fittest vector encountered by all agents in the swarm) the new particles' velocities are calculated via (Kennedy, 1995):

$$u_{in} = w \cdot u_{in} + c_1 \cdot rand \cdot \left(pbest_{in} - p_{in}\right) + c_2 \cdot rand \cdot \left(gbest_i - p_{in}\right), \tag{7}$$

where $n = 1,2,\ldots N$ represents each of the population size N of the swarm, w expresses the weight of the velocity of the previous iteration, c_1, c_2 are factors adjusting the tradeoff between the exploration of the search space and the exploitation of the possible optimal solution and $rand$ takes follows the uniform distribution over [0,1]. Finally, p_{in} stands for the position vector of element i of particle n.

Once the updated velocity of each agent has been calculated, its position is updated via:

$$p_{in} = p_{in} + u_{in}. \tag{8}$$

Simulation results and previous experience have shown that the values of the parameters that are suitable for solving MDM optimization problems may be regulated to $c_1 = c_2 = 2$ and $w = 0.5$ (Kapsalis, Kakarakis, & Capsalis, 2012).

Taking into account that the measurements are performed with tri-axial sensors (fluxgate magnetometers), initially a $(3 \times N_{MAG})$ matrix $\mathbf{B}_{meas}$ is introduced to accommodate the values of the measured magnetic field components $(B_{j_x}, B_{j_y}, B_{j_z})$.

As the proposed scheme unfolds, the agents of the PSO based method explore the search space in each iteration for the optimal solution and the algorithm yields the global optimum – *gbest* – parameters of the dipole sources (3 for each dipoles' position and 3 for its magnetic moment) that fit the measured magnetic field values $\mathbf{B}_{meas}$. The algorithm's goal is to minimize the function:

$$G = \sum_{j=1}^{N_{MAG}} \sum_{k=1}^{3} \left(\mathbf{B}_{gbest_{kj}} - \mathbf{B}_{meas_{kj}} \right)^2, \tag{9}$$

where k stands for the x, y, z components and the $(3 \times N_{MAG})$ matrix $\mathbf{B}_{gbest}$ contains the modeled magnetic field values corresponding to *gbest*. The stopping criterion of the algorithm is usually a small value of the function G (in the order of 1e-10) or alternatively, a maximum number of iterations. Additionally, a function is calculated to incorporate the measured magnetic field values from all measuring sensors:

$$F = \sum_{j=1}^{N_{MAG}} \sum_{k=1}^{3} \left(\mathbf{B}_{meas_{kj}} \right)^2 \tag{10}$$

Finally, in order to define an evaluation criterion for the *gbest* optimal solution, a function G_{rel} is introduced; the function readily represents the relative deviation between the measured and modeled magnetic field values and is used to assess the resulting model (goodness of fit). Low values of G_{rel} (typically below 10%) indicate that the dipole model can virtually represent the magnetic field of the EUT.

$$G_{rel}\left(\%\right)=\sqrt{\frac{G}{F}}\cdot 100 \tag{11}$$

Figure 4 depicts the flowchart of the implemented optimization method on the MDM problem.

The resulting global optimal parameters of the PSO algorithm will therefore correspond to the model that best reproduces the measured magnetic field values and

Figure 4. Flowchart of the PSO stochastic technique

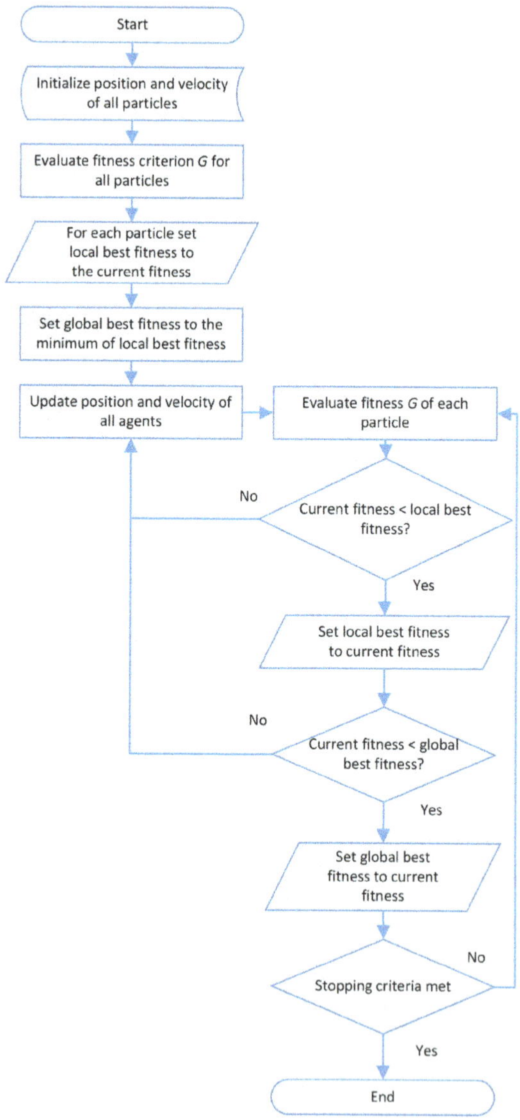

may be virtually used to represent the original unit or equipment. In the following section, the presented methodology is applied to real magnetic field measurements from dipole sources by employing a multi-magnetometer facility consisting of 12 fluxgate sensors.

DC MULTIPLE MAGNETIC DIPOLE MODELING

In this section, the proposed measurement setup and the test measurement methodology of the configuration that employs multiple magnetometers in order to simultaneously capture the magnetic signature of a unit is presented. Moreover, the demonstrated setup and methodology are validated via real measurements of simple magnetic dipole sources. The test cases that were measured and characterized include (i) one dipole with small magnetic moment and (ii) two dipoles with parallel magnetic moments, separated by approximately 20 cm. It should be noted that the presented results are indicative, in the sense that the aforementioned magnetic sources were repeatedly measured and characterized at various positions on the turntable.

Measurement Setup and Methodology

The measurements' configuration setup is shown in Figure 5. It consists of 12 sensorsthat simultaneously measure the components of static and low frequency magnetic fields. These sensors typically have a sensitivity ranging from DC up to a few KHz. Practical experience has shown that real spacecraft equipment may be modeled with up to 5 magnetic dipoles. As already mentioned, 6 parameters need to be determined by the algorithm for each dipole. Therefore, at least 30 independent equations are required in order to estimate accurately the parameters of the model. Since each sensor provides 3 measured values (2) – (5), the number of the sensors located at different positions (independent equations) should be at least 10. In the present facility, however, 12 sensors are used for redundancy purposes. As shown in Figure 5, the magnetometers form two circles above and below the turntable in order to capture the EUT's magnetic signature from several possible angles. In this configuration, both circles have a radius of r = 20 cm and are positioned at different heights (z =10and z =-10cm for the upper and lower circle respectively). The angle difference between two successive magnetometers is fixed at 60°. The choice of observation distance of the magnetometers is appropriate for capturing the magnetic signature of sources with weak magnetic behavior, but may be rescaled according to SNR demands and dimensioning of the unit limitations. Ideally, measurements on a sphere around the EUT would ensure the total coverage of the EUT's magnetic signature, but considerably additional measurement points are

Figure 5. Measurement setup consisting of 12 sensors

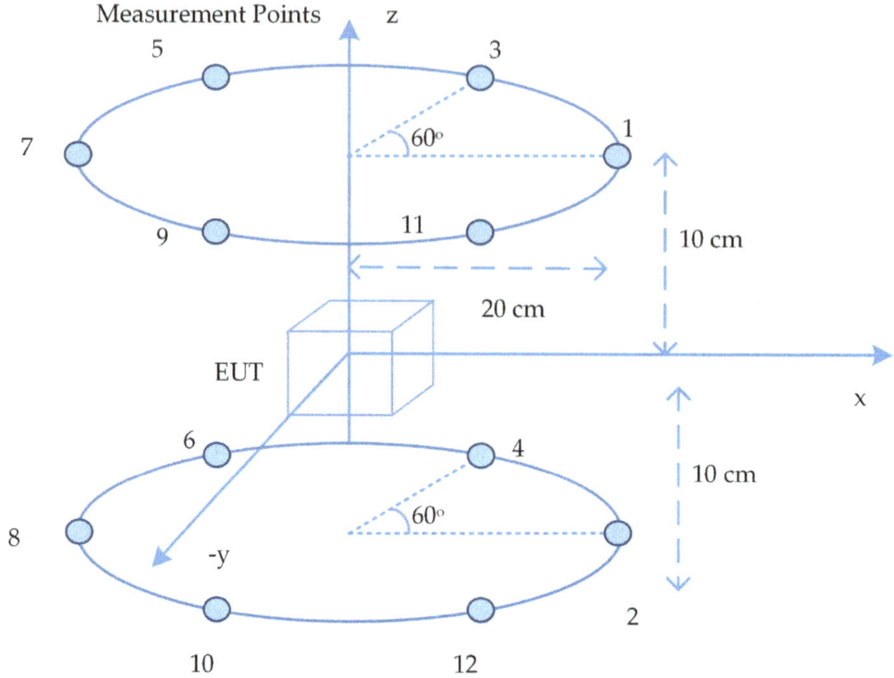

required. In the present work, the upper and lower planes of the configuration in Figure 5 indicate that the magnetometers are located provide better coverage of the magnetic signature of the unit from several possible angles. Other configurations may be implemented with the use of multiple magnetometers – for instance three circles, helix, etc. Finally, it is worth mentioning that no calibration procedure for the position of the 12 magnetometers is implemented.

The noticeableaspect of this configuration is the simultaneous capture from all 12 fluxgate sensors of the magnetic field of an EUT in the presence of Earth's magnetic field. Specifically, existing MCF facilities use a pair of Helmholtz coils to compensate the DC component of Earth's magnetic field that has approximate amplitude of around 50 μT. However, disturbances of the ambient field during the rotation of the turntable that the EUT is placed on (Earth's magnetic field variations, noise from power lines, nearly technical equipment, etc.) commonly require several repetitions.On the contrary, the test methodology of this measurement setup includes three successive measurements: (i) ambient field measurement before the placement of the EUT $(\mathbf{B}_{bef})$, (ii) capture of EUT's magnetic field $(\mathbf{B}_{EUT} + \mathbf{B}_{ambient})$, and (iii) removal of the EUT and ambient field measurement $(\mathbf{B}_{aft})$.

In principle, the total measurement time is very short (in most cases less than 1 min) in order to avoid significant environmental field disturbances. The average value of the ambient fields before and after the EUT measurement:

$$\frac{\mathbf{B}_{bef} + \mathbf{B}_{aft}}{2} \tag{12}$$

is then used to effectively compensate the ambient field during the EUT measurement.

In principle, the location of the real magnetic sources inside an EUT, as well as their magnetic moments, are a priori unknown. Consequently, in order to validate the proposed setup and modeling methodology a series of simple test cases are measured and characterized in the following paragraphs, namely a single dipole source and two dipole sources. It should be noted that the position of the dipole sources on the table is known with significant accuracy (in the order of mm), whereas their magnetic moments are a priori unknown. The accuracy of the prediction is mainly determined by the relative deviation between the estimated position of the test object and the real position of the dipole, as well as the goodness of fit criterion G_{rel}. Low values of the position deviation between the real and the modeled magnetic dipoles (typically less than 5 mm) and of the goodness of fit evaluation criterion (typically less than 10%) will ensure the accuracy of the prediction and validate the measurement configuration and the modeling process.

Validation With Magnetic Measurements of a Single Dipole Source

Initially, it is assumed that a simple EUT is composed of a single dipole source. The single dipole that exhibits a weak magnetic signature, is placed on the table and measured, following the test measurement methodology described in the previous subsection. For purposes of obtaining higher SNR, the circles' radii of the measurement configuration are reduced to 10 cm from the center of the turntable for both planes. Thereafter, the measured magnetic field values are delivered to the PSO modeling method in order to fit them with an equivalent modeled magnetic dipole. The validation procedure indicates that the parameters of the predicted and real dipoles are similar in terms of position and magnetic moment vectors, whereas the low value of goodness of fit evaluation criterion will eventually verify the fitting process. Depending on the parameters of the PSO stochastic technique, presented in previous section, the proposed methodology can be further generalized for purposes of estimating multiple dipole models. Extensive results for one measured test case are presented in the present subsection, focusing on the difference between the measured and the modeled magnetic fields, as well as the difference between the theoretical and predicted models. The dipole is positioned at (7.5, 2.5, 0.9) cm on the turntable and its magnetic moment is expected to be aligned only on one axis, while being negligible on the other two axes. The measured, as well as the modeled magnetic field values are tabulated in Table1.

Moreover, in Table 2 the estimated parameters of the model are presented and in Table 3 the calculated evaluation goodness of fit criterion G_{rel} is tabulated.

Table 1. Measured and Modeled magnetic field values of a single dipole source

	B_x [nT]		B_y [nT]		B_z [nT]		$\|B\|$ [nT]	
	Measured	**Predicted**	**Measured**	**Predicted**	**Measured**	**Predicted**	**Measured**	**Predicted**
Mag1	65.8	60.6	240.6	242.1	186.5	183.6	311.4	309.8
Mag2	40.4	41	-176	-176	-115.9	-115.9	214.6	214.7
Mag3	-62.1	-61.8	91.9	96.2	118.2	118.6	162.1	164.8
Mag4	-46.7	-47.1	-88.9	-88.9	-87.2	-85.7	133	132.2
Mag 5	-13.1	-13.4	-5.4	-6.7	33.9	33.7	36.7	36.9
Mag6	-12.9	-12.8	-0.5	-1.3	-30.8	-31.7	33.4	34.2
Mag7	-2.8	-1.8	-8.3	-9.1	18.3	19.1	20.3	21.2
Mag8	-2	-1.9	4.6	4.7	-19	-19.3	19.7	20
Mag9	3.7	4.3	-8.4	-9.2	20.5	21.5	22.5	23.8
Mag10	4.9	4.4	3.2	4.1	-20.9	-21.5	21.7	22.3
Mag11	24.4	24	-0.9	1.8	50.5	48.9	56.1	54.5
Mag12	22.4	21.9	-12	-12.2	-40.8	-43.3	48.1	50

Table 2. Model Prediction – Dipole with weak magnetic moment

	x [cm]	y [cm]	z [cm]	m_x [mAm²]	m_y [mAm²]	m_z [mAm²]	$\|m\|$ [mAm²]
Theoretical Model	7.5	2.5	0.9	unknown	unknown	unknown	unknown
Predicted Model	7.4	2.6	0.9	-0.003	-0.002	-1.94	1.94

Table 3.Goodness of fit (G_{rel}) – Dipole with weak magnetic moment

Total RMS of residuals [%]	**Total RMS of residuals [nT]**
2.1	1.531

The resulting parameters of the MDM for the EUT – dipole shown in Table 2 are similar and close to the theoretical values in terms of predicted position (x, y, z), since the relative deviation between the estimated position of the test object and the real position of the dipole is approximately 1.4 mm, ensuring the accuracy of the model's prediction. Moreover, the predicted magnetic moment of the EUT is oriented towards the z–direction, as excepted, and is positive when the magnet is placed upwards or negative when the magnetic is placed downwards, while its other vector components are negligible. Finally, the low value of the goodness of fit G_{rel} evaluation criterion tabulated in Table 3 (about 2%) confirms the effectiveness of the fitting process and thus, ascertains that the EUT can be efficiently represented by the predicted model. Finally, it should be noted that the magnetic field values are stronger in magnetometers 1-4, since these sensors are closer to the magnetic source and the magnetic dipole field follows the r^{-3} distance power law. Although the significant values of the magnetic field tend to dominate the fitting process, it is evident that the deviation between the measured and predicted magnetic fields is negligible for all fluxgate sensors. In Figures 6-8, the measured and modeled magnetic field values are compared for all three components of the field. Evidently, the model achieves an effective representation for all components of the magnetic field generated by the dipole.

Validation With Magnetic Measurements of Two Dipole Sources

In order to be able to characterize real spacecraft equipment, the developed test measurement methodology, the measurement configuration setup and the modeling method are required to manage even more challenging test cases, i.e. EUTs composed of more than one dipole. For this purpose, it is assumed that a more complex EUT is composed of two dipoles sources with parallel magnetic moments, separated by approximately 20 cm. The dipoles are positioned at (7.5, 7.5, 0.1) and (-12.5, 7.5, 0.1) on the turntable and their magnetic signature is obtained. The magnetometers are now located at two circles with radii 20 cm, since the generated magnetic field of the above EUT is stronger and the SNR is acceptable. In this case, 12 parameters of two dipoles (6 for each dipole) need to be determined by the method. Although the prediction of more than one dipole becomes increasingly challenging, the developed MDM methods must cope with these demanding requirements, since most spacecraft equipment consist of several magnetic sources. The measured and modeled magnetic field values are tabulated in Table 4, along with the model prediction and the goodness of fit criterion in Tables 5 and 6, respectively.

Firstly, it should be noted that although the measurement distance is increased, the measured magnetic field values at all magnetic sensors are significant in terms

Figure 6. Measured and modeled x – component of the magnetic field at the 12 sensors

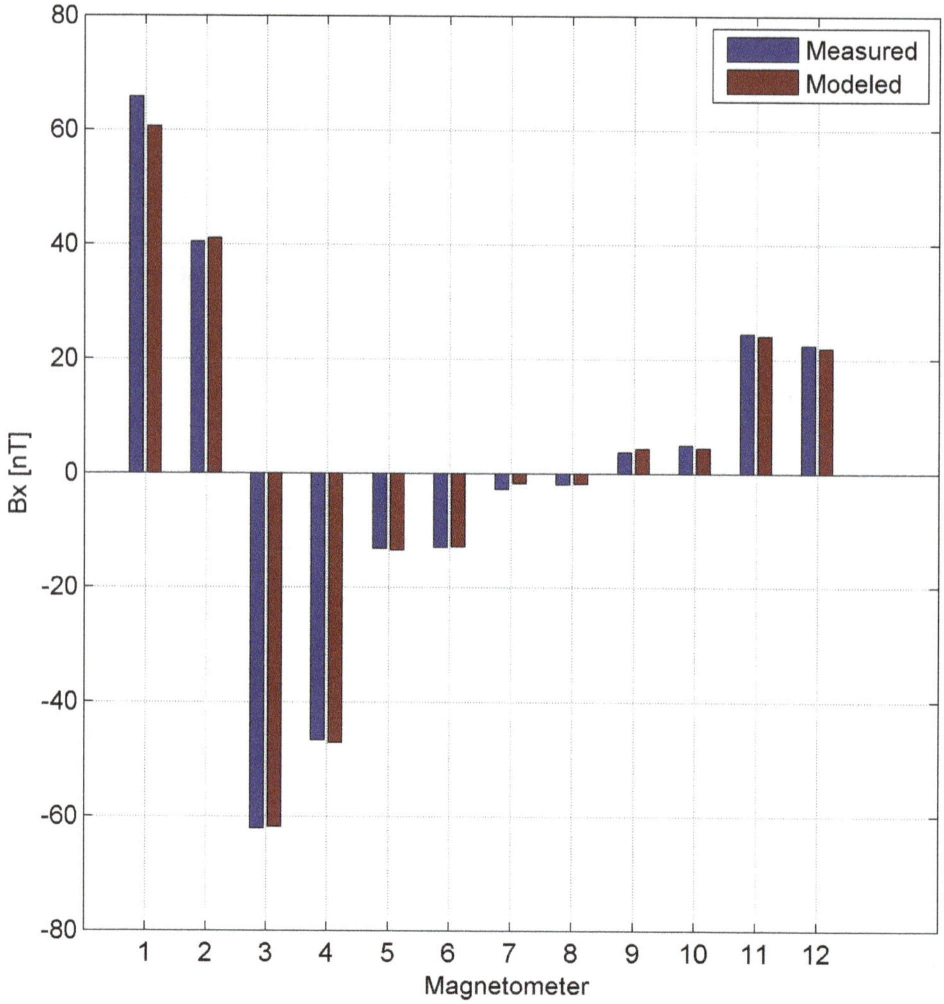

of SNR, since the magnetic dipole moments of the two sources are strong. The deviation between the measured and the modeled magnetic field values is negligible for all 12 magnetometers and for all three components of the magnetic signature, as tabulated in Table 4.Moreover, the relative deviation between the measured and modeled magnetic fields (less than 2%) and the maximum deviation between the theoretical and the predicted positions of the EUT (approximately 2.5 mm) readily confirm the capability of the developed method to overcome more challenging cases

Figure 7. Measured and modeled y – component of the magnetic field at the 12 sensors

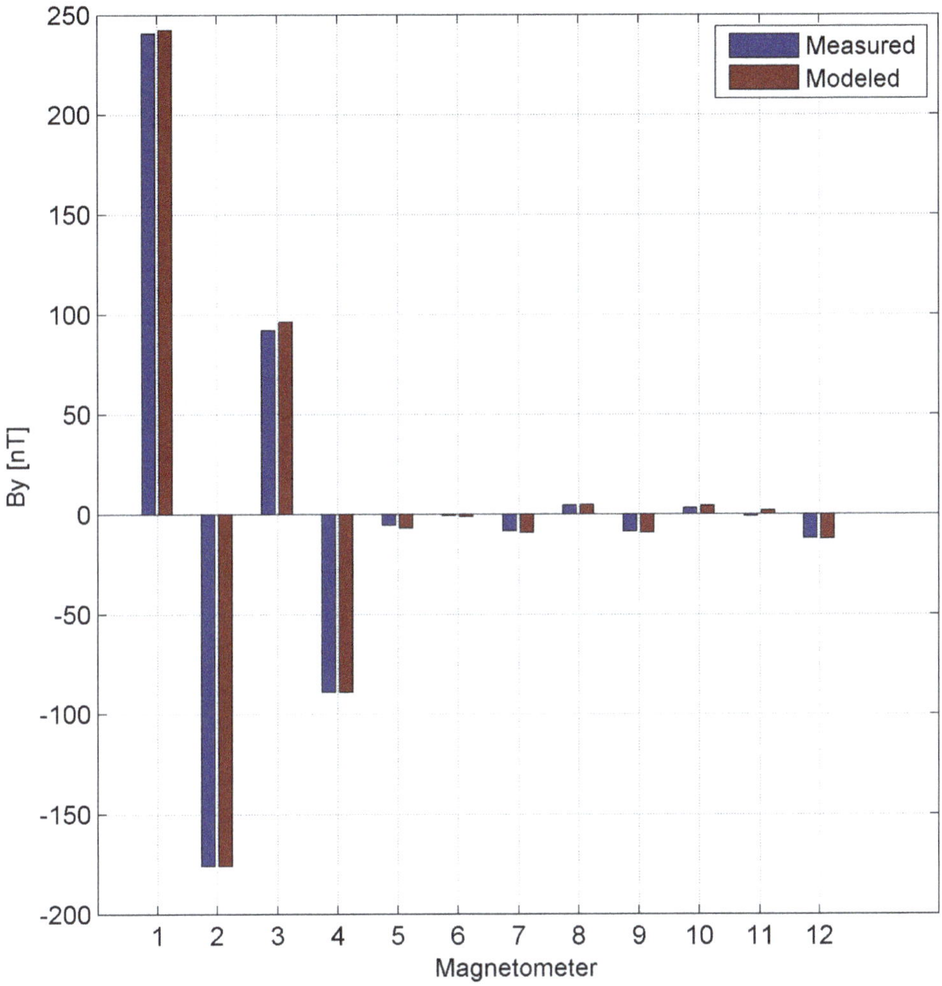

and effectively estimate the parameters of the two dipoles that can be employed to represent the EUT and reproduce its magnetic signature.

It is worth mentioning that the aforementioned modeling procedure may also be implemented on measurements of induced magnetic fields. In the following chapter, specific considerations are discussed regarding the test measurements of induced fields and the proposed methodology for modeling the interaction between different units of spacecraft equipment is described.

Figure 8. Measured and modeled z – component of the magnetic field at the 12 sensors

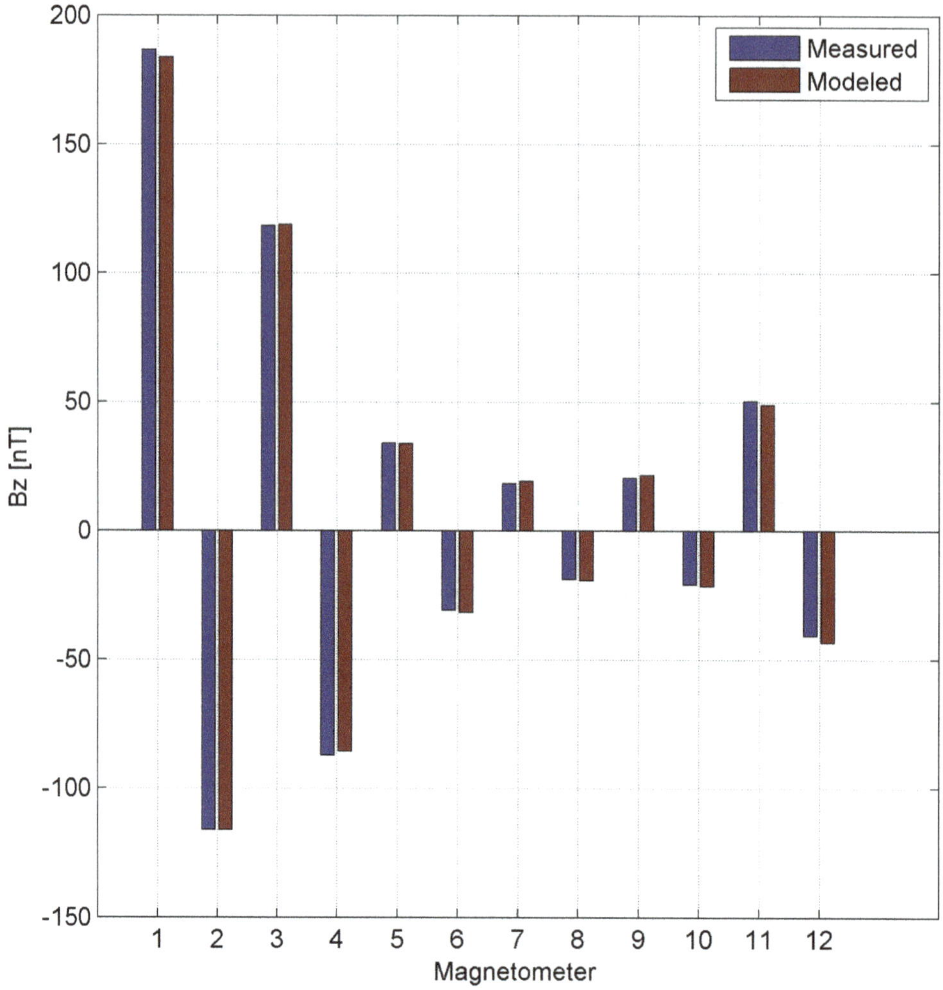

MODELING METHODS FOR INDUCED DC MAGNETIC FIELDS

The response of various EUTs to an applied external magnetic field may generate additional magnetic behavior. In the presence of external field(s), several EUTs exhibiting conducting and/or magnetic properties produce induced magnetic fields that may also cause disturbances at the sensitive instruments and contaminate their measurements.

Table 4. Measured and Modeled magnetic field values of two dipole sources

	B_x [nT]		B_y [nT]		B_z [nT]		$\|B\|$ [nT]	
	Predicted	**Measured**	**Predicted**	**Measured**	**Predicted**	**Measured**	**Predicted**	**Measured**
Mag1	107.8	103.4	-187.5	-201.1	415.1	417.4	468.1	474.7
Mag2	208.1	207.3	18.6	33.3	-487.7	-487.8	530.6	531.1
Mag3	13.8	15.9	3.8	9.6	989.5	989.6	989.6	989.8
Mag4	-155.9	-157.1	-277.2	-283.6	-1059.7	-1047.7	1106.4	1096.7
Mag 5	429.2	429.3	261.1	250.8	1153	1152.9	1257.7	1255.5
Mag6	368.9	360.6	-137.3	-122.6	-1037.1	-1047.1	1109.3	1114.2
Mag7	-497.7	-504.6	199.2	198.7	964.6	962.7	1103.6	1105
Mag8	-438.7	-449	-197.3	-199.9	-951	-945.1	1065.7	1065.2
Mag9	-55.9	-55.4	-139.6	-143.6	166.4	170.5	224.3	229.7
Mag10	-48.3	-47.2	137.5	147.8	-160.1	-153.8	216.5	218.4
Mag11	6.7	9.2	-132.1	-130.5	147.4	149.2	198	198.4
Mag12	38.3	37.2	122.4	129.1	-136.8	-136.3	187.5	191.4

Table 5. Model Prediction – Two dipoles with parallel magnetic moments

		x [cm]	y [cm]	z [cm]	m_x [mAm²]	m_y [mAm²]	m_z [mAm²]	$\|m\|$ [mAm²]
Theoretical Model	Dipole 1	7.5	7.5	0.1	unknown	unknown	unknown	unknown
	Dipole 2	-12.5	7.5	0.1	unknown	unknown	unknown	unknown
Predicted Model	Dipole 1	7.3	7.6	0.2	4.847	1.467	-28.11	28.57
	Dipole 2	-12.5	7.3	0.1	0.66	-1.123	-31.39	31.41

Table 6.Goodness of fit (G_{rel}) –Two dipoles with parallel magnetic moments

Total RMS of residuals [%]	**Total RMS of residuals [nT]**
1.6	5.92

Induced DC Measurements

Apart from a pair of Helmholtz coils to compensate Earth's magnetic field, existing MCFs typically employ another pair of Helmholtz coils that generates a (spatially) uniform external magnetic field at the volume of the EUT exhibiting conducting and/or magnetic properties, in order to measure and characterize the unit's induced magnetic behavior (ECSS, 2012). Typical levels of the DC applied magnetic field are +/- 50 μT and +/- 100 μT. It should be noted that several tests are conducted with different amplitude levels of the applied field for purposes of testing the linear response of the unit. In fact, various units are composed of magnetic materials (e.g. iron) and their response can be highly non-linear, exhibiting hysteresis effects. In these cases, a more detailed characterization of the unit is required and magnetic measurements are performed with several amplitude levels of the applied field. The tests are also conducted with applied external field in all 3 directions to measure the induced magnetization of the EUT along the 3 orthogonal axes.

Without loss of generality, it is assumed that the induced magnetic response can be modeled with a single equivalent induced magnetic dipole and its magnetic moment aligns to the direction of the external magnetic field. Thus, when an external field $(\mathbf{B}_{app}+)$ is applied vertically, the direction of the induced magnetic dipole moment stays constant during the EUT's rotation. If the external field is applied vertically in the opposite direction $(\mathbf{B}_{app}-)$, the remanent magnetic dipole's orientation stays constant. The induced dipole moments, however, will align to the new orientation of the field. In this way, the separation of the induced and remnant magnetic field contributions can be practically achieved. Specifically, the measured magnetic field values regarding the two configurations of the test are $(\mathbf{B}_{remn+ind}+)$ for positive and $(\mathbf{B}_{remn+ind}-)$ for negative orientation of the applied field.

The induced magnetic field measurements may be performed by employing the multiple magnetometer configuration depicted in Figure 5 and its associated test measurement methodology with the exception that a pair of Helmholtz coils must be used to generate the external magnetic field and control its uniformity. Thereafter, the measured induced magnetic field values can be modeled according to the well-established MDM methods and the PSO technique may be employed to estimate the magnetic moments $(\mathbf{m}_{\mathbf{B}_{app}}-)$ and $(\mathbf{m}_{\mathbf{B}_{app}}+)$, depending on the orientation of the magnetic field $(\mathbf{B}_{app})$, as described in a previous section. Then, vector analysis can be used to isolate the induced magnetic dipole moment of the EUT, as depicted in Figure 9.

Figure 9. Induced magnetic moment measurement and characterization

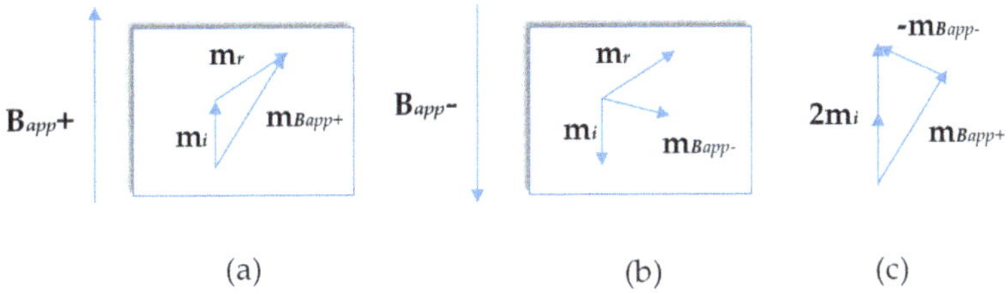

The difference between the moments $(\mathbf{m}_{\mathbf{B}_{app}}+)$ and $(\mathbf{m}_{\mathbf{B}_{app}}-)$ basically yields twice the induced magnetic moment, while the remnant magnetic moments cancel out:

$$\mathbf{m}_i = \frac{(\mathbf{m}_{\mathbf{B}_{app}}+)-(\mathbf{m}_{\mathbf{B}_{app}}-)}{2} \tag{13}$$

The characterization of the induced magnetic behavior of each EUT that will be mounted on the spacecraft is essential and should be taken into account in missions that demanding levels of magnetic cleanliness are required. Moreover, the induced magnetic properties will allow the accurate prediction of the in-flight performance of the equipment, since the total magnetic field generated by each EUT is the superposition of the individual contributions of all its magnetic sources. Finally, the generated magnetic field from surrounding equipment at the location of an EUT that exhibits conducting and/or magnetic properties is basically treated as external field that may additionally cause its induced response. The unit to unit interaction is consequently a critical issue in most space programs and the proposed modeling methodology is analyzed in the following subsection.

Unit – to – Unit Interaction

Except for the Low Earth Orbit (LEO) missions, where the spacecraft is located in Earth's vicinity and therefore inside the magnetosphere, several space missions exhibits the "so-called" zero-field conditions, i.e. the external magnetic field is in the order of pT. Nevertheless, external magnetic fields generated by surrounding equipment inside the vehicle are still present at the location of EUTs that possibly exhibit conducting and/or magnetic properties. Considering the interaction between two EUTs, let $\mathbf{B}_{1n}$ be the generated magnetic field of unit 1 at the location of unit 2, as depicted in Figure 10.

Figure 10. First order interaction between two units

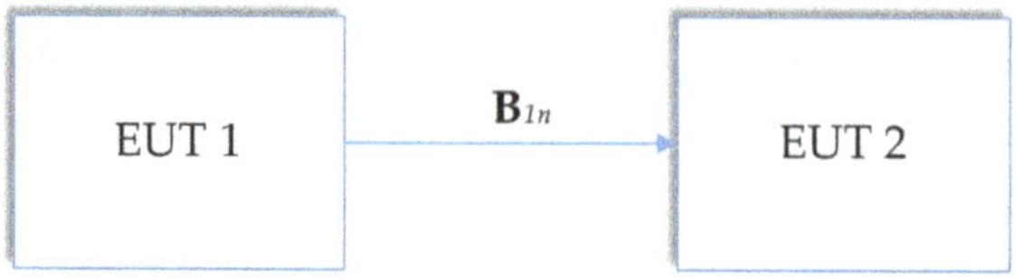

The magnetic field $\mathbf{B}_{1n}$ can be treated as external field at the location of unit 2, typically resulting in a two dipole model for EUT2, as shown in Figure 11. The magnetic dipole moment $\mathbf{m}_r$ represents the remnant magnetic properties of the unit, while the moment $\mathbf{m}_i$ represents the equivalent induced magnetic dipole moment of EUT2 in the presence of the applied external magnetic field $\mathbf{B}_{1n}$, originating from EUT1.

It is worth mentioning that the two magnetic moments $\mathbf{m}_r, \mathbf{m}_i$ are in general located at (x_1, y_1, z_1) and (x_2, y_2, z_2), i.e. different positions inside the volume of the EUT2. If the induced characterization test campaign has been successfully conducted, the estimation of the equivalent induced magnetic dipole's parameters $\mathbf{m}_2$ can be carried out, depending on the amplitude and direction of the external field $\mathbf{B}_{1n}$. Moreover, by employing the parameters of the induced magnetic dipole $(x_2, y_2, z_2, m_{x_2}, m_{y_2}, m_{z_2})$, the magnetic field originating from both dipoles can be calculated and summed up vectorially, allowing in-flight prediction of the EUT2's complete magnetic behavior.

The second order interaction from unit 2 to unit 1 is shown in Figure 12. EUT1 can be also modeled with two dipoles; dipole 1 represents the remnant magnetic properties of EUT1, while dipole 2 represents the induced magnetization of EUT1 in the presence of the external field originating from unit 2 $(\mathbf{B}_{2n} + \mathbf{B}_{2ind})$

Usually, the first order interactions are adequate to estimate the combined magnetic field originating from both remanent and induced magnetic dipoles (Junge & Marliani,

Figure 11. Two-dipole model for a unit, representing the remnant and induced magnetizations

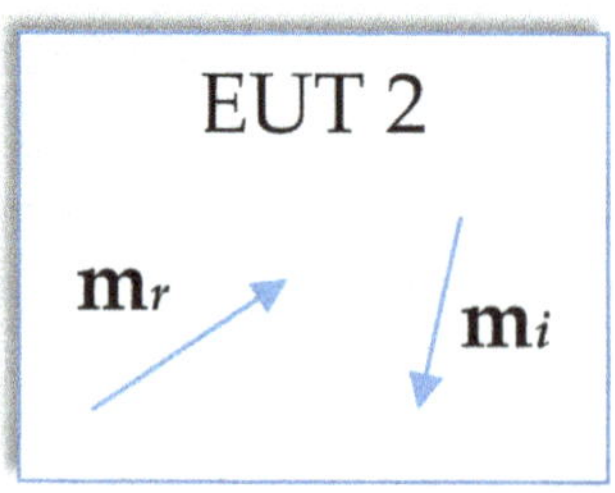

Figure 12. Second order interaction between two units

2011). Since the remnant magnetic properties of most spacecrafts and equipment on-board are consistently kept low or completely avoided in the spacecraft design, the contribution of induced magnetic dipoles in measurements' disturbance is expected to be significant. Practical experience has shown that the magnetic fields generated by induced dipoles can be at comparable levels to the remnant ones (e.g. Swarm). The demonstrated modeling method may be employed to accurately determine the unit-unit interactions and therefore allow precise in-flight predictions of the total magnetic field strength at the location of the magnetometers.

SOFTWARE BASED CALIBRATION ALGORITHM

As shown in the previous section, the parameters of the MDM can be accurately determined by employing near field measurements of the EUT's field and applying the PSO stochastic technique. It is a straightforward requirement that the location of the observation points (sensors) during the measurement test campaign of the EUT is precisely defined, since the estimation of the MDM's parameters involves the magnetometers' positions $\mathbf{r}_{0j}=\left(x_{0j},y_{0j},z_{0j}\right)$, $j=1,2,\ldots N_{MAG}$. In fact, the relative deviation of the position between the sensors and the EUT is one of the main contributors to the overall uncertainty of the model estimation (BIPM, IFCC, IUPAC, & ISO, 2008).

It is therefore critical to ascertain that the facility and the measuring configuration that are used to perform magnetic measurements of an EUT are calibrated in advance, in terms of magnetometers' position. Various calibration methods have been proposed in order to accurately determine the position of a test object, for instance calibration with laser or cameras (Hoang, Hernández, & Jo, 2014). Nevertheless, these methods are not always cost-effective.

In the present work, a software based calibration method is illustrated, employing the stochastic techniques. The method involves the snap-shot measurement of a well-defined magnetic source (current fed coil or permanent magnet) and then the PSO algorithm used for the MDM prediction, is modified and implemented on the calibration problem. The goal is to accurately determine the positions of the magnetometers in a facility before the EUT test campaign and thus, ensure the quality of the measurements and the estimation of the model. In this section, the proposed software based calibration technique is briefly described and is validated with real measurements. For this reason, a well-known calibration source (current fed coil) is employed, since its parameters are defined with significant accuracy. The MDM prediction problem entails the determination of the parameters of the source from near magnetic field measurements, utilizing the sensors at accurately specified positions. On the contrary, the proposed calibration methodology entails the determination of the position of the sensors, assuming that the parameters of the source are accurately known in advance.

Description of the Calibration Method

A measuring facility typically consists of one or several measurement points N_{MAG} that are positioned at $\mathbf{r}_{0j} = \left(x_{0j}, y_{0j}, z_{0j}\right)$, $j = 1, 2, \ldots N_{MAG}$. The position coordinates are expressed with respect to a reference point, typically the center of the turntable. Although these position coordinates are assumed to be well defined, their limited accuracy depends on various parameters (for instance the angular accuracy of the mechanical structure) and their resolution is in the order of mm or even cm. The precise magnetometer positions may thus deviate from (x_{0j}, y_{0j}, z_{0j}) by:

$$
\begin{aligned}
x'_{0j} &= x_{0j} + \Delta x_{0j} \\
x'_{0j} &= x_{0j} + \Delta x_{0j} \\
z'_{0j} &= z_{0j} + \Delta z_{0j},
\end{aligned}
\tag{14}
$$

where $(\Delta x_{0j}, \Delta y_{0j}, \Delta z_{0j})$ are displacement variables on each axis. Therefore, 3 parameters for each sensor need to be determined, namely its precise position coordinates $(x'_{0j}, y'_{0j}, z'_{0j})$.

The method involves the generation of a magnetic field from a magnetic dipole or quadrupole source with well-known specifications. Evidently, the test object's magnetic signature must remain unaffected of parameters such as temperature, humidity, etc. Moreover, its position (x, y, z) on the turntable must be precisely determined and its magnetic moment (m_x, m_y, m_z) well defined.

In contrast to the PSO method for model prediction, the position of each agent in the search space is now represented by a (1×3) vector $\mathbf{D}_j$ with elements (x_{0j}, y_{0j}, z_{0j}),

$j = 1,2,\ldots N_{MAG}$, since the3 parameters to be optimized are the position coordinates of each magnetometer. Furthermore, a $(3 \times N_{MAG})$ matrix $\mathbf{B}_{meas}$ is introduced to include the values of the measured magnetic field components generated by the ideal magnetic source at the measurement points $B_{j_x}, B_{j_y}, B_{j_z}$.

Additionally, the algorithm takes into account the parameters of the magnetic source - position (x,y,z) on the turntable and magnetic moment (m_x, m_y, m_z).The position coordinates of each magnetometer are originally randomly selected (initialization phase). After employing (2) – (5) using the magnetometer positions and the parameters of the magnetic source, the modeled magnetic field at any observation distance may be calculated and compared to the measured field values.

As the developed scheme unfolds, the particles explore the search space in each iteration for the optimal location of the sensor and the algorithm yields the global optimum – *gbest* – parameters (3 for each magnetometer's position) that fit the measured magnetic field values $\mathbf{B}_{meas}$. The algorithm's goal is to minimize the function:

$$G = \sum_{k=1}^{3} \left(\mathbf{B}_{gbest_k} - \mathbf{B}_{meas_k} \right)^2, \tag{15}$$

where k represents the x,y,z components and $\mathbf{B}_{gbest}$ is a $(3 \times N_{MAG})$ matrix containing the modeled magnetic field values corresponding to *gbest*.The algorithm stops when G is minimized or a maximum number of iterations is reached and outputs the precise positions of the sensors $(x'_{0j}, y'_{0j}, z'_{0j})$. Finally, the presented approach works iteratively for each magnetometer that is included in the measuring facility. For the validation of the proposed calibration method, a well behaved magnetic source with accurately determined parameters must be employed, as presented in the following subsection.

Validation of the Calibration Method With Real Measurements

A current-fed coil may be used as a calibration source, since its parameters can be precisely regulated. The current fed coil has the following characteristics:

1. Circular shape.
2. Inner diameter D_i =54 mm and outer diameter D_o = 57.4 mm.
3. Length < 4 mm.

4. Turns N =280 (+/- 1%).
5. The current is supplied by a twisted-wire pair.

When the coil is fed with I = 100 mA, the theoretical equivalent magnetic moment that is produced can be calculated:

$$m = \left|\mathbf{m}\right| = NI\pi\left(\frac{r_i^2 + r_i r_o + r_o^2}{3}\right), \tag{16}$$

where r_i and r_o are the inner and outer radii of the coil respectively. Evidently, when the coil is fed with I = 100 mA it generates a magnetic moment of about 68.7 mAm^2. It should be noted that the magnetic dipole moment of the coil source is oriented towards the z -direction when placed on the xy plane and is approximately negligible in the x,y directions. The coil is positioned at (0, 0, 2.1) cm on the table of the measuring configuration depicted in Figure 5 and its magnetic signature is captured according to the measurement test procedure. The configuration of the magnetometers is in two circles, having a radius of r = 10 cm and at different heights (z = 10 and z = -10 cm for the upper and lower planes respectively). Following the calibration method presented in the previous subsection, the calibrated magnetometers' positions are tabulated in Table 7, along with the Root Mean Square

Table 7.Calibrated and uncalibrated magnetometers' positions

	x' (cm)	**x (cm)**	**y' (cm)**	**y (cm)**	**z' (cm)**	**z (cm)**	**RMSE**
Mag 1	19.9532	20	-0.3944	0	10.1764	10	3.28e-14
Mag 2	20.1306	20	0.2211	0	-10.3486	-10	4.64e-14
Mag 3	10.0265	10	17.1281	17.3	10.0121	10	3.43e-14
Mag 4	10.0331	10	17.0639	17.3	-10.2764	-10	3.52e-14
Mag 5	-9.9753	-10	17.2381	17.3	10.2596	10	4.69e-14
Mag 6	-9.8716	-10	17.4494	17.3	-10.1948	-10	3.68e-14
Mag 7	-20.1491	-20	-0.3072	0	10.0561	10	4.64e-14
Mag 8	-20.0554	-20	-0.0728	0	-10.2295	-10	3.67e-14
Mag 9	-10.198	-10	-17.5206	-17.3	10.0551	10	3.67e-14
Mag 10	-10.019	-10	-17.3054	-17.3	-10.4682	-10	3.38e-14
Mag 11	10.0615	10	-17.1204	-17.3	9.937	10	4.69e-14
Mag 12	10.3681	10	-17.3144	-17.3	-10.2239	-10	3.42e-14

Error (RMSE) of the fitting process, i.e. the deviation between the measured and modeled magnetic field values at each sensor.

Evidently, the correction of the algorithm to the magnetometers' position is in the range of 0-5 mm. Consequently, it is worth mentioning that the software based calibration technique achieves significant accuracy levels in the prediction of the magnetometers' position that in practice is impossible to achieve with conventional calibration methods. Moreover, the insignificantly values of the RMS error confirm the validity of the calibration procedure for all magnetometers.

In order to illustrate the importance of the calibrated sensors' positions in measuring facilities before the spacecraft equipment testing, the current fed coil is positioned at a different location (2.5, 2.5, 2.1) cm on the turntable and measured again. Afterwards, its MDM parameters are estimated by employing the PSO technique, as mentioned in the previous section, both for the calibrated and the uncalibrated positions of the sensors. The resulting model parameters are shown in Table 8.

The resulting parameters for both cases are similar and close to the theoretical values in terms of predicted position (x,y,z) and magnetic moment (m_x,m_y,m_z). However, the goodness of fit evaluation criterion is minimized in the case that the calibration procedure has been realized, ensuring the accuracy of the model's prediction. Evidently, when high levels of magnetic cleanliness are required the in advance implementation of a calibration procedure on the measurement facility should be realized, since the accuracy of the estimated models is significantly enhanced.

In reality, the remote points where the sensitive equipment is placed and magnetic cleanliness is required are located outside the main structure of the spacecraft, mounted on a long boom (typically in the order of meters). Therefore, it is crucial to guarantee that the estimated model accurately represent the EUT at remote extrapolation distances. For these purposes, the MDMs of Table 8are used to predict the magnetic field at a circle of radius 20 cm with height $z = 1$m. The magnitude of the magnetic fields generated by the theoretical and the predicted model (with uncalibrated and calibrated positions) is depicted in Figure 13. Evidently, the deviation of the magnetic field that is generated by the estimated model with uncalibrated positions from the theoretical cannot be considered negligible when stringent magnetic cleanliness requirements are considered and hence, the calibration of the measuring setup prior to measuring and characterizing an EUT is necessary.

Finally, every calibration procedure has an associated calibration uncertainty, emerging both from the uncertainty properties of the reference quantity (calibration coil), but also from the method itself (Kakarakis, Spantideas, Kapsalis, Capsalis, & Junge, 2017). The uncertainty related to the calibration coil can be quantified:

Table 8. Model Prediction of the current fed coil with and without calibration of the magnetometers' positions

	x [cm]	y [cm]	z [cm]	m_x [mAm²]	m_y [mAm²]	m_z [mAm²]	$\|\mathbf{m}\|$ [mAm²]	Total RMS of residuals [%]
Theoretical Model	2.5	2.5	2.1	0	0	68.7	68.7	
Predicted Model with uncalibrated positions	2.5415	2.5012	2.1423	0.1236	0.1569	68.1912	68.19149	2.36
Predicted Model with calibrated positions	2.473	2.3389	2.1051	0.0244	0.0054	68.7573	68.7573	0.78

Figure 13. Estimation of the magnetic field at a remote circle of radius 20 cm with height z = 1m, generated by the models of Table 8

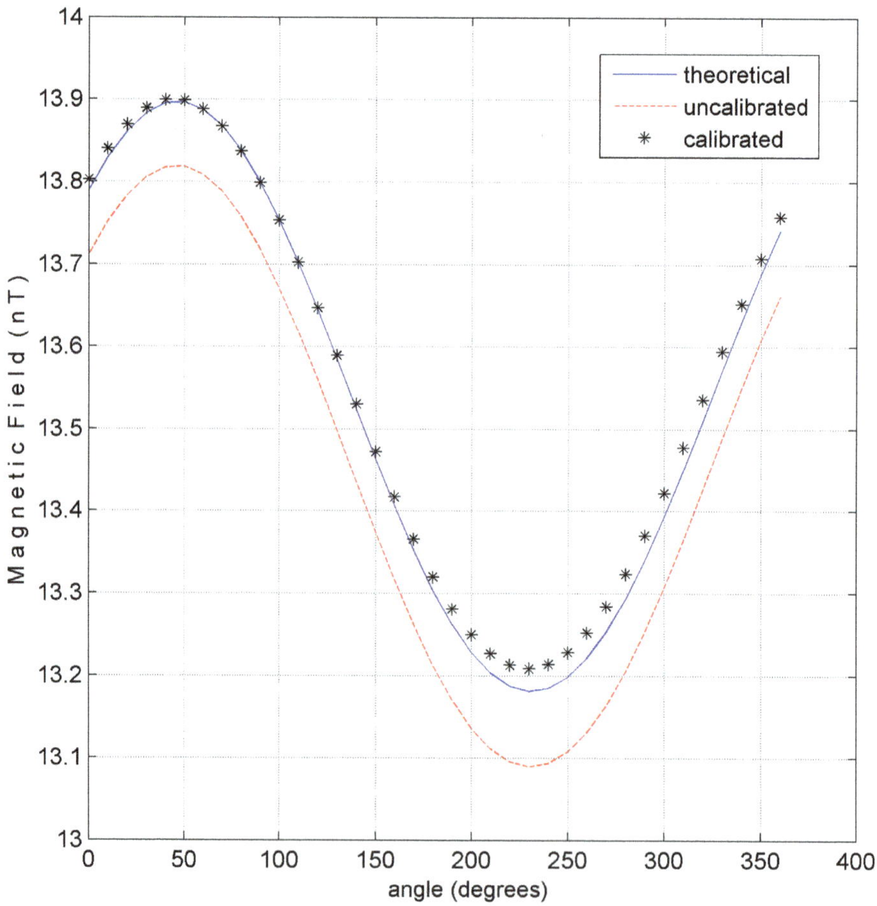

- The number of turns N is known with accuracy (+/- 1%).
- The current I is measured with 0.001 mA resolution.
- The imperfection of the windings results in expected uncertainty $\leq$ 1.5% in the coil's magnetic moment.
- The facility's accuracy of the position of coil on the turntable is $\leq$ 0.1 mm.

The total uncertainty associated with the calibration coil is estimated to be in the order of 2% in terms its magnetic moment and approximately 0.1 mm in terms of its position on the turntable. Therefore, the introduced uncertainty related with the calibration source results in uncertainty levels (deviation) of maximum 1 mm in the accuracy of the predicted magnetometers' positions, by employing the presented calibration method.

AC MAGNETIC FIELD VERIFICATION METHODS

Recent and upcoming space missions (Solar Orbiter, JUICE) have also stringent magnetic cleanliness requirements on AC time varying magnetic fields (Müller, Marsden, Cyr, & Gilbert, 2013; Spantideas, Kakarakis, Kapsalis, & Capsalis, 2016). The frequency range of interest coincides with the sensitivity of the on-board sensors and extends from sub-mHz to several 10 KHz. These strict requirements on low frequency magnetic fields involve measurements and characterization of time varying magnetic fields generated by the equipment. Thus, well-established MDM methods for static fields have to be complemented by additional techniques for AC magnetic field verification in space missions. In the present section, a proposed verification method for low frequency magnetic fields generated by spacecraft equipment is illustrated and validated via simulated results. The technique employs simultaneous time domain measurements from multiple sensors (configuration setup of Figure 5) and similar test methodology context to the static case.

Method Description

In order to comply with the magnetic cleanliness requirements on slowly time varying magnetic fields, the frequency range under investigation is limited to several 10 KHz. Without loss of generality, it is assumed that a unit can be represented by magnetic dipole sources with magnetic dipole moments that vary sinusoidally with time:

$$\mathbf{m}\left(\mathbf{r'},t\right)=\mathbf{m}\left(\mathbf{r'}\right)e^{-i\omega t} \quad (17)$$

Since the oscillating frequency $f = 2\pi\omega$ of the magnetic moment is expected to not exceed the upper limit of the requirements (usually 200 KHz), it is safe to assume that the radiated magnetic fields from the virtual dipole sources are quasi-static (Knoepfel, 2000). In fact, since the wavelength of the emitted magnetic fields is in the order of km or longer, the tests that are performed at small observation distances around the unit (typically in the order of m), involve near-magnetic field measurements. The time dependent magnetic field generated by an oscillating magnetic moment at an observation point $\mathbf{r}$ can be expressed (Jackson, 1999):

$$\mathbf{B}\left(\mathbf{r},t\right) = \frac{e^{-i\omega t}\mu_0}{4\pi}\left[\frac{3\left(\mathbf{r}-\mathbf{r}'\right)\left[\left(\mathbf{r}-\mathbf{r}'\right)\cdot\mathbf{m}\right]}{\left|\mathbf{r}-\mathbf{r}'\right|^5} - \frac{\mathbf{m}}{\left|\mathbf{r}-\mathbf{r}'\right|^3}\right] \tag{18}$$

where $\mathbf{r}'$ is the position of the magnetic moment $\mathbf{m}$ and μ_0 is the magnetic permeability of free space. The time varying magnetic field is similar to the DC static case (2) – (5), with the exception of its time oscillation with the frequency of the dipole moment ω. It can be further assumed that each dipole i, $i = 1,2,\ldots M$ corresponds to the oscillating frequency ω_i and thus, each frequency of the magnetic field spectrum can be represented by the contribution of a single magnetic dipole. Therefore, the frequency domain components of the radiated magnetic field may be separated for each frequency f_i, handled individually and treated DC-like, similarly to phasor representation. The magnetic field spectrum that is produced from the M magnetic dipoles at the j observation point can be constructed as the superposition of each individual contribution in the frequency domain:

$$\tilde{\mathbf{B}}_j = \sum_{i=1}^{M} Bx_{ij}\,\hat{\mathbf{x}} + \sum_{i=1}^{M} By_{ij}\,\hat{\mathbf{y}} + \sum_{i=1}^{M} Bz_{ij}\,\hat{\mathbf{z}} \tag{19}$$

Referring to Figure 1, the observation points $j = 1,2,\ldots N_{MAG}$ are tri-axial sensors that measure the time varying components of the magnetic field $B_{x_j}\left(t\right), B_{y_j}\left(t\right), B_{z_j}\left(t\right)$. The measured data can be stored in a matrix $\mathbf{B}_t$ with dimensions $(3 \times N_{MAG} \times t)$. Fast Fourier Transform (FFT) can be applied in order to transform the measured magnetic field time series of each sensor in the frequency domain, resulting in $\tilde{B}_{x_j}\left(f\right), \tilde{B}_{y_j}\left(f\right), \tilde{B}_{z_j}\left(f\right)$.

The magnetic field spectra components at a single frequency f_i can be selected, examined individually and treated DC-like. A $(3 \times N_{MAG})$ matrix $\mathbf{B}_{meas_i}$ is introduced in order to accommodate the values of the measured magnetic field spectra for the frequency f_i. This procedure can be performed iteratively for several frequencies under investigation in case of broadband emissions, or for selected peak frequencies if the EUT generates narrowband magnetic field emissions. Moreover, the measured magnetic field spectra of the EUT may be compared to the threshold values of the magnetic cleanliness requirements and thus, the identification of the contaminated frequencies is possible.

The measured values $\mathbf{B}_{meas_i}$ are then fitted with a single dipole source oscillating at f_i. The dipole's parameters that need to be determined include its position (x_i, y_i, z_i) and its magnetic moment $(m_{i_x}, m_{i_y}, m_{i_z})$, taking into account that each frequency is modeled with a single dipole. The presented method can be extended to estimate the parameters of several dipoles oscillating at the same frequency, similarly to the well-established DC modeling methods. The PSO stochastic technique can therefore be employed to generate a model of the EUT's magnetic behavior for each frequency f_i.

The developed PSO algorithm employs the equations (2) – (5) in order to estimate the 6 parameters of the dipole source that can be used to represent the EUT for the selected frequency f_i. As the algorithm proceeds, the global optimum – *gbest* – parameters of the dipole source are calculated to fit the measured magnetic field spectra values. The objective function that needs to be minimized can be expressed:

$$G_i = \sum_{j=1}^{N_{MAG}} \sum_{k=1}^{3} \left(\mathbf{B}_{gbest_{i_{kj}}} - \mathbf{B}_{meas_{i_{kj}}} \right)^2, \tag{20}$$

where k stands for the x, y, z components and the $(3 \times N_{MAG})$ matrix $\mathbf{B}_{gbest_i}$ contains the modeled magnetic field values for the frequency f_i corresponding to *gbest*. The algorithm stops when the deviation between the measured and modeled magnetic spectra values (G_i) is minimized, or when the maximum number of iterations is reached. Low values of G_i indicate that the fitted parameters of the dipole source can virtually reproduce the measurements of the EUT at the frequency f_i. The above modeling procedure can be then reiterated with the selection of a different frequency, resulting in frequency domain MDMs. The flowchart of the proposed method is depicted in Figure 14.

Consequently, the resulting global optimal parameters of the PSO algorithm may be employed to recreate the measured magnetic field values for each frequency and the magnetic field spectra can be constructed. In the following subsection,

Figure 14. Description of AC magnetic field verification method

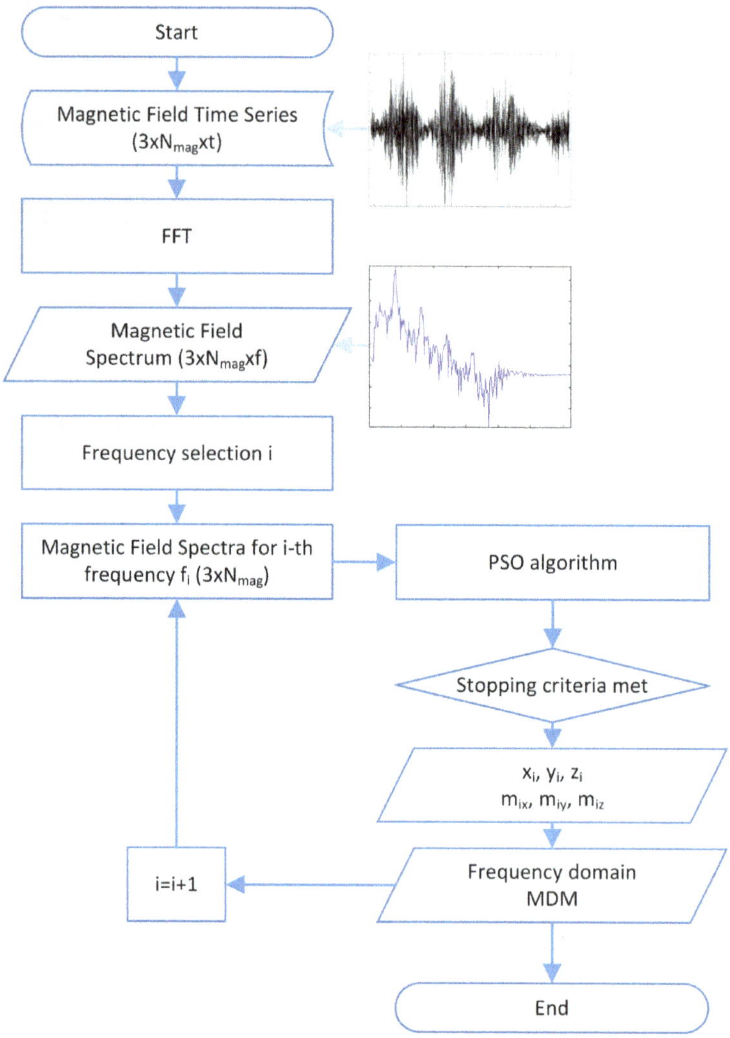

the proposed methodology for low frequency AC magnetic field verification is implemented on "so-called" simulated values generated by virtual dipole sources in a multi-magnetometer facility consisting of 12 sensors.

Validation With Simulated Measurements

In order to validate the presented method, two virtual dipole sources are considered as EUT. The time varying combined magnetic field of these simulated sources at the measuring configuration of Figure 5 is calculated according to equation (12) and then transformed into the frequency domain via FFT at each observation point. The spectrum values of the magnetic field are then distorted for purposes of introducing relative noise and uncertainty in the measurements. The magnetic field components are randomly distorted by maximum percentage of distortion q and create the "so-called" simulated measurements.

$$\tilde{B}_{x_j} = \tilde{B}_{x_j} \cdot q \cdot \left(rand - rand \right) + \tilde{B}_{x_j} \tag{21}$$

and similarly for the y,z component's spectra. The function $rand$ takes random values inside the interval [0, 1] following the uniform distribution and a maximum distortion of $q = 5\%$ is considered adequate to simulate magnetic field disturbances. The two dipoles are assumed to be located at different positions and, without loss of generality, exhibit sinusoidal magnetic moments aligned to the z-axis. Their parameters can be described:

- A dipole located at (5, 5, 1) cm having a magnetic moment of:

$$m_z = -30 \sin\left(2\pi f_1 t\right) \tag{22}$$

- A dipole located at (-8, 0, 0) cm having a magnetic moment:

$$m_z = 50 \sin\left(2\pi f_2 t\right) \tag{23}$$

where the frequencies $(f_1, f_2) = (7, 130)$ Hz and the magnetic moment is expressed in mAm². The overall time domain magnetic field at an observation point, which is produced by the contribution of both dipoles, is shown in Figure 15, using sampling frequency of 16384Hz, while the resulting distorted magnetic field spectrum at the same observation point is shown in Figure 16.

The peak frequencies of the magnetic spectra at the observation point may be identified in advance and the proposed method may be used to estimate the parameters of the MDMs only at these frequencies $(f_1, f_2) = (7, 130)$ Hz. The modeling procedure is carried out by employing the simulated measurements from the sensors' configuration of Figure 5 and the resulting models are tabulated in Table 9.

Evidently, the resulting parameters for both cases are close to the theoretical values in terms of predicted position (x, y, z) and magnetic moment (m_x, m_y, m_z).

Figure 15. Time varying components of the magnetic field generated by the virtual dipole sources

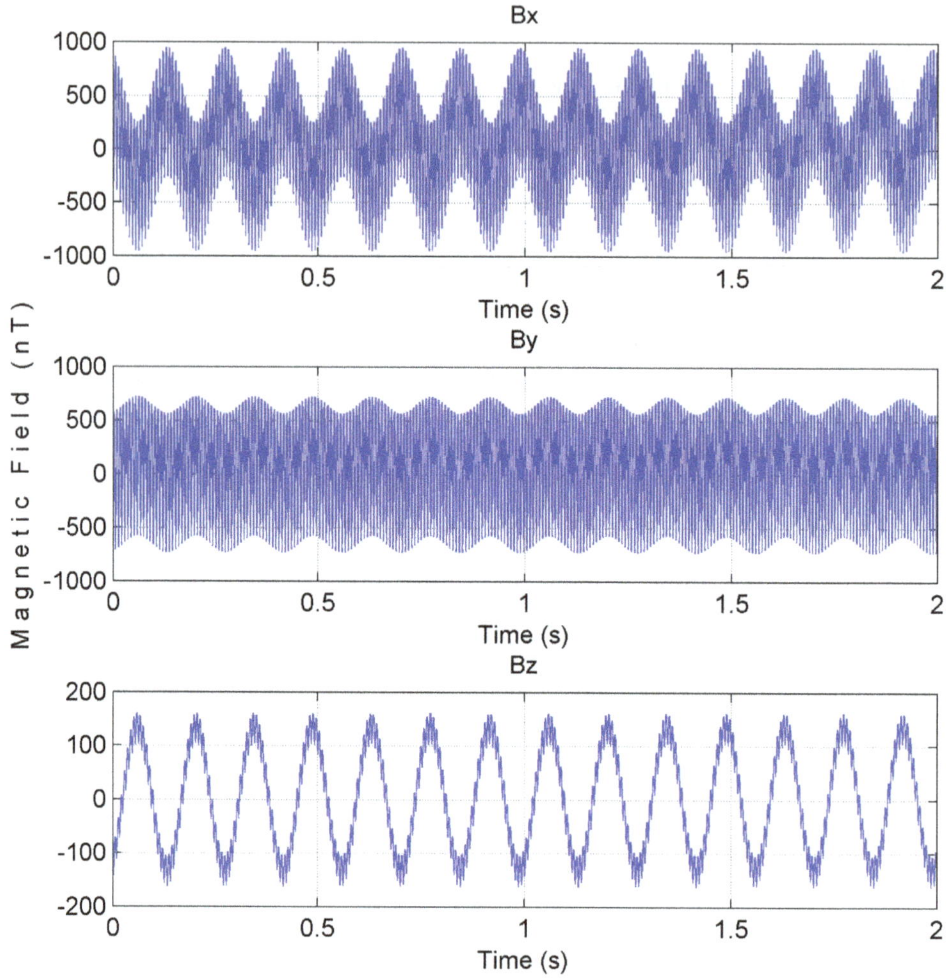

Additionally, the low values of the goodness of fit evaluation criterion (approximately 1-2%) ensure the accuracy of the predicted models.

The aforementioned modeling methodology may also be applied on measurements of AC induced magnetic fields. In the following section, magnetic measurements of induced AC magnetic field are discussed and a modeling methodology is proposed.

Figure 16. Spectra of the magnetic field's components generated by the virtual dipole sources

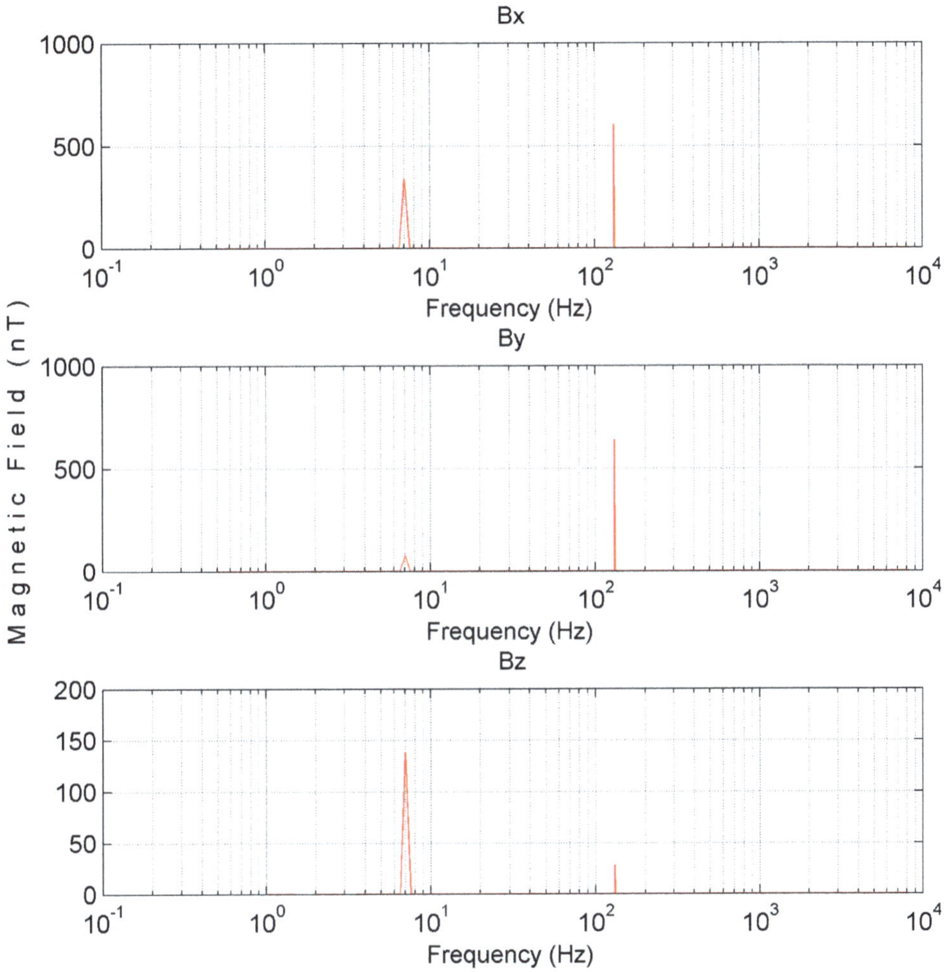

Table 9. Model Prediction – Frequency domain MDMs

Frequency [Hz]	x [cm]	y [cm]	z [cm]	m_x [mAm²]	m_y [mAm²]	m_z [mAm²]	$\lvert m \rvert$ [mAm²]	G_{rel} [%]
7	4.9808	5.0248	0.9639	0.1552	0.0586	49.9432	49.94348	1.0071
130	-7.9601	-0.0835	-0.0574	-0.1108	-0.1618	-29.9984	29.99904	1.981

VERIFICATION METHODS FOR INDUCED AC MAGNETIC FIELDS

Measurements and Modeling of a Unit's AC Induced Magnetization

In the present subsection, a baseline methodology is illustrated that may be used to carry out the AC induced magnetic dipole modeling analysis. Magnetic polarizability, also called AC magnetic susceptibility, is the dipolar response per unit volume from the unit under test to an external applied field $\mathbf{H}_a$. The induced dipole moment can be expressed (ECSS, 2012):

$$\mathbf{m}_{AC} = V\chi_{AC}\mathbf{H}_a \tag{24}$$

where V is the volume of the EUT and χ_{AC} is the AC magnetic susceptibility. In principle, most of the EUTs consist of conducting materials and exhibit different responses to an applied magnetic field depending on their shape, their conductivity σ and the frequency and amplitude of the external field (Chen & Gu, 2005). Moreover, at high frequencies, the induced (eddy) currents on the conducting surfaces of the unit possibly produce an AC induced magnetic behavior that tends to oppose the external field.

The response of these EUTs may be modeled and represented by an AC induced dipole moment $\mathbf{m}_{AC}$ and linearly expressed with respect to the external field via the frequency dependent 3×3 magnetic polarizability tensor χ_{AC}. For each direction of the applied field $\mathbf{H}_a$, an induced magnetic dipole model may be generated, according to the AC magnetic field verification method described in the previous section. The unit's response to an applied field of arbitrary orientation can be therefore estimated as a linear combination of responses corresponding to each principal axis x, y, z.

The measurement procedure is quite similar to the DC induced magnetic measurements. The external magnetic field should be as uniform as possible (spatially) in the volume of the EUT. Three sets of measurements of the EUT's magnetic signature are performed with external applied field $H_{a_x}, H_{a_y}, H_{a_z}$, as shown in Figure 17. These measurements shall be performed for several frequencies of the external field from DC to several 10 KHz. Furthermore, the typical levels of the external field are about 50 μT and 100 μT peak-to-peak in order to investigate the linearity of the EUT's response with respect to the amplitude of the applied field.

Figure 17. AC induced magnetic measurement and characterization

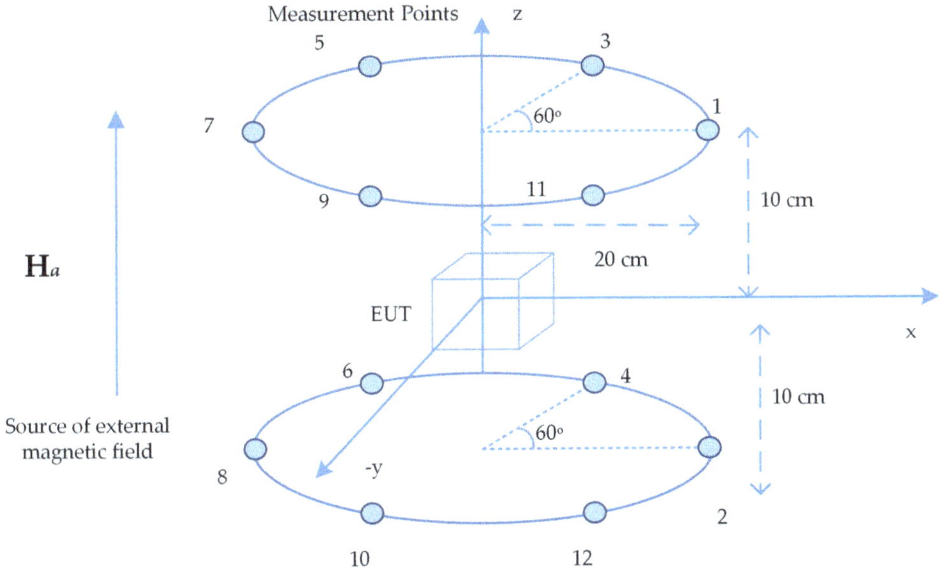

The induced magnetic field $\mathbf{B}_{ind}$ of the EUT is expected to emerge at the same frequency of the respective external magnetic field. The AC induced magnetization of the EUT may then be modeled in the frequency domain with a single magnetic dipole source oscillating at the frequency of the external applied field by employing the developed PSO based method, as described in the previous section.

FUTURE RESEARCH DIRECTIONS

Magnetic cleanliness is a critical aspect in space missions, but also in aircraft electromagnetic compatibility programmes. The levels of desired magnetic cleanliness become increasingly demanding and the development of new methods and techniques to cope with the stringent requirements is imperative. Although several discussed techniques are already included in the standard procedure of EMC testing, complementary methods that are presented in this chapter need to be fully validated. The illustrated test methodologies, after detailed and extensive evaluation, may be used as standard test methods for purposes of magnetic measurements and modeling of spacecraft equipment both for DC and AC magnetic field disturbances in magnetic cleanliness applications. Furthermore, the analysis should be supplemented to include magnetic field verification methods of spacecraft equipment that cannot be treated as magnetic dipoles, for instance solar arrays, harness routing, flywheels, etc. In fact, the abovementioned sources radiate significant magnetic fields and the characterization and modeling of their contribution to the system level magnetic cleanliness is essential.

CONCLUSION

In this chapter, modeling methods of DC and low frequency magnetic fields that are produced by unit or equipment mounted on a spacecraft are discussed for purposes of magnetic cleanliness in space missions. In the first part of this chapter, specific issues in already existing methods for measurements and modeling of DC magnetic fields are investigated, including special concerns for the induced DC magnetic behavior. The analysis involves the implementation of PSO stochastic technique on the magnetic dipole modeling inverse problem and aims to the reproducibility of the test results and elimination of uncertainty issues during the measurement procedure. The presented method is validated with real magnetic measurements of dipole sources. Moreover, a baseline methodology is presented concerning induced DC magnetic measurements and modeling, as well as unit-to-unit interaction inside the spacecraft. The analysis will allow the accurate estimation of dipole models that in turn will be employed for in-flight prediction of the units' magnetic behavior and interaction.

In the second section of the chapter, a calibration method that employs the PSO technique, targeting to accurately define the position of the magnetometers in a measuring facility and reduce the uncertainty in the modeling procedure, is presented. The calibration method is evaluated with real magnetic field measurements of a current fed coil source. The test results clearly indicate that the estimated model accurately represents the unit at remote extrapolation distances when a calibration procedure has been implemented prior to performing the magnetic measurements.

In the last part of this chapter, existing DC magnetic cleanliness methods are extended to include time varying field emissions for on-ground verification of low-frequency magnetic fields. The analysis employs the PSO stochastic technique in order to predict the parameters of virtual magnetic dipole sources that may be employed to reproduce the measured field data and represent the EUT in terms of radiated low frequency magnetic field. Additionally, specific considerations about the AC induced magnetic measurements and modeling are discussed.

REFERENCES

Acuna, M. H. (2002). Space-based magnetometers. *The Review of Scientific Instruments*, *73*(11), 3717–3736. doi:10.1063/1.1510570

BIPM, I., IFCC, I., IUPAC, I., & ISO, O. (2008). Evaluation of measurement data—guide for the expression of uncertainty in measurement. JCGM 100: 2008. *Citado en las*, 167.

Boschetti, D., Gervasio, G., & Marziali, I. (2012, May). Montecarlo approach for magnetic cleanliness evaluation on spacecraft. In *Aerospace EMC, 2012 Proceedings ESA Workshop on* (pp. 1-3). IEEE.

Carrubba, E., Junge, A., Marliani, F., & Monorchio, A. (2012, May). Particle swarm optimization to solve multiple dipole modelling problems in space applications. In *Aerospace EMC, 2012 Proceedings ESA Workshop on* (pp. 1-6). IEEE.

Chen, D. X., & Gu, C. (2005). AC susceptibilities of conducting cylinders and their applications in electromagnetic measurements. *IEEE Transactions on Magnetics*, *41*(9), 2436–2446. doi:10.1109/TMAG.2005.855450

Dumond, O., & Berge, R. (2012, May). Determination of the magnetic moment with spherical measurements and spherical harmonics modelling. In *Aerospace EMC, 2012 Proceedings ESA Workshop on* (pp. 1-5). IEEE.

ECSS, C. (2012). *European Cooperation for Space Standardization. ECSS-E-HB-20-07A. Space engineering – Electromagnetic compatibility Handbook*. ESA Publication.

Güttler, C., Hillenmaier, O., & Auster, U. (2009). Measurement of magnetic field emissions at low frequencies. *ESA Workshop on Aerospace EMC*, 30.

Hoang, V. D., Hernández, D. C., & Jo, K. H. (2014, April). Simple and efficient method for calibration of a camera and 2D laser rangefinder. In *Asian Conference on Intelligent Information and Database Systems* (pp. 561-570). Springer. doi:10.1007/978-3-319-05476-6_57

Jackson, J. D. (1999). Classical Electrodynamics. John Wiley & Sons.

Junge, A., & Marliani, F. (2011, August). Prediction of DC magnetic fields for magnetic cleanliness on spacecraft. In *Electromagnetic Compatibility (EMC), 2011 IEEE International Symposium on* (pp. 834-839). IEEE. doi:10.1109/ISEMC.2011.6038424

Kakarakis, S. D. J., Kapsalis, N. C., & Capsalis, C. N. (2015). A semianalytical heuristic approach for prediction of eut's multiple dipole model by reducing the number of heuristics. *IEEE Transactions on Electromagnetic Compatibility*, *57*(1), 87–92. doi:10.1109/TEMC.2014.2358712

Kakarakis, S. D. J., Spantideas, S. T., Kapsalis, N. C., Capsalis, C. N., & Junge, A. (2017). A Software-Based Calibration Technique for Characterizing the Magnetic Signature of EUTs in Measuring Facilities. *IEEE Transactions on Electromagnetic Compatibility*, *59*(2), 334–341. doi:10.1109/TEMC.2016.2615132

Kapsalis, N. C., Kakarakis, S. D. J., & Capsalis, C. N. (2012). Prediction of multiple magnetic dipole model parameters from near field measurements employing stochastic algorithms. *Progress In Electromagnetics Research Letters*, *34*, 111–122. doi:10.2528/PIERL12030905

Kennedy, J. (2011). Particle swarm optimization. In Encyclopedia of machine learning (pp. 760-766). Springer US.

Knoepfel, H. E. (2008). *Magnetic fields: a comprehensive theoretical treatise for practical use*. John Wiley & Sons.

Mehlem, K. (1978). Multiple magnetic dipole modeling and field prediction of satellites. *IEEE Transactions on Magnetics*, *14*(5), 1064–1071. doi:10.1109/TMAG.1978.1059983

Mehlem, K., Wiegand, A., & Weickert, S. (2012, May). New developments in magnetostatic cleanliness modeling. In *Aerospace EMC, 2012 Proceedings ESA Workshop on* (pp. 1-6). IEEE.

Müller, D., Marsden, R. G., St. Cyr, O. C., & Gilbert, H. R. (2013). Solar Orbiter. Exploring the Sun-Heliosphere Connection. *Solar Physics*, *285*, 25–70. doi:10.1007/s11207-012-0085-7

Musmann, G. (Ed.). (2010). *Fluxgate magnetometers for space research*. BoD–Books on Demand.

Paul, C. R. (2006). *Introduction to electromagnetic compatibility* (Vol. 184). John Wiley & Sons.

Primdahl, F. (1989). *A Pedestrian's Approach to Magnetic Cleanliness*. Danish Space Research Institute.

Spantideas, S., Kapsalis, N. C., Kakarakis, S. D. J., & Capsalis, C. N. (2014). A method of predicting composite magnetic sources employing particle swarm optimization. *Progress In Electromagnetics Research M*, *39*, 161–170. doi:10.2528/PIERM14092902

Spantideas, S. T., Kakarakis, S. D. J., Kapsalis, N. C., & Capsalis, C. N. (2016). Theoretical methods for studying distance and frequency scaling for AC magnetic fields in satellite missions. *IEEE Transactions on Magnetics*, *52*(4), 1–5. doi:10.1109/TMAG.2015.2496901

Weikert, S., Mehlem, K., & Wiegand, A. (2012, May). Spacecraft magnetic cleanliness prediction and control. In *Aerospace EMC, 2012 Proceedings ESA Workshop on* (pp. 1-5). IEEE.

KEY TERMS AND DEFINITIONS

Equipment Under Test (EUT): Any instrument or unit related to the space mission that produces a magnetic signature. These units can be inside (e.g., Power conditioning and distribution unit, RF switch, battery, etc.) or outside the main body of the spacecraft (e.g., solar arrays, boom, etc.) and need to be tested and magnetically characterized.

Induced Magnetization: Magnetic properties of a material that purely depend on the presence of an applied external magnetic field and directly vanish in its absence.

Magnetic Cleanliness: The generated magnetic field from the spacecraft is required to not exceed specific thresholds at the location of the magnetic sensors and therefore, disturb their measurements.

Magnetic Sensors: Instruments that are designed to measure magnetic fields (fluxgate magnetometers, search coil magnetometers, etc.). For instance, the fluxgate magnetometers have a magnetically susceptible core wrapped by two coils and measure the induced electric currents from a background magnetic field.

Magnetic Signature: The magnetic field generated by an EUT typically depends on the distribution of magnetic material inside the unit, as well as the electronics and the current paths that are incorporated in its design. Therefore, each unit is expected to emit an individual distinctive magnetic field.

Magnetic Source: Any source that generates a magnetic field. These sources typically include magnetic materials (e.g., iron, steel, etc.) and flowing electric currents.

Multiple Magnetic Dipole Modeling: The representation of a unit's generated magnetic field with a dipole model able to recreate and reproduce its magnetic signature.

Remnant Magnetization: Magnetic properties of a material that are always present, even in the absence of external fields (e.g., ferromagnetic materials).

Chapter 4
Electromagnetic Compatibility for Space Applications:
Test Specifications, Standards, and Methods

Stylianos Tsatalas
Emtech Space P.C. (CY) Ltd., Greece

ABSTRACT

In a space system or equipment, electromagnetic compatibility is a critical aspect affecting its operation, function, and performance. In the beginning of this chapter, an overview of the electromagnetic compatibility main considerations is provided. Then the main electromagnetic compatibility standards and relative useful handbooks for space applications are briefly presented. Testing methods and procedures to assess electromagnetic compatibility of space systems and units are discussed briefly covering all kinds of measurements. Moreover, information regarding the instrumentation and the testing facilities is also provided. Mission testing requirements and limits are also presented as examples for every electromagnetic compatibility test. Recent developments in test methods are also discussed where applicable.

INTRODUCTION

The main tool to assess electromagnetic compatibility of any electronic or electrical system for all industries, including space industry, is a variety of tests and measurements which represent actual phenomena in a controlled laboratory environment. What is very important when testing in laboratory environment, is to

DOI: 10.4018/978-1-5225-5415-8.ch004

ensure reproducibility and repeatability of test and/or measurements results. This is achieved primarily by the use of test standards produced by expert's committees. In these standards measurements and instrumentation principles apply together with the theoretical background relative to the nature of the electromagnetic phenomenon.

The objective of this chapter is to provide an overview of the main aspects of the electromagnetic compatibility such as the technical standards available (or suitable for the space sector) and the main tests typically performed to on-board equipment in spacecrafts or satellites. Effort was made to include as many information as possible in the frame of this context, like instrumentation requirement, general test arrangements, experienced difficulties but also alternative test solutions. The information provided is dealing only with unit testing; system testing in the space sector utilizes similar test methods but tailoring is expected in some cases.

General measurement principles and requirements are also discussed.

BACKGROUND

According to the definition found in the new and the older EMC EU Directive, but also in other relative contexts, "Electromagnetic compatibility" is the ability of equipment to function satisfactory in its electromagnetic environment without introducing intolerable electromagnetic disturbances to other equipment in that environment. This definition is valid for all systems and devices regardless of the operating environment (industrial, commercial, military or space).

Another useful definition is regarding "electromagnetic disturbance" which is any electromagnetic phenomenon which may degrade the performance of equipment or a system and may be electromagnetic noise, unwanted signals or changes in the propagation medium itself.

An important aspect of electromagnetic compatibility of systems or equipment is their "susceptibility" (immunity) which defines a measure of the ability to perform as intended without degradation in the presence of an electromagnetic disturbance.

In general, the assessment of the electromagnetic compatibility of a system or unit is performed by making specific tests based on methods described in relevant standards. Depending on the field of application different kind of standards applies. Some tests are similar or even identical. However, limits or performance criteria could be different for different application areas like between space and commercial systems or units.

Electromagnetic compatibility for space applications utilize test methods similar or identical to military ones, like MIL-STD-461-F/G. Differences are encountered in test limits or specific parameters of tests like for example frequency range or, in some case, in instrumentation. For systems used in commercial applications, like for example household devices, electromagnetic compatibility is assessed using test standards prepared from International Electrotechnical Committee. The test methods are similar; however, limits and performance criteria are in most cases different.

The most important difference in electromagnetic compatibility assessment between space and commercial systems is that in space systems the interaction between electrical, electronic, electromagnetic and electromechanical equipment within a space vehicle may cause performance degradation or problems in the functionality that could compromise the success of the whole space mission. Therefore, all equipment shall be characterized carefully and in detail to obtain the necessary data to allow the analysis of the interactions and to define mitigation measures, especially for sensitive or critical, for the space vehicle, equipment. In such cases, strict electromagnetic interference safety margins are defined. This matter is defined as intra-system EMC. On the other hand, in commercial systems these, interactions are less frequent as electronic or electrical parts are less sensitive. And, if such problems arise, these cause marketing or sales issues only.

Intersystem EMC is defined as interactions between external natural, induced or man-made electromagnetic environment and the space vehicle. An example of natural electromagnetic environment is the voltage building-up of a space vehicle when entered in plasma. External, man-made interference might result from radar beams during ground operations or in launching sequence. Interference could be also applied between the launcher and the space vehicle.

ELECTROMAGNETIC COMPATIBILITY TEST SPECIFICATIONS, STANDARDS AND METHODS

General Considerations for EMC Tests and Measurements

The essential measurement principles apply also to EMC; these are briefly discussed below:

- **Accuracy and Precision:** The accuracy of a measurement system (or an instrument) describes the nearness of the measured quantity to the quantity's true value. It is a measure of a statistical bias which can be systematic or random. Precision provides the degree to which repeated measurements of the same quantity in the same conditions provide the same results. It is a measure of statistical variability. A graph providing the definitions of accuracy and precision is given below (Figure 1). It is noted that precision is closely related to reproducibility and repeatability.

Figure 1. Accuracy versus Precision

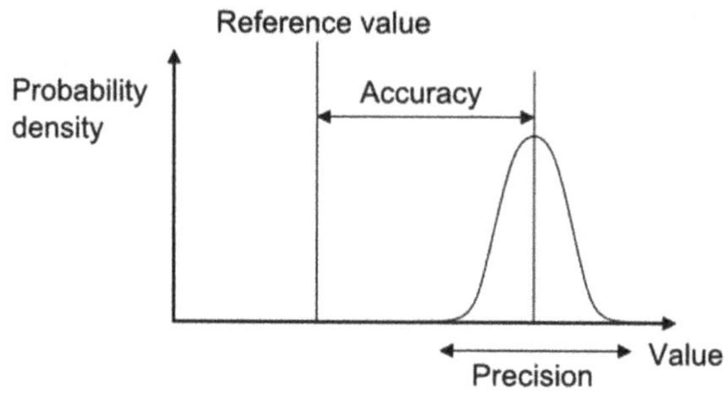

- **Repeatability and Reproducibility:** Repeatability is the variation in measurements performed on replicate items, by the same method, instrumentation and operator in the same conditions, in the same location, and within a short period of time. Various parameters may affect repeatability such as the instrumentation and the stability of the measured item but also the operator. Reproducibility is, in a broader sense, the degree of agreement between measurements conducted on replicate items, in different locations, by different operators. Both these terms are strongly related to precision; they are the main contributors to the precision of a measurement system or method.
- **Uncertainty:** It is a measure to reflect the incomplete knowledge of the quantity value. It characterizing the dispersion of the measured values of a quantity. Uncertainty should be accompanying a measurement result in EMC tests. The uncertainty budget of a measurement is typically defined and/or calculated following specific guidelines either defined in test standards or in relative handbooks such as the "Guide to the expression of uncertainty in measurement", issued by BIPM, 2008. Its evaluation is performed by Type A and Type B evaluation of the uncertainty. Type A is estimated by repeated measured values and it is assumed that the probability distribution that best describes the measured quantity is the Gaussian one. Type B is estimated by scientific judgment of the measurement conditions and test instrumentation or other available information. In this case it is assumed that the measured quantity lies within an interval [a,b] so generally the rectangular probability distribution can characterize the knowledge of the quantity. However, other probability distributions may apply if other information is available.
- **Complexity/Measurement Time and Facilities:** Electromagnetic Compatibility test and measurements require special Test Facilities so as to provide controlled environment. This is required both to minimize external

effects (i.e. interference that is not required for testing) but also to minimize the interference coming from EMC testing to other nearby activities. The necessary facilities are:

- ◦ **Anechoic Chambers:** An anechoic chamber is actually a shielded room which is lined with radio frequency absorbing material to reduce reflections from all internal surfaces. The anechoic chambers are categorized to full anechoic chambers which are completely covered with such material on all internal surfaces, wall, ceiling and floor. Another type of anechoic chamber is the semi- anechoic chamber which is a shielded room which is lined with absorbing material on all surfaces except the floor.
- ◦ **Shielded Rooms:** A shielded room is a room made by solid metal or any shielding material designed specifically to provide isolation of the internal electromagnetic environment from the external one. The purpose is to prevent ambient electromagnetic fields outside the shielding room from affecting test execution. Also, shielded rooms protect nearby activities from interference coming from electromagnetic compatibility test execution.
- ◦ **DC Magnetic Fields Test Facility – Magnetic Cleanliness**: The main characteristic of such test facility is to provide a test volume where the magnetic field of Earth is compensated so as the DC magnetic field of a unit can be measured in zero-field conditions.

Overview of Available Standards and Handbooks

- **MIL-STD 461E/F:** This is a standard applicable to electronic, electrical, and electromechanical equipment and subsystems procured by the US Department of Defense (DoD). It establishes test methods and verification requirements for the control of the electromagnetic interference (EMI) emission and susceptibility characteristics of such equipment and subsystems. It describes test setups and methods but also instrumentation requirements and characteristics for every test. It is applicable to equipment that may be used independently or as an integral part of other subsystems or systems with some limitations. The equipment, in a general form, should not be larger than a rack; it may include interconnections to other equipment and could be powered from prime power sources (mains power ac or dc).
- **ECSS EMC Standard (ECSS-E-ST-20-07C):** ECSS are standards prepared as an output of collaboration of the European Space Agency, national space agencies and European industry associations. It includes a description of all test methods and setups, measurement instrumentation and measurement

data reporting requirements. Moreover, it includes specific requirements regarding EUT operation and test setup specific preparations. Its philosophy is to define how and what should be accomplished. General test limits are also presented for some tests, however these could be different depending on the type of the space mission.

- **ECSS EMC Handbook (ECSS-E-HB-20-07A):** It is a standard which is prepared to be used to support the use of ECSS EMC Standard from engineers dealing with EMC issues. It provides rationale for unit level test requirements but also deals with design techniques for EMC both in unit and system level providing general rules and guidelines for several issues. Moreover, it is dealing with system level verification issues and with troubleshooting and retrofit techniques. For some of the test methods this handbook goes into detail into the test techniques and the relevant instrumentation. EMC analyses methods are also encountered together with their relative computational models.
- **Marshall Space Flight Center Electromagnetic Compatibility Design and Interference Control (MEDIC) Handbook:** This is an EMC handbook used by the National Aeronautics and Space Administration (NASA). Like the ECSS EMC handbook its purpose is to provide practical and helpful information in the electrical design of payload equipment and subsystems for electromagnetic compatibility. However, it doesn't go deep into the measurement methods but it deals with design techniques to reduce EMI risks, provides EMI suppression techniques and mitigation measures.

All the above-mentioned EMC standards and handbooks are specialized in space units and systems but analysis methods, design techniques and retrofit fixes for EMC might be found in other sources which could not be related to space applications as electromagnetic compatibility is a general issue of all industries and the main philosophy remains the same: to control EMC issues as soon as earlier in the design phases to reduce time, resources and possibly cost needed to resolve them after the implementation of a unit or a system.

Test and Measurement Techniques and Instrumentation for Space Applications

In this section, an overview of the EMC tests required to assess the electromagnetic compatibility of a unit used in a spacecraft will be provided. Test methods and test setups will be discussed together with instrumentation requirements and test facilities needed to perform the tests. Moreover, test conditions will be highlighted like ground planes, EUT mode of operation, arrangement and bonding. It is noted that in every test a reference to the test standards and handbooks mentioned above will be provided so the reader may find detailed information about the conduction of the test.

All the tests mentioned below are performed to identify the emissions emanating from the unit (Equipment) Under Test (EUT). The main categories of emissions testing are:

1. Conducted, where the parasitic noise coupled to cables is measured,
2. Radiated electric field, where the unintentional release of electromagnetic energy is measured,
3. DC magnetic field emissions for magnetic cleanliness requirements,
4. Low frequency radiated magnetic field emissions for magnetic cleanliness requirements.

The following tests are conducted in order to evaluate the functionality and the performance of a unit (Equipment) Under Test (EUT) in the presence of an electromagnetic disturbance. This determines the level of susceptibility of the unit in various electromagnetic phenomena. The main categories of susceptibility tests are:

1. Conducted susceptibility, where the degradation of the performance of a unit is assessed in the presence of interference on the EUT associated cables,
2. Radiated susceptibility, where the degradation of the performance of a unit is assessed in the presence of electric or magnetic field interference on the EUT and its associated harness,
3. Susceptibility to electrostatic discharges, where the EUT operation and performance is evaluated when an electrostatic discharge pulse is applied to the EUT.

Conducted Emissions Testing

The purpose of Conducted Emissions tests is the quantification of the noise coupled to input power lines of the equipment, including return ones, but also to the signal/data lines within a spacecraft. The test conditions differ between power and signal cables. These tests are performed, in general, in the frequency range of 30 Hz to 100 MHz and they involve the measurement of the noise currents induced in the lines of interest. The tests are divided in two frequency sub-ranges one concerning the audio and VLF frequencies up to 100 kHz and the other the RF ones, up to 100 MHz, or more.

Conducted Emissions tests are usually performed via a non-intrusive method; a suitable current probe is used to measure noise currents drawn on a power line.

The purpose of the tests is to measure the EUT contribution to voltage ripple to the power or data bus or signal leads.

In general, conducted emissions tests are performed in shielded rooms to avoid the interference from other sources. The instrumentation used usually includes the following:

- Measurement Receiver
- Current Probe
- Signal Generator with Amplifier
- Oscilloscope
- Data Recording Device
- LISNs
- Accessories:
  - Coaxial "T" connector and coaxial to bifilar transition
  - 1 Ω and 10 Ω power metal film resistors with inductance lower than 100 nH (for lower frequencies testing)

Before the beginning of the test a verification of the test setup should take place in order to verify the correct operation of all instrumentation, from the current probe to the data recording equipment. For this purpose, a suitable calibrated signal generator is required. The relevant procedures can be found in detail in the relevant ESA EMC Standard (ECSS-E-ST-20-07C, rev1 para 5.4.2 & 5.4.3).

The Equipment Under Test (EUT) has to be properly installed in the test setup, according to the applicable standard description but also to special conditions apply for the specific equipment and spacecraft. Its operation has to be stabilized before starting the measurement. If multiple modes of operation are possible it might be necessary to measure the current noise for each mode.

The usual representation of Conducted Emissions test results includes a graphic output of noise current versus frequency. Important test parameters shall be also reported, like test conditions, EUT operation mode and other test related information that might be useful for test reproducibility like, for example, EUT current consumption.

The most common conducted emissions tests are described below (Figure 2, Figure 3).

a. Test on Power Leads, to measure the differential mode noise currents. The frequency range of application is usually between 30 Hz and 100 kHz.

Figure 2. Measurement system calibration setup (Source: ESA ECSS-E-ST-20-07C, rev 1)

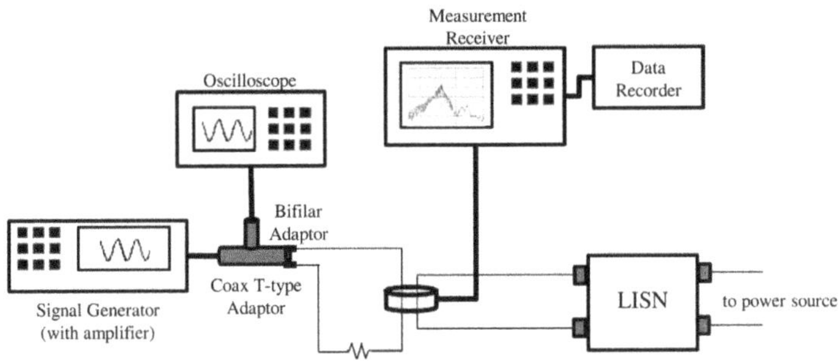

Figure 3. Measurement system test setup (Source: ESA ECSS-E-ST-20-07C, rev 1)

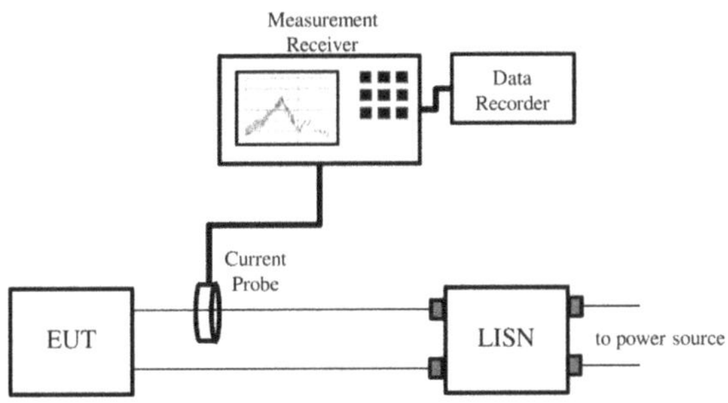

b. Test on Power and Signal leads, to measure the differential and common mode noise currents. The frequency of interest is between 100 kHz to 100 MHz (Figure 4).

In order to reject the common mode current and measure the Differential Mode current separately from the Common Mode one, the "true differential" method can be used. In this arrangement two complementary current probe setups can be used to measure respectively the "true" differential mode and the common mode noise

Table 1. Reference Standards and frequency range of test application, conducted emissions on power lines

Reference Standard	Frequency Range of Application
ECSS-E-ST-20-07C rev1 (5.4.2)	30 Hz to 100 kHz
MEDIC Handbook (2.2), CE01	30 Hz to 20 kHz
MIL-STD 461F (5.4), CE101	30 Hz to 10 kHz

Figure 4. Measurement system test setup, common mode (Source: ESA ECSS-E-ST-20-07C, rev 1)

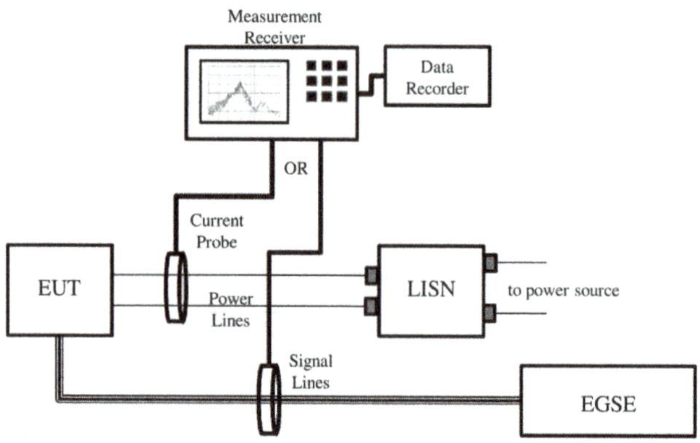

on the lines of interest. When the noise current is measured on each wire separately then DM and CM noises are combined and cannot be identified.

In general, these measurements are performed in the frequency domain but in some cases time domain measurements maybe preferred depending on the measurement sensitivity requirements.

An example of conducted emissions requirements from a real ESA mission (PLATO) is provided in the following figure (Figure 5).

Figure 5. Conducted Emissions Limit for ESA PLATO mission

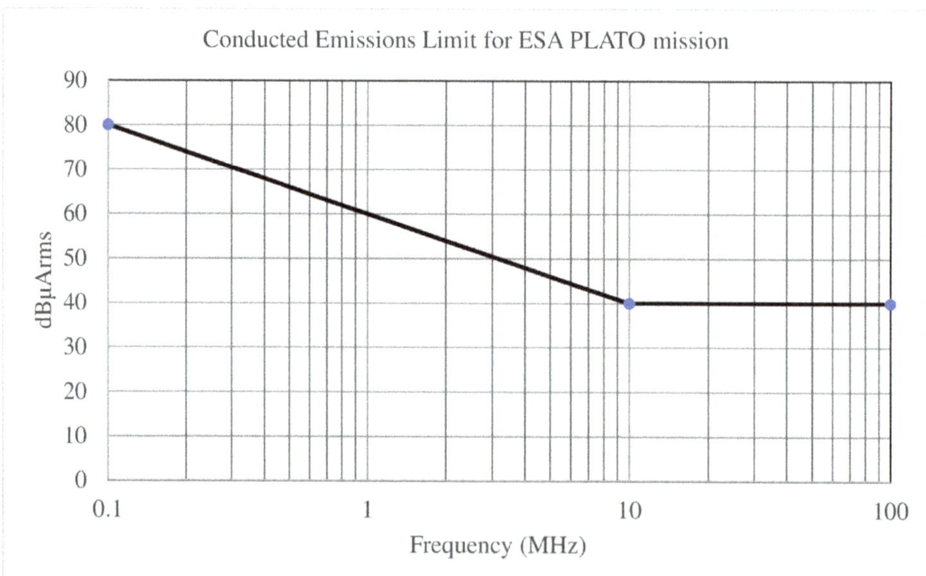

The measurement frequency range of the conducted emissions common mode measurements for the PLATO spacecraft's units is defined to be between 30 Hz and 50 MHz.

c. Inrush current test on power lines. This test quantifies the inrush (switch-on) current which is the momentary burst of current that flows into the power lines whet the equipment is turned on. This test additionally requires the use of a spike generator for the calibration of the test system before the measurement. Detailed information may be found in the relevant ESA EMC Standard (ECSS-E-ST-20-07C, rev1, 2012, para 5.4.4) and in the NASA EMC Handbook (NASA MEDIC Handbook, 1995, para 2.4), test is named CE07, or TT01.

Electric Field Radiated Emissions

Radiated emissions testing involves measuring the electromagnetic field strength of the emissions generated unintentionally by the equipment under test. These parasitic emissions are inherently come out from an electronic device due to the normal operation of any digital circuit which normally includes switching voltages and currents. The purpose of Radiated Emissions tests is to quantify the noise emanating from the equipment and the associated cables connected to it.

Measuring Radiated Emissions from an EUT is most of the times a complex process as the electromagnetic noise emanating from the Equipment Under Test has not the form of a homogeneous spherical pattern and it is usually involving a large range of Radio Frequencies. Parasitic emissions are tending to be directional and consequently the measurement test setup has to be flexible to identify them completely and measure them accurately. Due to the large frequency range that is required to be covered the measuring equipment includes various kinds of antennas. Moreover, depending on the measurement limits, it is required to use sensitive receivers to measure weak parasitic emissions in the presence of environmental noise.

Especially for Space applications, Radiated Emissions tests are performed, in general, in the frequency range of 14 kHz 30 MHz to 18 GHz. The test is divided

Table 2. Reference Standards and frequency range of test application, conducted emissions on power and signal lines

Reference Standard	Frequency Range of Application
ECSS-E-ST-20-07C rev1 (5.4.3)	100 kHz to 100 MHz
MEDIC Handbook (2.3), CE03	15-20 kHz to 50 MHz
MIL-STD 461F (5.5), CE102	10 kHz to 10 MHz

in two (or more) frequency sub-ranges one concerning the lower frequencies up to 30 MHz and the other the remaining range up to 18 GHz. The tests are applicable to any type of equipment.

The basic test site for measuring Radiated Emissions is the Open Area Test Site (OATS). The advantage of such a facility is the avoidance of the unknown reflections except the ones coming from the ground plane which have to be taken into account and normalized. However, the use of this kind of facility is limited due to weather conditions and, possibly, ambient electromagnetic noise. Radiated emissions tests are commonly ή typically performed in anechoic rooms to avoid the external interference from radio sources. Reflections of the electromagnetic waves on the anechoic rooms are reduced below a defined level due to their design and construction. This is achieved through measurements during the anechoic room construction. In general, the philosophy of such facilities is to provide reproducible results. The instrumentation used usually includes the following:

- Measurement receiver,
- Data recording device,
- Linearly polarized antennas; the following antennas are commonly used:
    - 14 kHz to 30 MHz, rod antenna,
    - 30 MHz to 200 MHz, biconical antenna,
    - 200 MHz to 1 GHz, double ridge horn, or log-periodic antenna,
    - 1 GHz to 18 GHz, double ridge horn antenna,
- Signal generators,
- Stub radiator,
- LISN (optional).

As mentioned also in Conducted Emissions measurements, before the beginning of the test a verification of the test setup must ή should take place in order to verify the correct operation of all instrumentation, from the antennas and the cables to the data recording equipment. For this purpose, a suitable calibrated signal generator is required. The relevant procedures can be found in detail in the relevant ESA EMC Standard (ECSS-E-ST-20-07C, rev1, 2012, para 5.4.6).

Special attention should be given so as the Equipment Under Test (EUT) is carefully installed in the test setup together with the associated cabling. The critical aspect in Radiated Emissions testing is to determine the test boundary of the EUT (together with its cabling) in order to define the details of the test conditions like i.e the measuring points around the EUT. The relevant standards provide general guidance for this purpose. Usually, the EUT is placed in a test table and the measuring antennas are placed in 1m distance from the front edge of the EUT boundary. Before the measurement, a survey to define the surface of the EUT that produces the maximum emissions have to be performed and the final measurement has to be made for this surface. As in Conducted Emissions testing, EUT operation has to be stabilized before starting the measurement. If multiple modes of operation are possible it might be necessary to measure the current noise for each mode.

The representation of Radiated Emissions test results is similar to the ones produced from Conducted Emissions measurements and includes a graphic output of electromagnetic noise versus frequency. The electric field emission from the equipment is expressed in units of dB above 1 μV/m and the measurement distance is 1m. Test relevant details are crucial to be reported, like test conditions (EUT position, cable type and layout), EUT operation mode and other test related information that might be useful for test reproducibility. Radiated Emissions tests usually provide results less reproducible comparing to other EMC tests and that is why is important to maintain test conditions information in detail.

The Radiated Emissions tests for the Electric Field, used for Space applications are described below.

a. Electric Field Radiated Emissions (low frequencies). The frequency range of application is typically between 14 kHz and 30 MHz. In this frequency range, the rod antenna is used to measure the electric field emanate from the EUT boundary. It is noted that this test is performed in order to essentially characterizing extreme near electric field emission. Such measurement can be useful to:

    ◦ Protect a victim inherently sensitive to purely electric field;
    ◦ Prevent capacitive crosstalk between cables.

b. Electric Field Radiated Emissions (high frequencies). The frequency range of application is typically between 30 MHz and 18 GHz. Several antenna types are used to cover the whole frequency range (measurement setup in Figure 6).

Table 3. Reference Standards and frequency range of test application, electric field radiated emissions

Reference Standard	Frequency Range of application
MEDIC Handbook (2.5), RE02	14 kHz to 10 GHz, lower part of frequency range
MIL-STD 461F (5.17), RE102	10 kHz to 18 GHz, lower part of frequency range

Figure 6. Electric Field measurement system test setup (Source: ESA ECSS-E-ST-20-07C, rev 1)

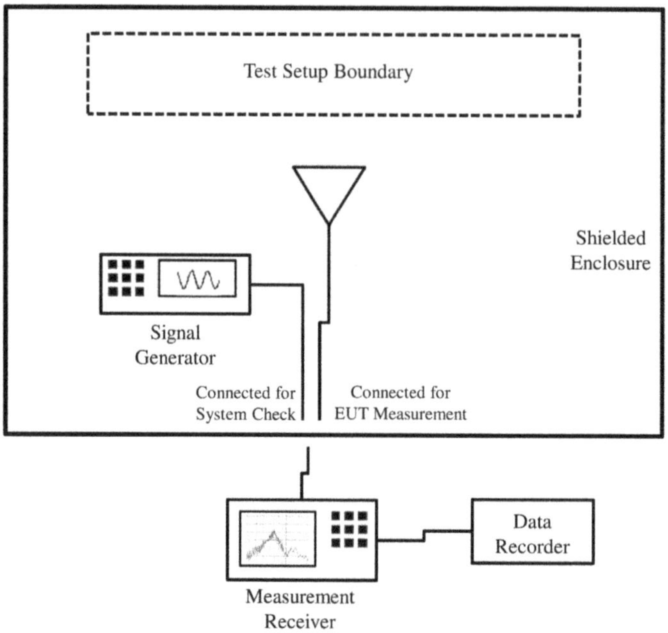

It has to be noted that the upper frequency range can be reduced to gain time in cases where results are well below the test limit between 500 MHz and 1 GHz or when all rising times of digital circuitry in the equipment are longer than 35ns. However, depending on mission requirements some frequency notches have to be tested above 1 GHz.

An example of radiated emissions (electric field) requirement from a real ESA mission (PLATO) is depicted in Figure 7. The typical frequency range of the requirement is between 30 MHz and 18 GHz. However, it is noted that depending on the unit under measurement type (i.e RF equipment, non-RF equipment) the upper frequency could be reduced significantly. Also, there might be discrete frequencies

Table 4. Reference Standards and frequency range of test application, electric field radiated emissions

Reference Standard	Frequency Range of Application
ECSS-E-ST-20-07C rev1 (5.4.6)	30 MHz to 18 GHz, upper part of frequencies
MEDIC Handbook (2.5), RE02	14 kHz to 10 GHz, upper part of frequencies
MIL-STD 461F (5.17), RE102	10 kHz to 18 GHz, upper part of frequencies

Figure 7. Test limits for radiated electric field emissions (Source: ESA PLATO EMC Specification Document, 2014)

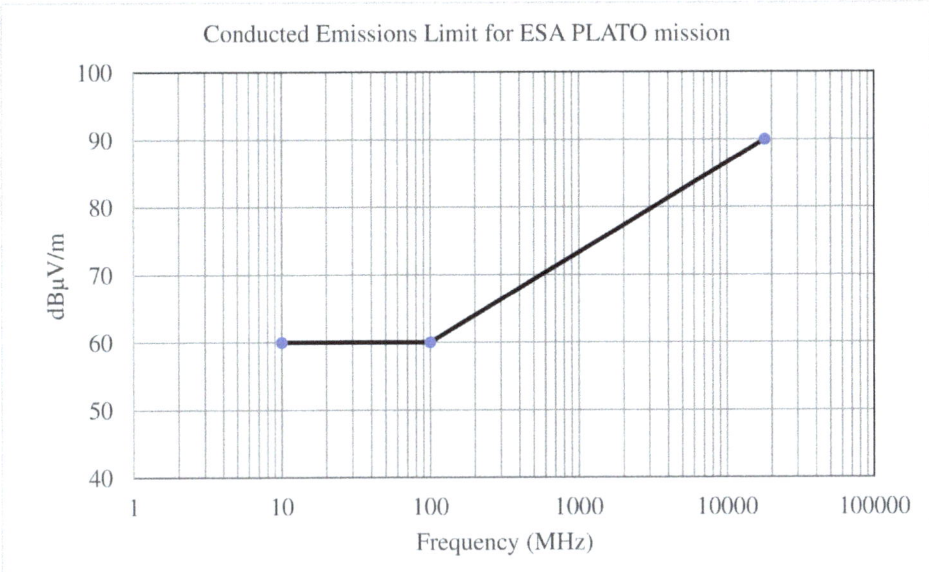

where the emission limit could be defined separately. In PLATO mission, for two discrete frequencies, for the S-band and X-band receivers the emission limit has been defined to be less than 30 dBμV/m (ESA PLATO EMC Specification Document, 2014).

Magnetic Field Radiated Emissions, 30 Hz to 50 kHz

Magnetic field emissions testing purpose is to measure the magnetic field strength coming out from the equipment under test and the associated cabling. These emissions are produced from equipment and subsystems, interconnecting cables, power and antenna transmission lines. Usually, the measurements are performed at equipment (unit) level and then system level verification can be performed through analysis.

Magnetic Field Radiated Emissions measurement from an EUT can be performed in several points around the EUT in order to fully characterize it. The general idea is to represent the EUT with multiple magnetic dipoles based on the measurement data collected and by using tools for magnetic dipole modelling. Having measurements in different distances for the EUT the accuracy of the magnetic dipole model can be ensured.

For magnetic field radiated emissions testing, it is not necessary, but still desired, to use a special facility like shielded or anechoic rooms. The measuring instrumentation to be used depends on the frequency range of interest and includes the following:

- Magnetic field instrumentation; the following types are commonly used:
  - mHz region to 20 Hz, Fluxgate magnetometers.
  - 1 Hz to 70 kHz, induction coil magnetometers (search coil magnetometers-SCM).
  - 10 Hz to 100 kHz, air coil loops.
- Measuring Receiver for Frequency Domain measurements or Oscilloscope for Time Domain measurements.
- Data Acquisition System.

As the above mentioned magnetic field measurement instrumentation are of different design philosophy the selection of the most suitable type to be used depends on the frequency range, noise level, sensitivity and dimensions. Some of them are mission specific parameters. Sometimes it could be reasonable to use more than one types of magnetometers to cover the frequency range of interest with adequate noise level and sensitivity. Overlapping frequency ranges could also exist and test results from these ranges can be used for validation of the measurement data.

Test setup verification procedures are not described in the relevant Space Standards. However, a typical procedure can be found in the relevant military standard (MIL-STD-461F, 2007). Such procedures can be designed prior to measurement to include all the measuring equipment based on the similar procedure exist for Radiated Electric Field Emissions test.

The test setup is similar to Radiated Emissions Testing. The Equipment Under Test (EUT) is placed in a test table and the measuring magnetometers are placed in 1m distance from the edges of the EUT boundary. The measuring points around the EUT have to be identified based on the model requirements. Depending on the mission requirements for model accuracy and quality the number of measuring points and the associate distance from the EUT can be decided in order to fulfill measurement validity. The relevant standards do not provide guidance for this purpose but techniques from the DC magnetic field testing can be partially utilized as described in the EMC ECSS standard.

As in other Emissions measurements, EUT operation has to be stabilized before starting the measurement. However, it might be necessary to collect test data during the turn-on period of the EUT to identify transient magnetic fields which might be necessary for modeling purposes. If multiple modes of operation are possible it might be necessary to measure magnetic field for each mode or magnetic field transients when the mode is changing.

The radiated magnetic field test results can be represented as a graphic output of the magnetic field expressed in units of dB above 1 pT versus frequency. Test details are important to be reported, like test conditions affecting magnetic emissions like EUT operation mode (and change of mode in case of transient fields), current consumption and other typical information like magnetometer orientation regarding to EUT orientation, cable types and layout that might be useful for test reproducibility.

As the frequency range of the magnetic field to be measured could be really low (within the Hz region for example) this measurement is facing, inherently, issues regarding the instrumentation noise floor, due to environmental noise, that affect the sensitivity of the measurement system. In recent missions, the sensitivity requirements have been lower than the noise floor. In such cases, the use of gradiometers to reject the common part of the ambient field should be considered to be used. It is also important that the measurement system must adequately filter human generated magnetic fields from nearby power lines. Experience shows that it is critical to effectively control the grounding of the test set-up. Also, dual (gradiometer) measurements could be required to remove geophysical and man-made noise affecting the measurement.

In cases where time domain measurements have been selected special attention should be drought to the Data Acquisition system performing the analog to digital conversion as it could contribute substantially to the noise floor.

Spectral density of spacecraft generated AC magnetic field measured at SCM sensor position, other than narrowband peaks described in currently candidate ESA mission, THOR, requirement (ESA THOR mission requirement document, 2015, para R-SC-2550), must not exceed the red curve connecting the points defined in the following table by a straight line (red curve in the attached plot, Figure 8).

Spectral density of spacecraft generated AC electric field noise measured by EFI sensors, other than narrowband peaks described in requirement R-SC-2570 (ESA THOR mission requirement document, 2015, para R-SC-2570), must not exceed the curve connecting the points defined in the following table (red curve in the attached plot, Figure 9).

Table 5. Reference Standards and frequency range of test application, magnetic field radiated emissions

Reference Standard	Frequency Range of application
MEDIC Handbook (2.6), RE04	30 Hz to 50 kHz
MIL-STD 461F (5.16), RE101	30 Hz to 100 kHz

Table 6. AC magnetic field power spectral density versus frequency limit

Power Spectral Magnetic Field Density	$2x10^{-3}$	$3x10^{-6}$	$2x10^{-8}$	$2x10^{-11}$	$2x10^{-11}$	$1x10^{-8}$	$1x10^{-7}$	nT^2/Hz
Frequency	0.1	1	10	300	2000	$1x10^{5}$	$2x10^{5}$	Hz

Figure 8. Test limits for radiated AC magnetic field emissions (Source: ESA THOR Missions Requirements Document, 2015)

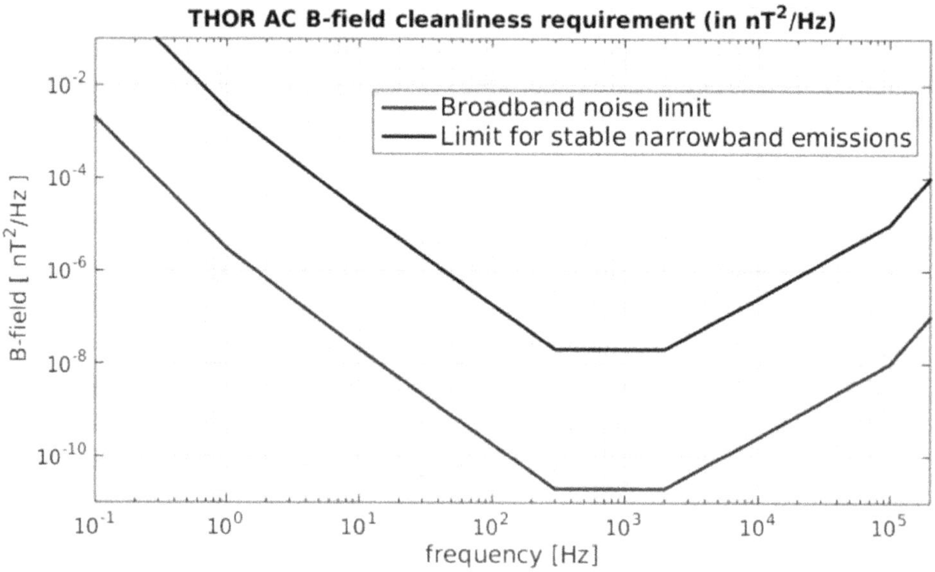

Table 7. AC magnetic field power spectral density versus frequency limit

Power Spectral Electric Field Density	$1x10^{-10}$	$1x10^{-12}$	$1x10^{-14}$	$2x10^{-17}$	$2x10^{-18}$	$2x10^{-18}$	$1x10^{-16}$	V/m^2
Frequency	1	10	100	1000	$1x10^{4}$	$1x10^{5}$	$2x10^{5}$	Hz

DC Magnetic Field Emission, Magnetic Moment

In various exploration missions, spacecrafts carry magnetometers in order to reveal the structural composition of planets and to study the complex interactions between the solar wind and the planetary environments. The Magnetic field levels to be measured during the mission are usually extremely low, so the spacecraft must be

Figure 9. Test limits for radiated AC electric field emissions (Source: ESA THOR Missions Requirements Document, 2015)

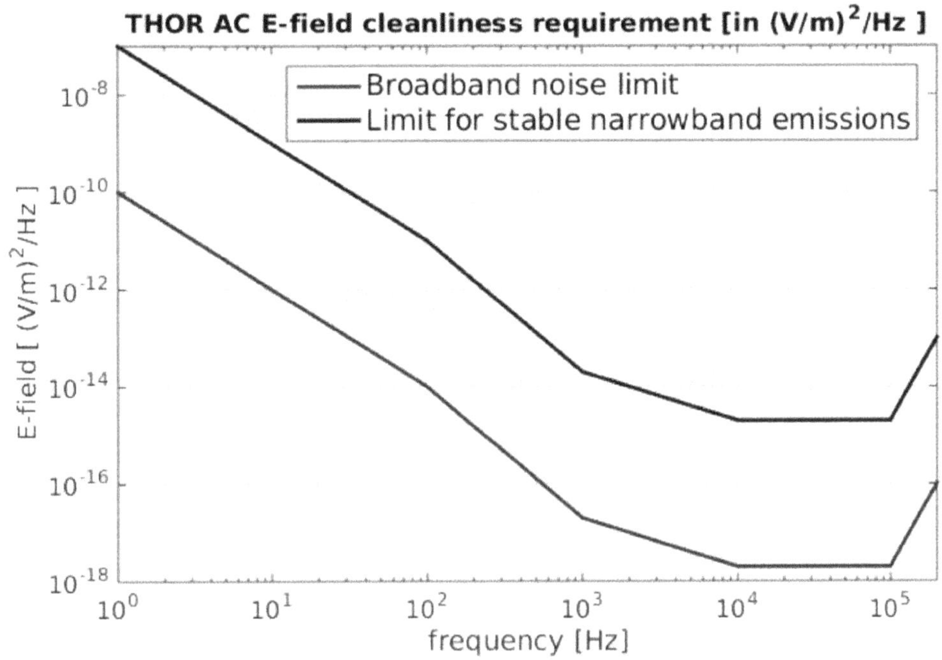

"magnetically clean". Magnetic cleanliness of a spacecraft is the purpose of DC magnetic field emissions testing. The target of the test is to obtain an estimation of the magnetic moment of the EUT.

The method involves the measurement of the DC magnetic field around the EUT at several points. The philosophy of the method is to represent the EUT with multiple magnetic dipoles based on the measurement data obtained. Then, by using numerical solvers for magnetic dipole modelling the magnetic signature of the EUT in terms of dipole(s) position and its magnetic moment(s). Having measurements in different distances and positions for the EUT the accuracy of the multiple magnetic dipoles model can be ensured.

For DC magnetic field testing, the EUT should be installed in an area where the Earth's magnetic field is compensated. A typical facility for this purpose is called Magnetic Coil Facility (MCF) and provides zero field conditions so as even weak EUT magnetic fields can be measured adequately.

The measuring instrumentation typically includes the following:

- Magnetic field measurement instrumentation, 3-axis Fluxgate magnetometers.
- Data Acquisition System.

The test setup is similar to Magnetic Field Radiated Emissions Testing but, as mentioned above, is performed in a Magnetic Coil Facility. The Equipment Under Test (EUT) is placed on a rotating table and the measuring magnetometers are located in well-defined positions around the turntable. Usually one to two magnetometers are used for this measurement. A data acquisition system collects the measurement data from the 3-axis magnetometers while the table is rotating in various step angles (i.e every 10^0). The measuring distance is determined based on sensitivity requirements coming from the mission objective. Such a facility is available at ESA ESTEC and relative information can be found at the hyperlink: http://sci.esa.int/lisa-pathfinder/45846-magnetic-coil-facility/

Test setup verification procedures are not described in relevant Standards but usually include the measurement of the residual field within the MCF test volume. It is important to measure the magnetic field within the test volume and how much this deviates from zero.

The measurement data are then used to model the EUT magnetic properties. Using Multiple Dipole Moment (MDM) solvers the EUT magnetic signature can be produced in the form of magnetic dipole(s) with known magnetic moment(s) and position(s). But, these solvers require measurement data of good quality to provide a good estimation of the magnetic dipoles properties. So, measurement distance should be selected carefully; data acquisition system selection is also of major importance as intrinsic noise may affect the measurement quality.

The relevant standards provide adequate guidance for this test. The test procedure includes measurements of the EUT in different angles on the rotating table in different test conditions:

- While EUT is not operating.
- After de-perming process (ECSS-E-ST-20-07C rev1, 2012, para 5.4.5.3, ECSS-E-HB-20-07A, 2012, para 7.1.4).
- After perming process (ECSS-E-ST-20-07C rev1, 2012, para 5.4.5.3, ECSS-E-HB-20-07A, 2012, para 7.1.4).
- Measuring the stray field while EUT is operating.

The DC magnetic field test results can be represented as a graphic output of the magnetic field expressed in units of nT versus angle (or, in general, versus sensor position). Test details to be reported include the test conditions (EUT operation, perm – deperm process).

In some cases, the sensitivity requirements of the measurement are low and it is important to handle measurement system noise issues. The use of gradiometers to reject the common part of the ambient field should be considered to be used. As in Radiated Magnetic Field testing it is important to adequately reject human generated magnetic fields existing in the measurement environment like nearby power lines. Detailed information may be found in the relative ESA EMC standard (ECSS-E-ST-20-07C, rev 1, 2012, para 5.4.5).

An example of typical DC magnetic field requirement requires that the total magnetic field of the spacecraft measured at the position of the mission magnetometer (MAG) outboard sensor shall be below 5 nT, (ESA THOR Mission Requirements Document, 2015).

Recent developments in the DC magnetic field measurement methodology, employs the use of multiple magnetometers around the EUT instead of rotating it and use one or two magnetometers. This methodology employs a "snapshot" like measurement of the magnetic field; the magnetic field is measured simultaneously through all magnetometers. A measurement of the ambient magnetic field in the area of measurement before and after the measurement of the EUT is performed to collect magnetic field data in order to cancel the effect of ambient field in the measurement of the EUT. In this way, there is no need for compensation of the Earth's field so as Magnetic Coil Facility is not required. Such a facility is currently in validation phase and is available at ESA ESTEC premises (Spantideas, Kapsalis, Kakarakis, Capsalis, 2016).

Conducted Susceptibility

Conducted immunity testing involves injecting RF voltages or currents into each of the cables associated with the equipment under test (EUT). As it is technically difficult to generate uniform fields in low frequencies in typical test facilities it is preferred to perform conducted testing as susceptibility problems at low RF frequencies are normally associated with cable coupling. Conducted susceptibility testing is an alternative, although not entirely equivalent, to radiated susceptibility testing at lower RF frequencies providing also very good reproducibility.

The purpose of these tests is to simulate the behavior of the EUT and its connected cables to RF noise in low frequencies present from various sources within the spacecraft. It is also to determine the susceptibility level of the EUT to these interference signals applied to EUT incoming and outcoming cables so as to ensure that EUT functionality and performance is not degraded from interference noise present on the power or signal lines.

To run Conducted Susceptibility tests could be a time-consuming process depending on the number of the cables attached to the EUT. Additionally, this test requires various types of equipment to cover the frequency range of interest starting from 30 Hz up to 100 MHz or more for Space applications. The test is divided in two frequency sub-ranges one concerning the lower frequencies up to 50-100 kHz and the other the remaining range up to 100 MHz. The tests are applicable to any type of equipment.

The tests for conducted susceptibility require a shielded room to be performed. This is mostly in order to avoid producing noise that may affect other activities nearby but also to ensure that ambient interference signals will not affect the test. The instrumentation that is required to perform the measurements includes the following:

- Signal generator with amplitude or pulse modulation capability
- Pulse (spike) generator, 1 kHz – 100 kHz, adjustable duty cycle (for voltage transient test); output characteristics should be in accordance with ECSS standard
- Power amplifier
    - Audio Frequencies up to 50 kHz
    - 50 kHz to 100 MHz or more
- Current probes
    - Current injection probe, 50 kHz – 100 MHz
    - Current measurement probe for lower frequencies
    - Current measurement probe, 50 kHz – 100 MHz
- Differential high voltage-probe
- Injection transformer
- Accessories
    - Power metal film resistor with high peak power capability
    - One or two calibration fixture(s)
- Oscilloscopes
- Data (waveform) recording device
- Spectrum analyzer (optional)
- LISN (optional)

As mentioned also in other measurements, it is required to verify the correct operation of the test setup before starting the test such as the current probes, the cables and the data recording equipment. For this purpose, a suitable calibrated signal generator is required. The relevant procedures can be found in detail in in the relative ESA EMC standard (ECSS-E-ST-20-07C, rev 1, 2012, para 5.4.7). It is noted that the current probes are used for monitoring the current applied to the line tested but also for the application of the interference signal for the upper frequency sub-range.

The Equipment Under Test (EUT) is installed in the test setup together with the associated cabling. It is important to determine the type of cables that will be tested (power, analog/digital signal) in order to identify a way to assess the correct functionality and the performance of the EUT. For, example, to assess the susceptibility of the power bus an oscilloscope might be used to measure the voltage ripple and compare with the applicable limit. Digital signal lines might need special equipment to identify susceptibility levels like instruments measuring Bit Error Ratio in a digital bus. The EUT operation has to be stabilized before starting the measurement. If multiple modes of operation are possible it might be necessary to repeat the test for each mode separately.

The representation of Conducted Susceptibility test results includes test data relative to the test system, like the level of the RF signal versus frequency, together with the metrics used to assess the performance or the functionality of the EUT. When the performance of the EUT is degraded during testing this should be recorded as detailed as possible (i.e the signal frequency when the degradation happened and information about the performance degradation). Other test relevant details could be reported, such as test conditions like cable types and layout, EUT operation mode and any other related information that might be useful for test reproducibility.

The typical Conducted Susceptibility tests used for Space applications are described below.

a. Conducted Susceptibility on power lines (audio frequencies). The frequency range of the interference application is typically between 30 Hz to 100 kHz. In this frequency range, an injection transformer is used to apply the sinusoidal interference signal to the EUT power leads. A current probe is used to monitor the applied current (Figure 10).

Figure 10. Measurement system test setup (ECSS-E-ST-20-07C rev 1)

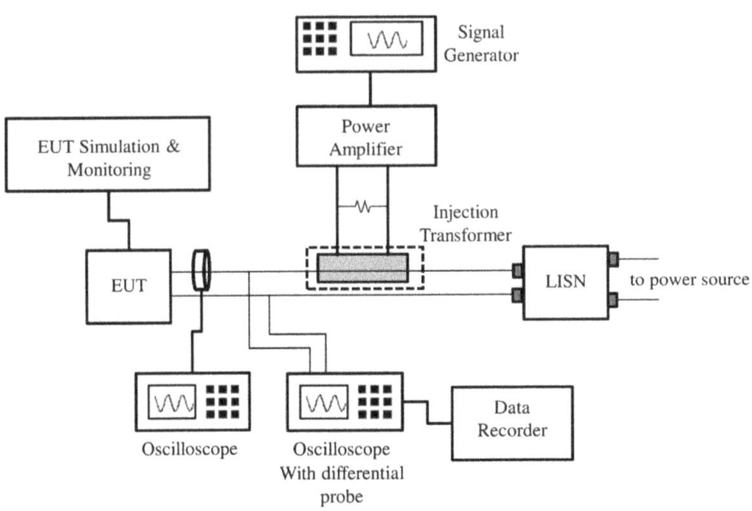

b. Conducted Susceptibility on power and signal/data lines, bulk current injection (upper frequency range). The frequency range of the interference application is typically between 100 kHz to 100 MHz. A current injection probe is used to apply the sinusoidal pulsed modulated interference signal to the EUT associated cables power or signal ones. As in the acoustic frequency range test, a current probe is used to monitor the applied current (Figure 11, Figure 12).

An example of conducted susceptibility requirement from a real ESA mission (PLATO), suggests that the levels of the applied voltage levels (measured at 1 Ω resistor), for the differential mode testing should be:

- $V_{injection} = 1\ V_{RMS}$, in the frequency range from 30 Hz to 50 kHz
- $V_{injection} = 1\ V_{RMS}$, in the frequency range from 50 kHz to 50 MHz, with a 50% AM envelope modulation 1 kHz square wave.

Table 8. Reference Standards and frequency range of test application, conducted susceptibility, power lines

Reference Standard	Frequency Range of application
ECSS-E-ST-20-07C rev1 (5.4.7)	30 Hz to 100 kHz
MEDIC Handbook (2.7), CS01	30 Hz to 50 kHz
MIL-STD 461F (5.7), CS101	30 Hz to 150 kHz

Figure 11. Bulk Current injection excitation pulse (Source: ECSS-E-ST-20-07C revision 1)

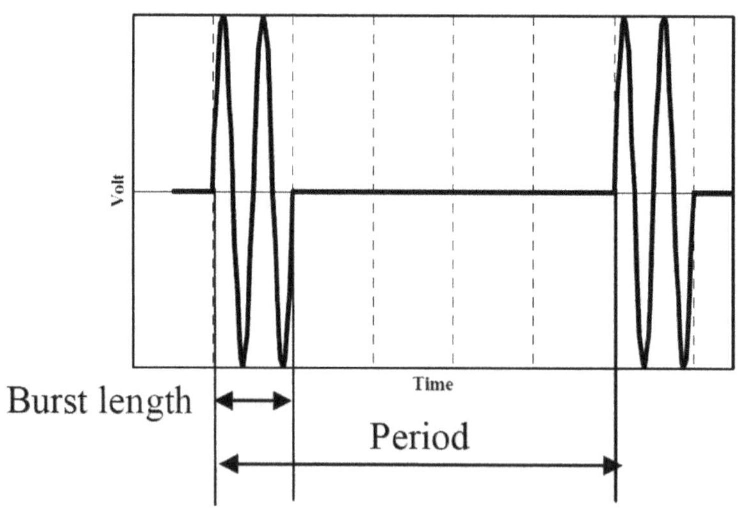

Figure 12. Bulk Current injection test setup (ECSS-E-ST-20-07C rev 1)

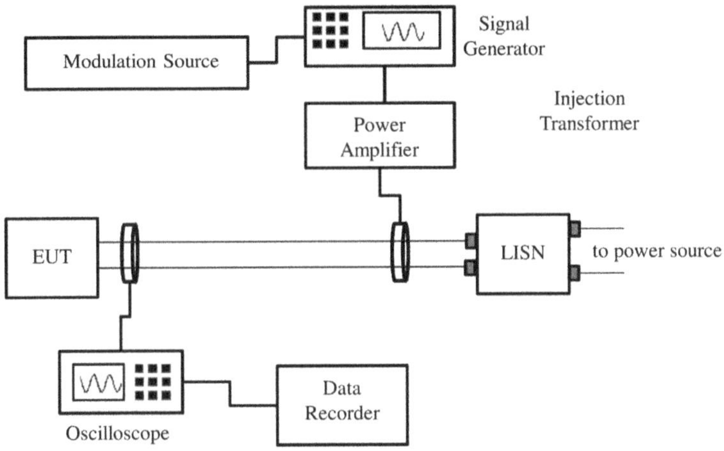

Additionally, the common mode applied levels should be as follows:

- $V_{injection}$ = 1 V_{RMS}, in the frequency range from 50kHz to 50MHz, with 50% AM envelope modulation 1 kHz square wave

Bulk Current Injection method is described in a lot of standards like the ESA EMC standard (ECSS-E-ST-20-07C, rev1, 2012, para 5.4.8) the ISO 11452-4 automotive standard, the MIL-STD-461-G, 2007 para 5.13) in which is named as test CS114 (Figure 13). It is also included in the test standard for commercial equipment testing, IEC 61000-4-6, where it is proposed as an alternative method to direct injection when a non-invasive injection of the interference signal is required such in the case of multi-conductor non-standard cables. As mentioned above, conducted susceptibility methods, like bulk current injection, has been employed as alternatives to radiated susceptibility testing which is considered to face various difficulties to be conducted in this frequency range especially regarding field generation.

Recent studies findings (Hill, 1992), (Adams, Cruz & Melquist, 1992) claims that radiation and probe injection cannot be considered as direct equivalent methods due to the fact that their nature is different; injection probes behave like lumped sources while radiated fields based on distributed coupling along the length of the cable. Interference signal injection with current probe can be considered as equivalent provided that the induced current can be correlated with the one developed when radiated field is applied. Recently, a new technique has been presented (Javor, 1997), (Klingler, Szelag & Heddebaut, 1994) known as Double Bulk Current Injection (DBCI) where two current probes are used for RF interference injection. Experimental

Table 9. Reference Standards and frequency range of test application, conducted susceptibility, power and signal lines

Reference Standard	Frequency Range of Application
ECSS-E-ST-20-07C rev1 (5.4.8)	50 kHz to 100 MHz
MEDIC Handbook (2.8), CS02	50 kHz to 400 MHz
MIL-STD 461F (5.13), CS114	10 kHz to 200 MHz

preliminary evaluations (Grassi, Spadacini, Marliani & Pignari, 2008), (Spadacini & Pignari, 2004) show that, even approximately, the injected radio frequency (RF) currents show correlation to the noise currents coupled to the EUT and the harness from direct electromagnetic field.

The test setup is depicted in Figure 13. The two current probes are fed separately with the specified signal but with amplitude and phase control for each one. There are some limitations in the method application but it has been evaluated in the recent years and is considered as ready to be implemented.

c. Conducted Susceptibility on power lines, voltage transients. These voltage transients are short and applied on the input power lines of the EUT. They are simulating voltage transients that may appear in the power lines as a consequence of other power bus users switching on and off other units or modules or when these units change operating mode. Moreover, voltage spikes may be produced when there is a short circuit on the power bus.

Figure 13. Measurement system test setup (ECSS-E-ST-20-07C rev 1)

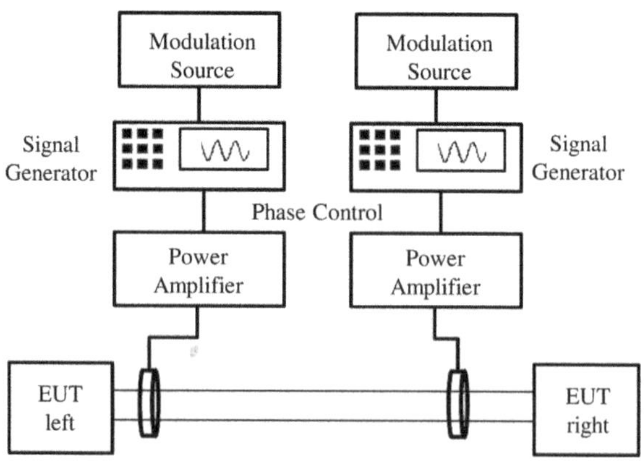

Table 10. Reference Standards and frequency range of test application, conducted susceptibility, power lines, voltage transients

Reference Standard	Frequency Range of Application
ECSS-E-ST-20-07C rev1 (5.4.8)	50 kHz to 100 MHz
MEDIC Handbook (2.8), CS02	50 kHz to 400 MHz
MIL-STD 461F (5.13), CS114	10 kHz to 200 MHz

The voltage spike injection is performed by using a suitable pulse generator either in series or in parallel with the power lines under test. The voltage transient pulse waveform to be applied is described also in the standards and in Figure 14 and Figure 15.

A verification procedure is available in the relevant standards. Typically, in both cases (in series or parallel), the injection shall be calibrated on a 5 Ω resistor. After the calibration procedure, the same settings shall be used for the injection.

Bulk Cable Injection test, as described in the relevant ESA EMC Standard (ECSS-E-ST-20-07C, rev1, 2012) includes an arrangement that utilizes sinusoidal signal pulse modulated only. However, there are also other BCI tests not included in the ECSS-E-S-20-07C standard which are considered to cover interference signals other than sinusoidal ones. These are described as CS115 and CS116 BCI tests in the relative military EMC standard (MIL-STD-461-F, 2007, para 5.14 & 5.15). These tests are also discussed in the ESA EMC Handbook (ECSS-E-HB-20-07A, 2012, para 4.3.9).

Figure 14. Conducted Susceptibility on power lines, voltage transients test setup, in series application (ECSS-E-ST-20-07C rev 1)

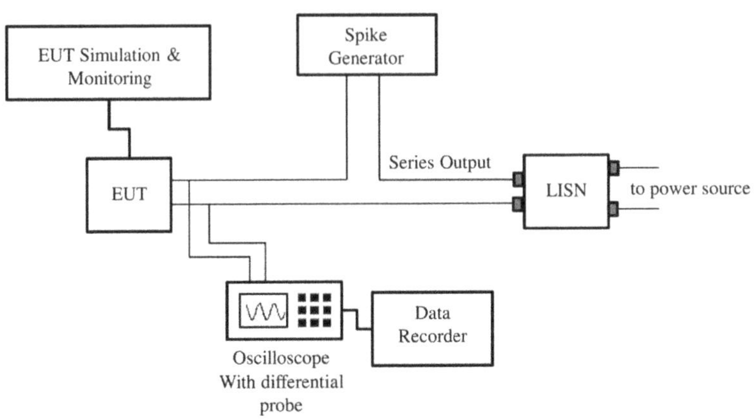

Figure 15. Conducted Susceptibility on power lines, voltage transients test setup, in parallel application (ECSS-E-ST-20-07C rev 1)

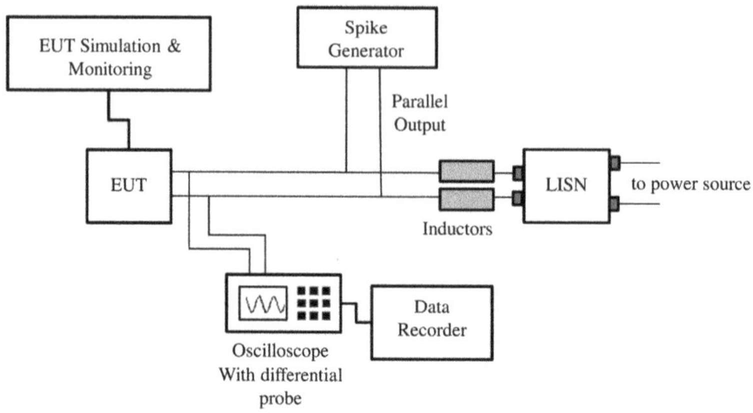

CS115 (Impulse Excitation, Figure 16) test can be performed in order to verify that the EUT functionality and performance is not affected by fast rise and fall time transients that may be present due to platform switching operations and coupled to EUT associated cables.

Figure 16. MIL-STD-461F/CS115 signal characteristics (Source: DoD MIL-STD-461F)

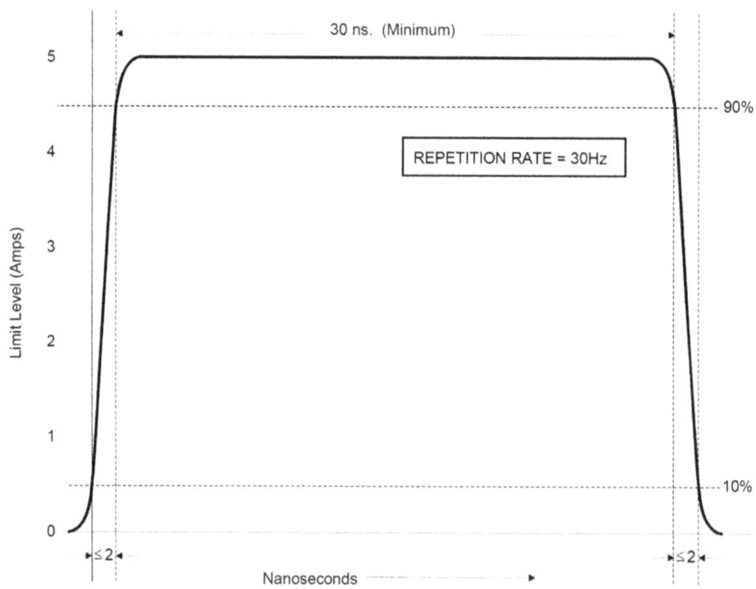

CS116 (Damped Sinusoidal Transients, Figure 17) test is performed to verify that the EUT functionality and performance is not affected when damped sinusoidal transients coupled onto EUT associated cables and power leads; the basic concept of this test setup is to simulate electrical voltage and current waveforms of a damped sine.

Radiated Susceptibility, Electric Field

A unit or a module within a spacecraft is subjected to radiated electric fields coming from nearby units especially the ones that intentionally utilize radio frequencies i.e for wireless transmission, or from external sources like the launch pad or the launcher. Radiated immunity testing involves the application of an RF electric field the equipment under test (EUT) boundary, i.e with its associated cables.

The purpose of these tests is to apply electric field (to simulate the RF noise) using transmitting antennas in RF frequencies and assess the EUT functionality and its performance. Additionally, the susceptibility level of the EUT to electric field applied should be determined in order to apply countermeasures, if required.

Radiated Susceptibility tests are also time-consuming as, typically, a large frequency range has to be covered. The typical frequency range is between 30 MHz to 18 GHz. The tests are applicable to any type of equipment.

The tests for radiated susceptibility should only be performed in a shielded room in order to avoid bothering nearby activities. Instrumentation required to perform the measurements includes the following:

- Signal Generators suitable to cover the whole frequency range
- RF Power Amplifiers
- RF power meter

Figure 17. MIL-STD-461F/CS116 signal characteristics (Source: DoD MIL-STD-461F)

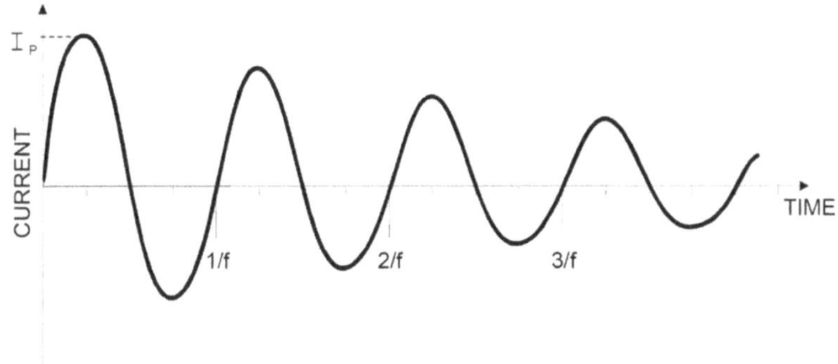

- Directional Couplers
- RF Attenuators
- Linearly polarized transmitting antennas; the following antennas are commonly used:
  - 30 MHz to 200 MHz, biconical antenna
  - 200 MHz to 1 GHz, double ridge horn, or log-periodic antenna
  - 1 GHz to 18 GHz, double ridge horn antenna
- Receiver Antenna(s) for 1 to 18 GHz for verification purposes
- Electric field sensors
- Measurement receiver
- Data recording device
- LISN (optional)

A verification procedure is described in detail in the relevant ECSS Standard in order to check the measurement system operation before starting the test such as the signal generator and the RF amplifiers and the transmitting antennas. This is achieved by using receiving antennas or electric field sensors, which are physically small – electrically short devices, in a specific test setup. A suitable calibrated signal generator and a measuring receiver is required to perform this system check. The relevant procedures can be found in detail in the ESA EMC Standard (ECSS-E-ST-20-07C, rev1, 2012, para 5.4.11).

The Equipment Under Test (EUT) is installed in the test setup together with the associated cabling and is left to stabilize after turned on (Figure 18). The EUT performance is assessed in the same way like in conducted susceptibility testing. Various instruments can be used for monitoring the EUT behavior depending on the type of the signal cables connected (power, analog/digital lines). So, instruments like oscilloscopes, protocol analyzers etc.

The level of the RF signal applied to produce the required electric field versus frequency, together with the EUT susceptibility test results relative to the assessment of the EUT performance should be reported. When degradation of the EUT performance is noted during testing, it is reported in detail to better aim the corrective actions required to overcome the problem. Other details are useful to be also reported, such as test conditions like cable types and layout, EUT operation mode and any other related information that might be useful for test reproducibility.

A radiated susceptibility test requirement from a real ESA mission (PLATO) is defined in the relevant document as follows:

The applied electromagnetic field in the frequency range 30MHz-18GHz in both vertical and horizontal polarizations should be applied with the following intensity (1 kHz 50% square wave modulated):

Equipment with harness routed:

Figure 18. Radiated Susceptibility test setup (ECSS-E-ST-20-07C rev 1)

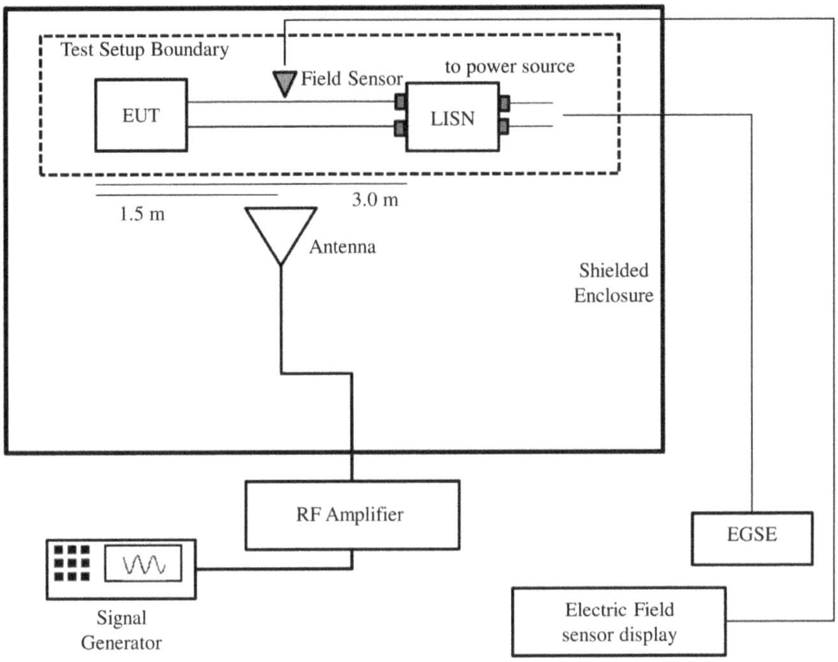

- Inside the S/C body: 2 V/m
- Outside the S/C body (even partially): 10 V/m", (ESA PLATO EMC Specification Document, 2014)

Radiated Susceptibility, Magnetic Field

Magnetic field radiated magnetic field testing involves the application of low frequency magnetic fields on the EUT. The test is focused to identify the EUT performance when submitted to low frequency magnetic fields. The purpose is

Table 11. Reference Standards and frequency range of test application, radiated susceptibility, electric field

Reference Standard	Frequency Range of Application
ECSS-E-ST-20-07C rev1 (5.4.11)	30 MHz to 18 GHz
MEDIC Handbook (2.11), CS02	14 kHz to 10 GHz
MIL-STD 461F (5.20), RS103	2 MHz to 40 GHz

to determine the susceptibility of the EUT to such fields. In general, this test is performed when there is estimation that the equipment is potentially sensitive to low frequency magnetic fields or when it is going to be installed nearby to a low frequency magnetic field source.

This test is presented in detail in the ECSS standard and it is not suitable to apply the magnetic field to EUT harness due to the size of the magnetic loops used. More specifically the magnetic field is applied around the EUT face or to the electrical interface connectors, parallel to the EUT faces and parallel to the axis of connectors. The frequency of the applied magnetic field is typically between 30 Hz to 100 kHz. The test instrumentation could include the following:

- Signal source
- Power amplifier
- 12 cm radiating loop having the following specifications
- 4 cm loop sensor
- Measurement receiver
- Calibration fixture in order to verify the level of the applied signal
- Current probe
- LISN

The verification procedure is also described in detail in the relevant ECSS-E-ST-20-07C rev1 (5.4.11) Standard. In this Standard, the applicable signal levels are verified using a specific test setup (Figure 19). The Equipment Under Test (EUT) is installed in the test setup together with the associated cabling and should be left to stabilized after turned on. The EUT performance is assessed in the same way like in other susceptibility testing.

Figure 19. Radiated Susceptibility, Magnetic Field test setup (ECSS-E-ST-20-07C rev 1)

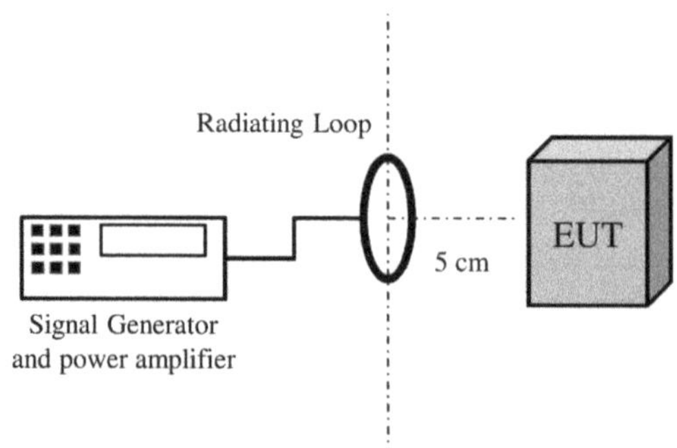

Regarding test results representation, it is required to report the level of the magnetic field applied to the EUT together with assessment results of the EUT performance. Any degradation of the EUT performance is noted during frequency scan. Other details are useful to be also reported, such as test conditions like cable types and layout, EUT operation mode and any other related information that might be useful for test reproducibility.

As mentioned, this test is performed using a continuous wave excitation signal as interference. Other conditions that may be present in a spacecraft, like transients coming from high current pulses, where the phenomenon is considered as broadband or when sensitive harness is subjected to magnetic fields coming from Power Distribution units (i.e from solar arrays), cannot be simulated using this test and suitable test setups and procedures would be prepared, in a case specific basis.

Susceptibility to Electrostatic Discharges

In some previous missions, satellites experienced in orbit anomalies due to electrostatic discharges. For this reason, a test setup and procedure is included in the relevant ESA EMC Standard (ECSS-E-ST-20-07C, rev1, 2012, para 5.4.12). to define a test method for application of electrostatic discharges to harness attached to equipment on a spacecraft. It is noted that it is identified that ESD test generators used in commercial products industry and defined in the commercially used for ESD testing, (IEC 61000-4-2, 2008, para 6) can be also used for space testing.

The test setup is proposed also in the relative ISO standard for Electrostatic Discharge testing (ISO 14302, 2002) and is widely used in several industries. It utilizes an injection circuit utilizing a wire for coupling the electrostatic discharge pulse to the harness under test. This injection circuit is an RC pulse forming network and can be connected to an ESD generator adjusted in contact discharge mode. It is considered as common mode injection by coupling a parallel wire to harnesses under test. The necessary equipment to perform the electrostatic discharge test should be as follows:

Table 12. Reference Standards and frequency range of test application, radiated susceptibility, magnetic field

Reference Standard	Frequency Range of Application
ECSS-E-ST-20-07C rev1 (5.4.10)	30 Hz to 100 kHz
MEDIC Handbook (2.10), RS02	Transient pulses only
MIL-STD 461F (5.19), RS104	30 MHz to 100 kHz

- ESD generator (Figure 21) as specified in IEC 61000-4-2 (or other high voltage source)
- The discharge primary circuit
- Monitoring devices:
  - Two current probes, 100 MHz bandwidth
  - One high-voltage probe, 1 MHz bandwidth
- Two-channels digital oscilloscope

Figure 20. Measurement system test setup (ECSS-E-ST-20-07C rev 1)

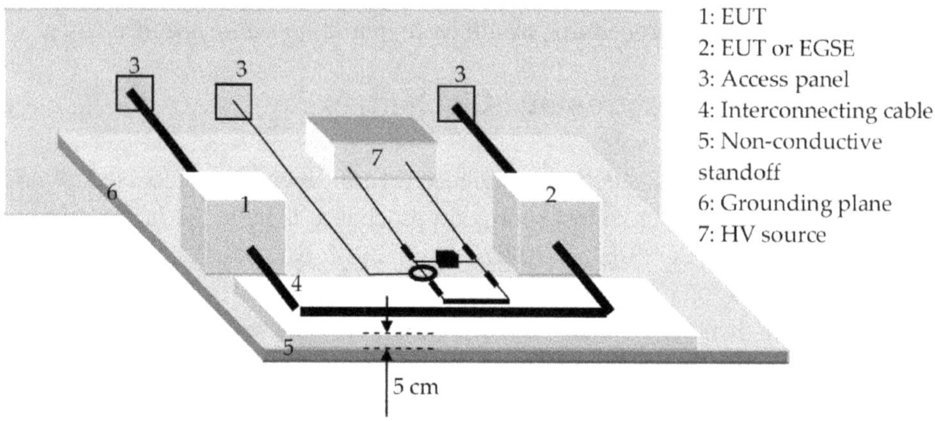

Figure 21. ESD Generator, pulse forming network (ECSS-E-ST-20-07C rev 1)

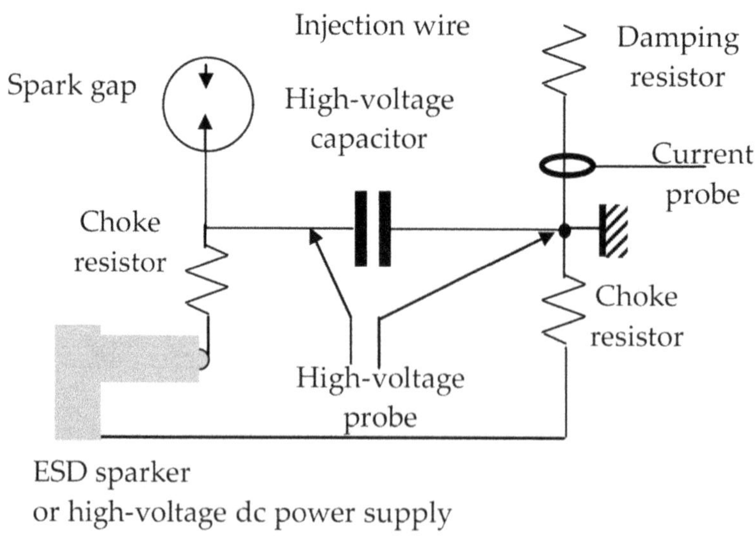

The test setup is depicted in Figure 20 and the application of the pulse is made by the ESD generator inducing the high voltage pulse through the pulse forming network to a wire which is in parallel with the harness under test.

In this way, the discharge pulse is applied by conductive coupling which also ensures that the pulse characteristics are controlled better. Moreover, the location of the injection is better defined.

However, there are still test conditions that may affect reproducibility like the length of the discharge loop in relation to the length of the cable under test and also the distance between the two affecting the coupling position. These conditions are described in the relative ECSS standard but leave some freedom to test operators. Another issue affecting the test conduction is the possibility to use different kind of high voltage sources except the typical ESD generator. The two main parameters of the test, the ESD generator and the coupling circuit affect, to some extent, the reproducibility.

In a real ESA mission, the applicable electrostatic discharge susceptibility levels are defined as follows:

The unit shall not exhibit any failure, malfunction, unintended responses, degradation of performance or deviation beyond its specified BOL performance tolerances when exposed to ESD on its interfaces as follows:

- ≥15 discharges
- ≥15 mJ / pulse
- Pulse voltage 15 kV
- Repetition rate 0.1-1 Hz", (ESA PLATO EMC Specification Document, 2014)

FUTURE WORK

Measurement reproducibility and repeatability was always issues of main concern in Electromagnetic Compatibility testing. It is evident that performing the tests in the same way in different test sites should be the target of all people involved in EMC test campaigns. One way to ensure good measurement reproducibility/repeatability is to follow the methodologies described in the relevant standards when possible. It is, however, proven by experience that in practice there are still reproducibility issues. On the other hand, in the Space sector there are several cases where, driven by the mission requirements, there is some need of tailoring of the EMC test setups and procedures. Moreover, there are some areas that the test setups and procedures need to be improved in terms of being more representative of the nature of the phenomena.

CONCLUSION

Electromagnetic Compatibility Testing for Space Applications is dealing with several physical phenomena coming out from the operation of the on-board equipment of a spacecraft, of the launch pad or the launch vehicle but also from the external environment as well. Consequently, several test setups are defined in relevant standards and used by companies, organizations and laboratories to identify potential non-conformities and aim the mitigation measures to be found and applied.

REFERENCES

Adams, J. W., Cruz, J., & Melquist, D. (1992). Comparison measurements of currents induced by radiation and injection. *IEEE Transactions on Electromagnetic Compatibility*, *34*(3), 360–362. doi:10.1109/15.155856

ESA. (2014). *PLATO EMC Specification, ESA-PLATO-ESTEC-MIS-SP-002*. ESA TEC-EEE.

ESA. (2015). *THOR Mission Requirements Document ESA-THOR-EST-MIS-RS-001, ESA-ESTEC*. ESA THOR Study Team.

Grassi, F., Spadacini, G., Marliani, F., & Pignari, S. A. (2008). Use of double bulk current injection for susceptibility testing of avionics. *IEEE Transactions on Electromagnetic Compatibility*, *50*(3), 524–535. doi:10.1109/TEMC.2008.926810

Hill, D. A. (1992). Currents induced on multiconductor transmission lines by radiation and injection. *IEEE Transactions on Electromagnetic Compatibility*, *34*(4), 445–450. doi:10.1109/15.179277

International Electro-technical Commission, IEC 61000-4-2. (2008). *Electromagnetic compatibility (EMC) - Part 4-2: Testing and measurement techniques - Electrostatic discharge immunity test*. Author.

International Standardization Organization (ISO). (2002). *Space Systems – Electromagnetic Compatibility Requirements*. Author.

Javor, K. (1997, August). On field-to-wire coupling versus conducted injection techniques. In *Electromagnetic Compatibility, 1997. IEEE 1997 International Symposium on* (pp. 479-487). IEEE.

Klingler, M., Szelag, M., & Heddebaut, M. (1994). Double bulk current injection: A possible substitute to field-to-wire coupling. *Proc. EUROEM*, 94, 1249-1256.

National Aeronautics and Space Administration. (1995). *Marshall Space Flight Center Electromagnetic Compatibility Design and Interference Control (MEDIC) Handbook*. Marshall Space Flight Center.

Secretariat, E. C. S. S. (2012). *ECSS-E-ST-20-07C Rev. 1, Space Engineering – Electromagnetic Compatibility, ESA-ESTEC*. Requirements & Standards Division.

Secretariat, E. C. S. S. (2012). *ECSS-E-HB-20-07A, Space Engineering – Electromagnetic Compatibility Handbook, ESA-ESTEC*. Noordwijk, The Netherlands: Requirements & Standards Division.

Spadacini, G., & Pignari, S. A. (2004). A bulk current injection test conforming to statistical properties of radiation-induced effects. *IEEE Transactions on Electromagnetic Compatibility*, *46*(3), 446–458. doi:10.1109/TEMC.2004.831896

Spantideas, S. T., Kapsalis, N. C., Kakarakis, S. D., & Capsalis, C. N. (2016, May). A novel technique for accurate extrapolation of complex magnetic sources. In *Aerospace EMC (Aerospace EMC), 2016 ESA Workshop on* (pp. 1-5). IEEE. doi:10.1109/AeroEMC.2016.7504599

US Department of Defense. MIL-STD-461-F. (2007). Requirement for the control of electromagnetic interference characteristics of subsystems and equipment. Author.

KEY TERMS AND DEFINITIONS

BCI: Bulk current injection.
BIPM: Bureau international des poids et mesures.
CM: Common mode.
DC: Direct current; as acronym also a synonym for "time-invariant."
DM: Differential mode.
DoD: Department of Defense (USA).
ECSS: European cooperation for space standardization.
EMC: Electromagnetic compatibility.
EMI: Electromagnetic interference.
ESA: European Space Agency.
ESTEC: European Space Research and Technology Centre.
EU: European Union.
EUT: Equipment under test.
LISN: Line impedance stabilization network.
MCF: Magnetic coil facility.

MEDIC: Marshall Space Flight Center electromagnetic compatibility design and interference control.

MDM: Magnetic dipole modelling.

NASA: National Aeronautics and Space Administration.

OATS: Open area test site.

PLATO: Planetary transits and oscillations of stars (ESA mission).

RF: Radio frequency.

S/C: Spacecraft.

SCM: Search coil magnetometer.

THOR: Turbulence heating observer.

Chapter 5
Numerical Assessment in Aeronautics for Electromagnetic Environmental Effects

Miguel David Ruiz-Cabello Nuñez
University of Granada, Spain

Sergio Fernández Romero
National Institute of Aerospace Technology (INTA), Spain

Marc Pous
Universitat Politècnica de Catalunya (UPC), Spain

Enrique Pascual Gil
AIRBUS, Spain

Luis M. Diaz Angulo
University of Granada, Spain

David Poyatos Martínez
National Institute of Aerospace Technology (INTA), Spain

Mireya Fernández Chimeno
Universitat Politècnica de Catalunya (UPC), Spain

Guadalupe G. Gutierrez
Airbus Defence and Space, Spain

Amelia Rubio Bretones
University of Granada, Spain

Manuel Añon Cancela
National Institute of Aerospace Technology (INTA), Spain

DOI: 10.4018/978-1-5225-5415-8.ch005

Ferran Silva
Universitat Politècnica de Catalunya (UPC), Spain

Jesus Alvarez
AIRBUS, Spain

Mario Fernández Pantoja
University of Granada, Spain

Borja Plaza Gallardo
Ingeniería de Sistemas para la Defensa de España S.A. (ISDEFE), Spain

Luis Nuño
Polytechnic University of Valencia, Spain

Rafael Gómez Martín
University of Granada, Spain

David Escot Bocanegra
National Institute of Aerospace Technology (INTA), Spain

Pere J. Riu
Universitat Politècnica de Catalunya (UPC), Spain

Rafael Trallero
National Institute of Aerospace Technology (INTA), Spain

Ricardo Jauregui Telleria
Universitat Politècnica de Catalunya (UPC), Spain

Salvador G. Garcia
University of Granada, Spain

ABSTRACT

Electrical and electronic systems on board air vehicles are susceptible to electromagnetic interference (EMI). This has made the topic of electromagnetic compatibility (EMC), a major concern for aircraft safety. The use of composite materials worsens this situation, for their poor shielding and low conductive capabilities. Some of the main experimental E3 certification scenarios used in aeronautics are revisited in this chapter. Guidelines to achieve simple, yet accurate, numerical models of them are provided, with appropriate tradeoffs between computational simplicity and accuracy. The numerical method, endowed with extended capabilities, has been chosen for this task for its ability and efficiency to deal with complex problems of arbitrary materials. The feature selective validation (FSV) IEEE standard procedure, commonly used to quantify the comparison of data in electromagnetic problems, is also revisited. The simulation of three different air vehicles in several certification scenarios is finally described and the numerical results compared to experimental data.

INTRODUCTION AND BACKGROUND

Modern aircrafts have become increasingly dependent on electronic equipment to control its systems. This has led to new safety concerns with respect to their immunity levels against EM hazards and, in consequence, their assessment by aircraft manufacturers. Besides, the variety of potential EMI sources are increasing dramatically due to the appearing of new artificial sources in addition to the natural ones. This situation is aggravated by the pervasive use of composite materials in aircraft structure: CFC, CFRC, CFRP, etc. These materials, lighter and stronger from the mechanical point of view, are poorer conductors than metals and therefore have lower shielding capabilities.

From the EMC point of view, the main EM threats for an AV can be summarized as follows:

- **Lightning Indirect Effects (LIE): 0 to ~50 MHz:** Indirect effects are caused by the electric current flowing through the structure and internal wiring as a consequence of the impact of a lightning strike. This is, undoubtedly, the most important threat to onboard electronic equipment, becoming critical for aircrafts mostly made of CFCs (Meyer et al., 2008) like, for instance, modern UAVs.

Figure 1. Typical EM Hazard in aeronautics. Reprinted from (Cabello, 2017d)

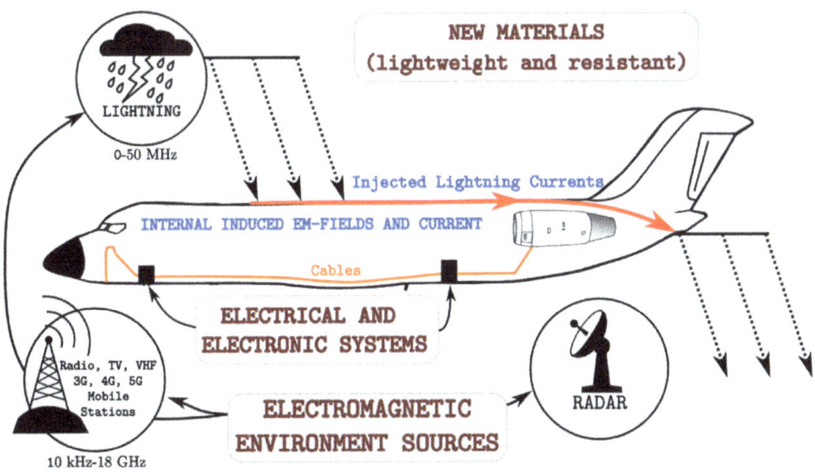

- **High Intensity Radiated Fields (HIRF): 0 to ~100 MHz:** This threat is caused by artificial intentional or unintentional, external or internal RF sources. These are constantly increasing in number: TV, mobile networks (3G/4G/5G), radars, navigation satellite systems, etc. and may couple to cables and equipment, potentially causing malfunction for high field levels.
- **Electromagnetic Pulses (EMP) of Nuclear or Non-Nuclear Origin: 0-100 MHz:** Most of current non-nuclear low-level electromagnetic pulse generators are not capable of radiating enough EM energy to produce a significant damage. However, novel devices are appearing, which are able to involve much higher power levels with extremely short durations. These modern weapons, also named E-bombs, are becoming cheaper and susceptible of being used in terrorist acts. They can wreak havoc on computers and networks, yielding temporary disruptive effects (Radasky et al., 2004; Radasky, 2014).

In all these cases, as a result of the exposition to the EM hazard, transient currents will flow along the aircraft surface, creating EM fields which penetrate into the fuselage through apertures such as windows, or by diffusion through parts made of poorly conducting materials (*Figure 1*). Inside the aircraft, these transients induce currents to the EWIS, which, in turn, couple to equipment, potentially compromising their safe operation or even creating permanent damage.

Several aviation incidents, reported in the last quarter of the past century, triggered the attention of aviation agencies to include strict EMC requirements for the airworthiness certification processes of AVs during their whole life-cycle. Before a newly developed AV model is permitted to operate, it must get a certificate of airworthiness issued by an aviation regulatory authority. For instance, in civil aviation, the regulatory authority is the EASA in the EU, or the FAA in the US. Within the EMC context, certification methods are mainly based on experimental tests and the guidelines provided in several standardized documents and certification guides: Eurocae ED-105/SAE-ARP5416; SAE-ARP5415; Eurocae ED-107/SAE-ARP5583; MIL-STD-202; MIL-STD-461;EUROCAE ED-14/RTCA/DO-160, MIL-STD-464, STANAG 4370, etc..

Classically, the EMC assessment of EMI in the aeronautic industry has relied on experimental campaigns, often carried out at late stages of the aircraft manufacturing when correction and redesign costs due to unforeseen EMI issues are prohibitive. Nowadays, the computing power of modern clusters is making it possible to employ CEM tools to simulate and predict EMC issues. At early stages of design, these allow engineers to make design decisions like the use of alternative cable routings, shielding materials, bonding components, material replacement, return network addition, etc. CEM tools are demonstrating their utility at all the design phases of an aircraft, to improve EMC safety aspects, and to ease and speed up the experimental certification process.

However, international aviation agencies are still reluctant to accept simulations as a replacement of experimental tests. To achieve this, strict procedures need to be consolidated, and standard engineering guidelines must be agreed. These must describe a systematic methodology to achieve *validated* computational models, suitable to produce data which is acceptable for certification purposes. A formal declaration about this topic was stated by the EASA and constituted one of the main outcomes of the HIRF-SE project, funded by the European Union under the FP7 program between 2008 and 2013, with a 23 M€ budget. This project gathered over 40 partners from 11 EU countries, including all major universities, airframers, test-houses, and software-houses. The main target was to build a synthetic framework to be used for the validation of numerical solvers with experimental data in the assessment of HIRF effects on AVs.

During HIRF-SE, the concept of a *validated* model for a specific AV was settled down. Such a model must include, with full traceability and reproducibility, the whole path necessary to achieve the input required by the computer solver; starting from the CAD data of the AV and the experimental setup, with the steps followed to clean, simplify, and mesh it, resulting in the actual input file fed to the solver. It must also include all the mathematical models (material treatment, cables, sources, post-processing…) used by the solver, the methods employed for its computer implementation, the executable version of all the tools used in the process, and the solver itself used for the simulation. Finally, a sufficient set of experimental and simulation data agreeing according to some consolidated validation criterion must also be provided. The typical workflow of the whole procedure is given in *Figure 2*.

Several other European partners, including all major aeronautic stakeholders, agencies and public research entities, have also pointed in the same direction[1]. Some of these comments have been included in the EUROCAE certification guides for numerical methods (Eurocae ED-107A, 2010), as a support for experimental campaigns.

At Spanish national level, the HIRF-SE project was continued through several other projects (UAVEMI, UAVE3, Alhambra-UGRFDTD…) funded by the Ministry of Economy, the Andalusian regional govt., and by AIRBUS. These projects gather

Figure 2. CEM for EMC workflow in aeronautics

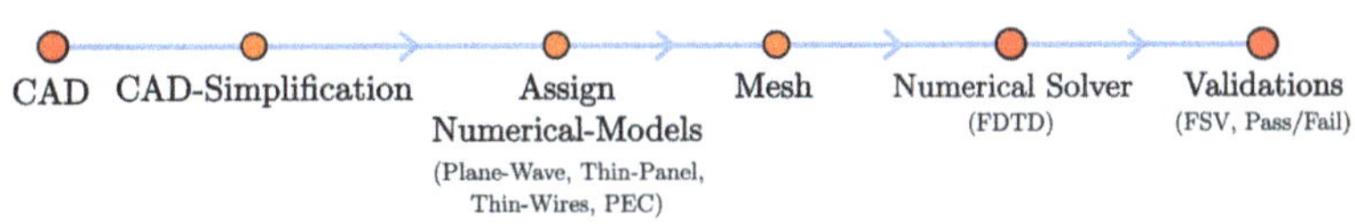

a military national airworthiness agency (INTA), a major airframer (AIRBUS), and three top-ranked universities (UGR, UPC and UPV). A representative set of researchers of these institutions are co-authors of this work, responsible for measurement campaigns, CAD treatment, simulations, validations, etc. Among other achievements, these projects have served to build and validate[2] the SEMBA-UGRFDTD solver (Garcia *et al.*, 2013), used for the simulations presented in this chapter. This solver is capable of treating the complexity found in EMC test-setups and its maturity has been achieved after broad validations, which include cross-comparisons with commercial and in-house time and frequency-domain solvers, and with experimental data (Gutierrez et al., 2012; Gil & Gutierrez, 2011).

The main objectives of this chapter are to provide stones that pave the way for such a roadmap. It is divided in two parts. In the first one, the authors revisit some of the most common procedures employed in E^3 certification in aeronautics. Next, a short description of the basic FDTD method is given. References of the required extensions and models of the basic method to cope with the complexity found in certification setups are also pointed out, paying special attention to the material and cable modeling. The guidelines to build computationally affordable models, retaining all the relevant physics that are present in the experimental setup are also described and put in relation to CAD geometrical cleaning, simplification and meshing. The FSV standard (IEEE Std 1597.1, 2008) is also presented as a tool for an automatic validation procedure to assess the matching between experimental and numerical results. Such expert-like tools are definitely required in this complex discipline where the incertitude in parameters is often not well controlled.

The second part of the chapter provides application examples of the procedures discussed in the first part. Three examples of experimental EMC certification scenarios are described: a radiated test-setup for the AIRBUS C-295 aircraft, a reverberating chamber setup for the central fuselage of the INTA MILANO UAV, both for HIRF analysis; and a conducted current setup for the INTA SIVA UAV for LIE assessment. Most of the experiences described here in the context of aircraft EMC analysis, can be fully extended to general space systems, which are also susceptible to some of the threats described above, apart from some other ones intrinsically due to being without the protection of the earth atmosphere and magnetosphere.

TESTS FOR AERONAUTICAL E^3 COMPLIANCE

Let us now revisit three common test setups for aircrafts, considered in the (Eurocae ED-107A, 2010) and (Eurocae ED-105A, 2013) aircraft certification guides[3]:

1. **Direct Drive (DD) (or Direct Current Injection (DCI)) Tests:** The aim of this test is to find the current which couples to the internal cables of the aircraft when a current is injected and drained on different points of the fuselage (*Figure 3*). The combination of injection and drain points must be such that it represents the worst cases of operation. To this end, the standard recommends the design and construction of a coaxial return wire network around the aircraft to minimize the radiation to the exterior (Eurocae ED-107A, 2010). This configuration maximizes the ratio between induced currents and source power and extends the high-frequency applicability of the test setup. These tests are significant only up to the first aircraft resonances which are associated to its largest dimension. They are considered in (Eurocae ED-107A, 2010) for HIRF assessment, and in (Eurocae ED-105A, 2013) for LIE. They are referred to as LLDD in (Eurocae ED-107A, 2010, section 6.4.2), when low levels of currents are injected, and as High Level Direct Drive (HLDD) in (Eurocae ED-107A, 2010, section 6.3.4), for high levels of currents (see also RTCA/DO-160 G Section 22). A representative example of a DCI LLDD test is given later in this chapter for the INTA SIVA UAV.
2. **Radiated Tests:** In this setup, typically used in HIRF analysis (Eurocae ED-107A, 2010), the aircraft is located, ideally, in the far field region of an antenna. The tests are usually conducted in anechoic or semi-anechoic chambers, or in an OATS. The specific position and the number of antennas depend upon the aircraft widespread, PoE (apertures, joints, slots…), and the location of the system under test. It is important to ensure that all leakage points of the fuselage are illuminated with the field, which means the transmit antennas should be located far enough to ensure that all these leakage points are included in the antenna beamwidth. The antenna positions and polarizations must be such that the system under test is illuminated in the worst cases of coupling. It is

Figure 3. Typical DCI test in nose/tail layout. Reprinted from (Cabello, 2017d)

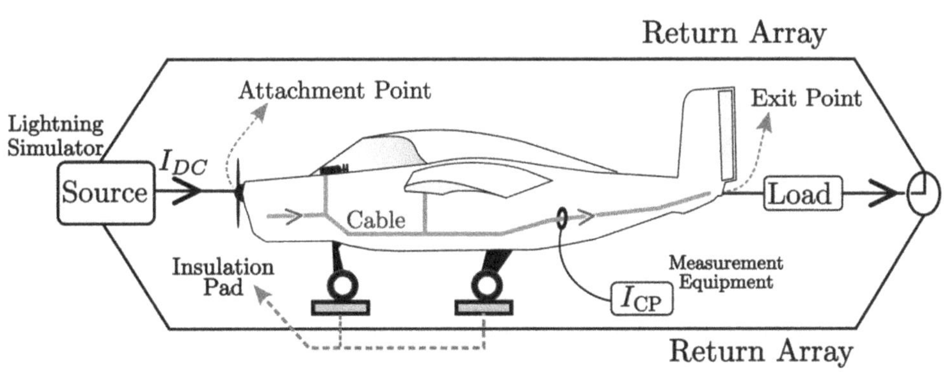

essential that the fields within the fuselage are measured in a statistical manner. They can be performed with low field levels, either to assess the cable currents (LLSC tests) between 1 MHz and 400 MHz (Eurocae ED-107A, 2010, section 6.4.3), or the fields inside the aircraft (LLSF tests) above 100 MHz and up to 18 GHz, or even 40 GHz if needed, (Eurocae ED-107A, 2010, section 6.4.4). They can also be performed with high field levels (Eurocae ED-107A, 2010, section 6.3.5). A LLSF radiated test performed in an OATS will be described for an AIRBUS C-295 aircraft in this chapter.

3. **Tests in a Reverberation Chamber:** The use of RC is considered in (Eurocae ED-107A, 2010, section 6.3.6) as an alternative to High Level radiated tests for HIRF analysis in OATS. The RC creates a good statistically independent distribution of EM fields which are equivalent to illuminate the object with all directions and polarizations. Hence, they allow us to take into account the effect of all apertures (slot, holes and penetrable materials) simultaneously. Results are found by averaging the measurements made after stirring the chamber paddles. In this chapter, RC tests are used to find the field in critical points inside the central fuselage of the INTA MILANO UAV.

As previously described, direct drive and radiated tests can be categorized as high level tests and low level tests. The high level tests are aimed to evaluate directly the impact of high amplitude RF fields on the immunity of aircraft systems. On the other hand, the low level tests are used to determine the internal aircraft environment through the estimation of the transfer function on a test point, $H_T(f)$. Low level tests, usually non-destructive, are mainly preferred (Eurocae ED-107A, 2010), and will be the ones used of the experimental setups of this chapter.

In this regard, $H_T(f)$ is defined as the ratio of the requested magnitude $X(f)$ (induced cable voltages or currents, internal fields or surface currents) to the magnitude of the source $S(f)$ (RF fields or injected currents):

$$H_T(f) = \frac{X_{requested}(f)}{S_{source}(f)} \tag{1}$$

In some cases, this transfer function must be corrected for different reasons; for instance, to remove the amplitude and phase characteristics of the probes, wires, and amplifiers or to create a correction factor using surface currents on the ground versus in the air to compensate the test measurement made on the ground (Eurocae ED105A and ED107A). The corrected transfer function $H_{TC}(f)$, is:

$$H_{TC}\left(f\right) = \frac{H_T\left(f\right)}{H_C\left(f\right)} \tag{2}$$

where $H_C(f)$ is the correction factor.

Several low level tests and their commonly used transfer functions (Rierson, 1997; G. Rasek & S. Loos, 2008; Eurocae ED-107A, 2010; Romero et al., 2012) are presented below:

1. The objective of the LLDD DCI tests is to obtain the transfer function relating the aircraft surface current to the current induced on the cable bundles[4]. The data processing consists of, first to measure the skin currents and the induced current on cables normalized by the injected current; and second, to calculate a correction factor using three dimensional models to derive the aircraft's skin current pattern for the applicable external HIRF environment. This technique is used in the EMC certification process from 10 kHz to the first resonance frequency of the aircraft.

$$H_{T\,LLDD1} = \frac{I_{cable}}{I_{injected}} \tag{3}$$

$$H_{T\,LLDD2} = \frac{J_{test}}{I_{injected}} \tag{4}$$

$$H_{C\,LLDD} = \frac{J_{in\ flight}}{J_{on\ ground}} \tag{5}$$

where $J_{in\,flight}$ and $J_{on\ ground}$ are skin current densities determined through simulations and J_{test}, I_{cable} and $I_{injected}$ are currents measured during tests.

2. The objective of the LLSC tests is to obtain the transfer function relating the external electric field to the current induced on the cable bundles. The data processing consists on the calculation of the ratio between the induced current in the wire bundle and the illuminating antenna field strength, and to normalize

this ratio to 1 V/m. This technique is used over the first aircraft resonance (up to 400 MHz).

$$H_{T\ LLSC} = \frac{I_{cable}}{E_{calibrated}} \tag{6}$$

where I_{cable} is the induced wire bundle current and $E_{calibrated}$ is the field strength used for illuminating the aircraft.

3. The objective of the LLSF tests is to measure the transfer function relating external RF fields to internal fields from 100 MHz to 18 GHz. Thus, the test consists of two phases according to (Eurocae ED-107A) for OATS, and (IEEE 299.1, 2014) for RC. First, the field from the transmitting (*Tx*) antenna is measured at the required location and the forward power fed to the antenna is recorded. This measurement is called the reference ($E_{reference}$). After that, the aircraft is placed inside the test volume, the same forward power is fed to the radiating antenna and the fields inside different bays or cavities are measured (E_{inside}). The outcome of the LLSF tests is the fuselage attenuation or SE, defined as the ratio between the reference measurement and the measurements made inside the aircraft.

$$H_{T\ LLSF} = \frac{E_{reference}}{E_{inside}} \tag{7}$$

Transfer functions tend to have a common behavior, mainly depending on the aircraft size and well described in certification guides (Eurocae ED-107A, 2010) and (AC 20-158A, 2014). For instance, in DCI tests, the currents along cables inside a typical-size aircraft present the generic profile given in *Figure 4* which has been obtained as the result of years of testing experience. We can distinguish three frequency bands which depend on the electric size of the aircraft and its apertures. EM fields in each band produce different electromagnetic coupling effects inside and on the fuselage (Rasek & Gabrisak, 2011; Schroder et al., 2014, Rasek et al., 2015; Parmantier et al., 2012):

- **Low Frequency** $\left(10\text{kHz}, 50\text{MHz}\right)$. In this range, the field penetration through the apertures is low and will mainly depend on the conductive

Figure 4. Generic Transfer Function qualitative behavior in a log-log scale. X-axis represents frequency in GHz, and Y-axis a generic transfer function.

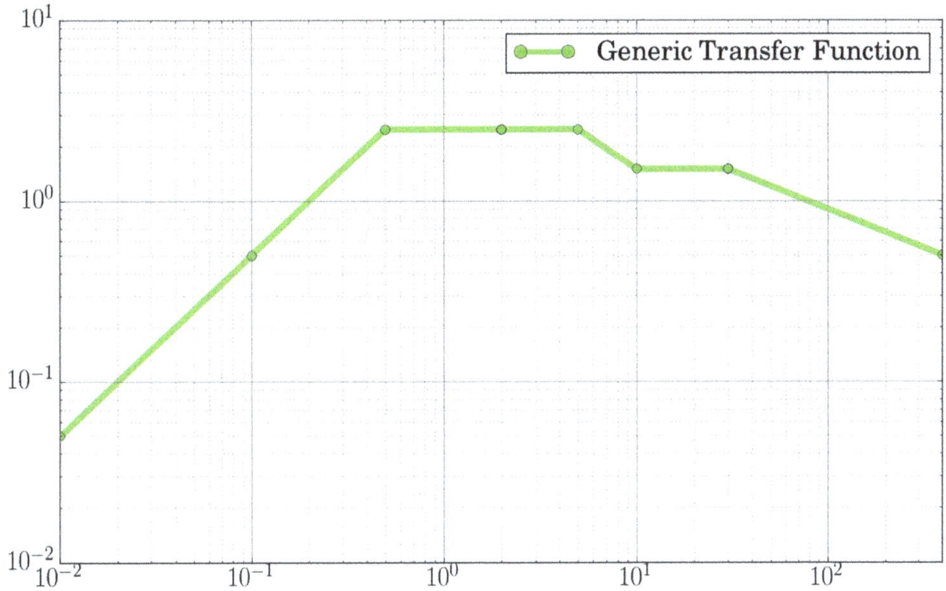

properties of the fuselage. There are no resonances inside the aircraft, which behaves as a cavity under its cutoff frequency. The external EM fields induce strong surface currents flowing along its skin. These currents are inductively coupled to the internal cables which can be modeled by a simple *LR* series circuit in this range of frequencies. The characteristics of this circuit will depend on the geometry and position of the cables with respect to the apertures of the aircraft. When modeled in this way, the cutoff frequency of the transfer function is given by

$$f_{\text{cutoff}} = \frac{R}{2\pi\ L} \tag{8}$$

where the resistance R depends on the conductivity and dimensions (length and section) of the cable and its loads, and the inductance is given by the cable dimensions (length and radius) and the distance of the cable to the current-return (e.g., the fuselage). Typically, the inductive part can be masked by low-resistive behavior due to low-loss cables, or cables being directly grounded to the fuselage.

- **Medium Frequency** $(10\text{MHz}, 100\text{MHz})$: The apertures of the aircraft become more penetrable as frequency increases while the fuselage presents similar penetration characteristics to the low-frequency regime. The maximum coupling is reached at the firsts resonances of the aircraft structure or that of the wiring. This normally occurs when their size is a quarter or half the wavelength.
- **High Frequency** $(100\text{MHz}, 40\text{GHz})$. In this band, the penetration through apertures of the EM energy into the fuselage is strong, dominating over the penetration trough the conductive skin. The surface current (jump between internal and external tangential magnetic field) on its skin diminishes. The field that penetrates into the fuselage may create destructive and constructive interferences, behaving as a reverberant cavity, which may even exceed the external field with very high peaks. This may suppose a major potential risk of this band, which affects the aircraft electronics devices directly.

COMPUTATIONAL ELECTROMAGNETICS IN E^3

CEM solvers have proven for years to be able to reproduce EMC experimental scenarios including full aircrafts (Gutierrez et al., 2016). Numerical simulations allow the engineer to take into account the whole geometrical and physical complexity of a problem. Parametrical studies can be carried out to draw conclusions on the fundamental parameters and the role of their uncertainty in the sensitivity of the results during the design phase. This helps to minimize the rework activities in order to found weaknesses and mitigate EM vulnerabilities. This reduces the cost of development and keeps time and costs associated to testing under control.

At the certification/qualification testing phase, CEM algorithms allow to cross validate measurements with simulations, improving the confidence in tests. They can also complement experimental campaigns, helping in their design and optimization. Finally, CEM can simulate situations that cannot be measured with experimental setups, for instance:

- Real in-flight conditions, to remove the ground effects of experimental tests performed on ground (Eurocae ED-107A, 2010).
- Real far-field conditions which are often difficult to reproduce experimentally even in large and well controlled OATS. Indeed, lots of tests demand a more controlled environment, like that found in RCs or in anechoic chambers. However, these are often expensive and not large enough to accommodate a

whole aircraft, but just parts of it, not allowing the verification of the complete system.

A wide variety of CEM algorithms and methods exists, capable of addressing the electrical sizes and complexity encountered in the aircraft industry. Among them, the FDTD method, proposed by Yee in 1966 (Yee, 1966), is today one of the most well-known and popular ones.

FDTD is formulated in time domain in an explicit manner and, hence, it can find transient and broadband responses in a single run. It can deal with arbitrarily complex dispersive and non-linear materials. It can be easily programmed using parallel techniques (MPI, OpenMP) in state-of-the-art computer cluster architectures, to solve very complex and large problems. These reasons make FDTD one the most valuable methods for EMC analysis.

FDTD

FDTD and its main numerical features are well known in literature and they are part of most syllabuses in undergraduate studies. Some good references are the books (Taflove & Hagness, 2005; Yu, Mittra, Su, Liu & Yang, 2006; Sullivan, 2000; Kunz, Steich & Luebbers, 1993; Mittra, 2014). There is a vast literature exploring every single feature, formulating it to handle all kind of materials and boundary conditions, seeking for improved stability reformulations, searching for innovative meshing methodologies, etc. Particularly, hybridizations with other techniques allow the analysis of cable bundles and harnesses, an especially important topic in the context of EMC.

Many of the published works are actually successful *key advances* in this matter. However, some of these methods are just not robust enough to be used beyond pure proofs of concept for academic examples and in no manner can be employed in problems of the complexity found in EMC in aeronautics. For this reason, just the basic fundamentals of FDTD are described, together with robust and accurate models to address EMC problems. Special attention is paid to our own ideas which allow handling aeronautic materials and cable bundles as they play an important role in the energy coupling inside an aircraft.

FDTD Fundamentals

Maxwell's curl equations in time domain can be written in a symmetric manner as

$$-\nabla \times \vec{E} = \vec{M}_T + \mu_0 \frac{\partial \vec{H}}{\partial t}$$
$$\nabla \times \vec{H} = \vec{J}_T + \varepsilon_0 \frac{\partial \vec{E}}{\partial t} \tag{9}$$

where ε_0 , μ_0 are the free-space permittivity and permeability, $\vec{J}_T$, $\vec{M}_T$ are the total electric and magnetic current densities, $\vec{E}$ and $\vec{H}$ the electric and magnetic field vectors; all of them function of time and space $(\vec{r}, t)$. The total current densities include: the independent source current densities $\vec{J}_{\text{source}}$, $\vec{M}_{\text{source}}$, the polarization/magnetization current densities $\vec{J}_{\text{p}}$, $\vec{M}_{\text{p}}$, in general dispersive and anisotropic; and the ohmic conduction current density terms $\vec{J}_{\text{c}} = \sigma \vec{E}$, $\vec{M}_{\text{c}} = \sigma_M \vec{H}$

$$\vec{M}_T = \vec{M}_{\text{source}} + \vec{M}_c + \vec{M}_p$$
$$\vec{J}_T = \vec{J}_{\text{source}} + \vec{J}_c + \vec{J}_p \tag{10}$$

These equations require, to be complete, the constitutive relationships relating the polarization and magnetization current densities with their causes. Typically, these are found in frequency domain as

$$\vec{J}_p(\omega) = j\omega\varepsilon(\omega)\vec{E}(\omega) \ , \ \vec{M}_p(\omega) = j\omega\mu(\omega)\vec{H}(\omega) \tag{11}$$

with $\varepsilon(\omega)$, $\mu(\omega)$ being the constitutive parameters of the material. The frequency domain expressions can be translated into time domain either if differential or convolutional form. A proper way of doing this is found in (Han, Dutton & Fan, 2006) where arbitrary dispersion is handled by expanding the constitutive parameters in arbitrary-order partial fraction pole/residue series found by vector fitting techniques (Gustavsen, 2006).

The classical FDTD method (Yee, 1966) employs centered operators to solve Maxwell's equations (and the time-domain constitutive relationships) in Cartesian coordinates, by replacing the space and time derivatives by

$$\frac{\partial f(\nu,\ldots)}{\partial \nu} \approx \frac{f(v + \Delta v, \ldots) - f(v - \Delta v / 2, \ldots)}{\Delta v} \tag{12}$$

This kind of discretization requires the space to be sampled in unit cells of sides $(\Delta x, \Delta y, \Delta z)$, in a not necessarily uniform grid, with the EM field-components placed in the well-known Yee's cube staggered arrangement (*Figure 5*). The E and H-field components are also shifted in time by a $\Delta t / 2$ factor, yielding a final discrete explicit marching-on-in-time algorithm which finds the solution at a given time instant as a function of the solution at previous time steps.

Just for illustration purposes, let us write the 1D FDTD equations for plain ohmic media

$$\begin{aligned} E_i^{n+1} &= C_{a,i}\, E_i^n + C_{b,i}\left(H_{i-1/2}^{n+1/2} - H_{i+1/2}^{n+1/2}\right) \\ H_{i+1/2}^{n+1/2} &= D_{a,i+1/2}\, H_{i+1/2}^{n-1/2} + D_{b,i+1/2}\left(E_i^n - E_{i+1}^n\right) \end{aligned} \tag{13}$$

with

$$\begin{aligned} C_{a,i} &= \frac{2\tau_i - \Delta t}{2\tau_i + \Delta t},\ C_{b,i} = \frac{2\Delta t \tau_i}{\varepsilon_i \Delta_i \left(2\tau_i + \Delta t\right)},\ \tau = \varepsilon / \sigma \\ D_{a,i+1/2} &= \frac{2\tau_{M,i+1/2} - \Delta t}{2\tau_{M,i+1/2} + \Delta t},\ D_{b,i+1/2} = \frac{2\Delta t \tau_{M,i+1/2}}{\mu_{i+1/2} \Delta_{i+1/2} \left(2\tau_{M,i+1/2} + \Delta t\right)},\ \tau_M = \mu / \sigma_M \end{aligned} \tag{14}$$

where Δ_i is the cell size at the space position i, and $\Delta_{i+1/2}$ is the cell-size of the dual mesh defined by $\Delta_{i+1/2} = \left(\Delta_i + \Delta_{i+1}\right) / 2$. The subscript i denotes the space position ($x = i\Delta$), and the superscript n the time instant ($t = n\Delta t$). To find the

Figure 5. Position of the EM fields in Yee's cell (Reprinted from Cabello, (2017d))

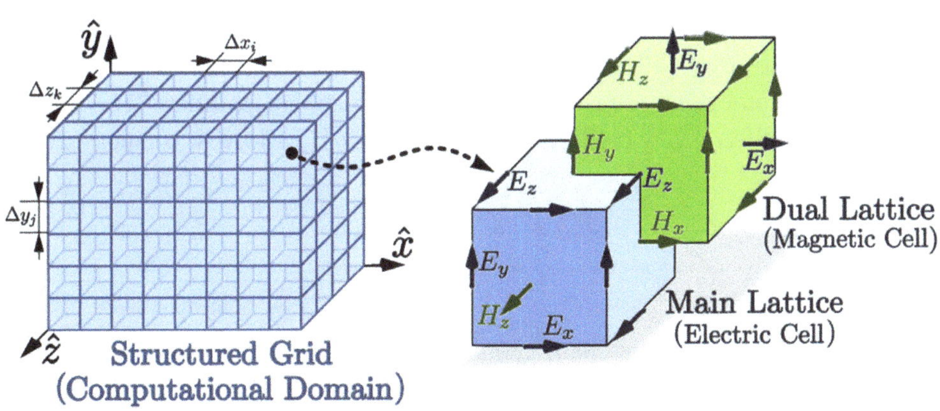

constants in, a second-order[5] centered average operator has also been employed to approximate the identity operator, so as to render a numerical scheme compatible with Yee's space-time arrangement.

In order to simulate open problems, proper reflectionless absorbing boundary conditions are required. A major breakthrough advance in FDTD was introduced in (Berenger, 1994) with the PML technique, capable of attaining reflections under 120 dB. PMLs have been extended to truncate almost any kind of material: dispersive, anisotropic, bianisotropic, metamaterials (Garcia et al., 1999, Taflove et al., 2013; Hao et al., 2009). FDTD, endowed with PMLs, and dispersive material treatment, can handle almost any complex problem found in electrical engineering.

A main restriction of the marching-on-in-time explicit algorithm is that it requires the time step to be minored by the smallest space step of the spatial mesh to ensure its numerical stability. This is given by the well-known CFL or Von-Neumann criterion, which requires the so-called CFL Number to fulfil in 3D (Taflove & Hagness, 2005)

$$\mathrm{CFLN} \equiv \frac{c\Delta t}{\underset{\forall i}{Min}\left\{\left(\sqrt{\frac{1}{\Delta x_i} + \frac{1}{\Delta y_i} + \frac{1}{\Delta z_i}}\right)^{-1}\right\}} \leq 1 \tag{15}$$

This reduces to $\mathrm{CFLN} \equiv \frac{c\Delta t}{\sqrt{3}\Delta} \leq 1$ for a homogeneous isotropic mesh. The condition relaxes in 1D to

$$\mathrm{CFLN} \equiv \frac{c\Delta t}{\underset{\forall i}{Min}\left\{\Delta_i\right\}} \leq 1 \tag{16}$$

Hence, a fine space discretization does not only imply an increase in computer memory, but also larger CPU simulation times[6] due to the subsequent reduction in the time step. For homogeneous materials, typically a spatial resolution of 15-20 PPW at the maximum frequency provides enough accuracy[7] (Taflove & Hagness, 2005)

$$\Delta = \frac{c}{f_{\max}\ \mathrm{PPW}} \tag{17}$$

However, when there are small geometrical features (edges, curvatures, slots, wires, thin panels…), all of them require being properly discretized-in-space, resulting into a global time step minored by the smallest space increment. This brute-force approach to deal with multiscale problems can imply huge-memory setups and long-lasting simulations, unaffordable even for modern computers. For this reason, sub-cell equivalent models are introduced. Some of them are:

- To handle curvature in a properly accurate manner without resorting to a dense mesh, conformal algorithms have proven to be useful. They are based on the use of the integral forms of Ampère and Faraday's laws just locally in cells traversed by material interfaces. Refer to (Dey & Mittra 1997; Cabello et al., 2016) for more details. Other approaches include subgridding techniques combined with local time stepping schemes to mix meshes with different densities in the same simulation (Zivanovic et al., 1991).
- To deal with thin materials, like the CFC skin of an aircraft, NIBC can be used (Sarto, 1999). Another alternative referred to as SGBC, proposed by the authors in (Cabello et al., 2017a), is briefly described in the next sub-section.
- To handle cable bundles, once again brute-force meshing is not the solution. In the last sub-section of this section, the authors present a hybrid MTLN based in (Berenger, 2000) hybrid method, that can also be extended to cables not aligned with the Cartesian grid (Guiffaut et al., 2010).

Though further details will not be provided here, there are many other situations also requiring sub-cell treatment in FDTD: thin slots, junctions, gaskets, etc. All of them are important in EMC, as they are responsible of coupling EM energy inside the aircraft.

Subcell Models of Lossy Thin Panels

Conductive thin panels (metallic, CFC, etc.) are part of the skin of an aircraft. These can be very complex, usually multilayered, anisotropic and frequency dispersive. Typically they are much thinner than usual mesh sizes, and hence they cannot be resolved by the usual FDTD scheme as this would imply using unaffordable large meshes. In this section, the authors briefly describe the SGBC technique proposed in (Cabello et al., 2017a, Cabello et al., 2017b), which hybridizes a 1D Crank-Nicolson FDTD scheme to deal with the field propagation inside the thin-panel, with the usual 3D Yee- FDTD scheme for the rest of the problem. SGBC has been used in all the simulations presented in the last part of this chapter to model the composite parts of the aircrafts.

SGBC starts from the same assumption made by NIBC (Sarto, 1999). It postulates that plane waves impinging on a conductive planar thin-panel with oblique incidence, refract at a close-to-normal angle regardless of the actual angle of incidence, as far as the refractive index is much higher inside the thin-panel than outside. It can be easily proven that for grazing incidence $\theta_i \to \pi / 2$ (the worst-case), the wave refracts at an angle $|\theta_t| < 10^{-2}$ with the normal if

$$f\,[\mathrm{GHz}] < 1.8\,\sigma\,[\mathrm{kS/m}] \tag{18}$$

This means that for the typical conductivities ($\sigma > 10^4\ \mathrm{S/m}$) of the panels found in automotive or aeronautics, the model holds up to 18 GHz regardless of the panel thickness.

Let us briefly describe the SGBC method for a simple isotropic planar ohmic thin panel, of thickness th, aligned with a Cartesian plane[8], having a constant

Figure 6. Cross section of a FDTD cell with a SGBC boundary and details of the thin-panel subgridding
(Reprinted from (Cabello et al., 2017b) ©2017 IEEE, with permission from IEEE).

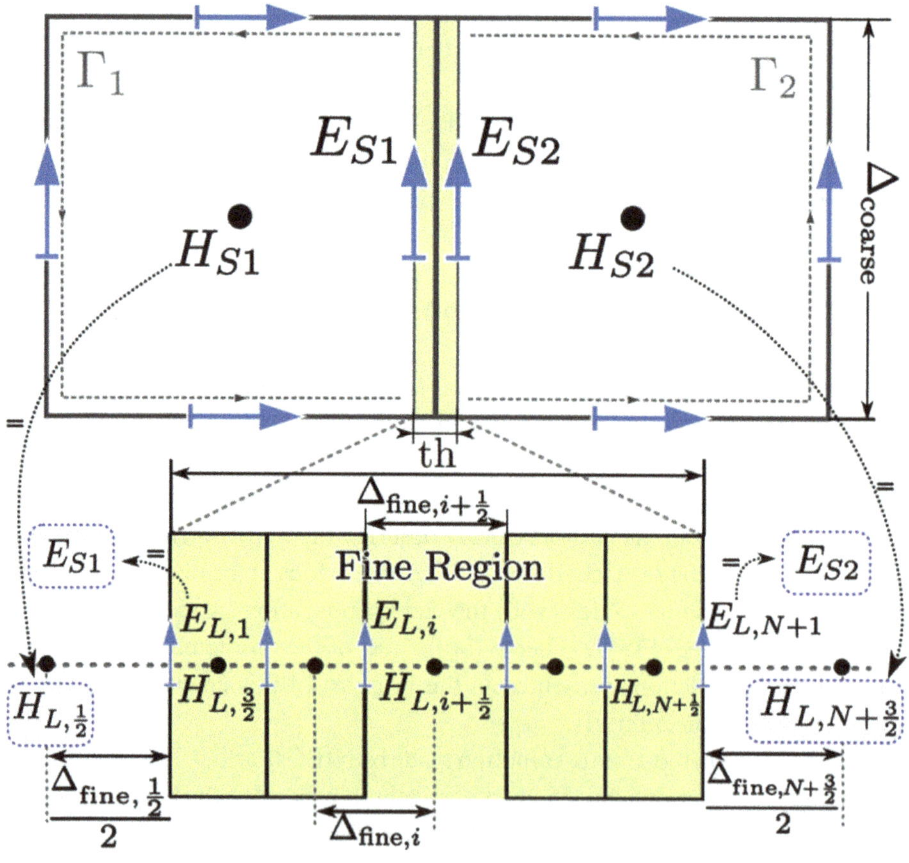

conductivity, embedded in free space. First, the thin-panel is sub-gridded into N 1D FDTD-cells of size Δ_{fine}, according to the setup of *Figure 6*. The outer 3D FDTD cell size is $\Delta_{\text{coarse}} > \Delta_{\text{fine}}$. Next, the tangential E-fields on each side of the thin-panel E_{S1}, E_{S2} are duplicated at the usual 3D Yee position, to account for each face value, which are employed accordingly to update the magnetic field component at each side H_{S1}, H_{S2} by using the integral form of Ampère's law. The SGBC algorithm continues as follows:

1. The fields inside the panel domain denoted by $E^n_{L,i}$, $H^n_{L,i+1/2}$ are updated by a Crank-Nicolson 1D scheme (CNTD) (Yang et al., 2006), which is just an unconditionally stable variation of the Yee FDTD scheme, maintaining is the usual space staggering, but co-locating the E and H-fields in time by means of an average operator

$$\begin{aligned} H^{n+1}_{L,i+1/2} &= D_{a,i+1/2} H^n_{L,i+1/2} + D_{b,i+1/2} \frac{E^n_{L,i} - E^n_{L,i+1} + E^{n+1}_{L,i} - E^{n+1}_{L,i+1}}{2} \\ E^{n+1}_{L,i} &= C_{a,i} E^n_{L,i} + C_{b,i} \frac{H^{n+1}_{L,i-1/2} - H^{n+1}_{L,i+1/2} + H^n_{L,i-1/2} - H^n_{L,i+1/2}}{2} \end{aligned} \quad (19)$$

Unlike for 1D Yee-FDTD, the resulting marching-on-in-time scheme is now implicit, but it can be easily transformed to yield a simple tridiagonal system of equations in E, and an explicit equation for H. Refer to (Cabello et al., 2017a) for details.

2. E-fields and H-fields outside the thin-panel are advanced in the usual 3D Yee-FDTD manner.
3. The connection between the coarse and fine mesh is made through a hybrid implicit-explicit (HIE) method described in detail in (Cabello et al., 2017a).

SGBC has been introduced as a robustly stable replacement of NIBC which was often reported to exhibit late-time instabilities (Kobidze, 2010; Nayyeri et al., 2013). Some of its characteristics are:

- SGBC has exhibited a strong late-time stability for all the cases that we have simulated, whereas NIBC has often run into late-time instabilities, sometimes requiring heuristic time-step reductions of up to 1% that of the analytical limit .

- The unconditional stability of CNTD allows us to choose the time step just constrained by the exterior coarse discretization .
- The computational overburden of solving the 1D tridiagonal system of equations does not compromise the computational efficiency of the full problem.
- The extension of this method to anisotropic materials would no longer be affordable by 2D CNTD. Proper variations of it, like a 2D Alternating Direction Implicit FDTD (Garcia et al., 20007), also unconditionally stable, can be an alternative.
- Instead of a 1D CNTD method, a pure 1D Yee-FDTD method could also have been used inside the slab by employing an ETD method like that described in (Cabello et al., 2017c), not requiring dramatic reductions in the times step (50% that of Eq. for moderately conductive media). Hence, a 2D Yee-FDTD method could also be an affordable alternative to 2D CNTD to address anisotropic lossy panels.
- Often the internal nature of the multilayered lossy thin panel is not known, but just the S-parameters, either found from measurements, or from microscopic assumptions. In this case, SGBC could still be used by retrieving effective macroscopic constitutive parameters following the roadmap described in (Cabello et al., 2017b).

Subcell Models of Cables

Cables are the main characters responsible for coupling EMI inside the equipment that they interconnect. Their accurate simulation is a must, and again meshing them by brute in FDTD, is not computationally viable. The MTLN equations are the appropriate alternative to deal with cable-bundle waveguiding analysis. When they are coupled to Maxwell's space-time curl equations, they provide a way to analyze the cable-bundle behavior inside a complex inhomogeneous 3D EM environment.

A typical implementation of a coupled FDTD-MTLN method is based on (Berenger, 2000), which extends the thin-wire approach of (Holland, 1981), to many-wire bundles, by taking into account the cross-talk between the individual wires. It can also be generalized using the ideas of (Guiffaut, 2012; Ledfelt, 2001; Edelvik, 2002) to permit the treatment of wires not aligned with the Cartesian FDTD grid. Just as a reminder, let us recall the coupled FDTD-MTLN Holland equations for a single wire with a p.u.l. resistance R running along the x-direction in free space

$$\begin{aligned} &\partial_x I + C\partial_t V = 0 \\ &\partial_x V + L\partial_t I + RI = \langle E_x \rangle \\ &\qquad\qquad\qquad\qquad\quad \Uparrow\!\Downarrow \\ &\partial_y H_z - \partial_z H_y = \varepsilon_0 \partial_t E_x \end{aligned} \tag{20}$$

where the p.u.l. inductance L, and the capacitance C, fulfill the usual $LC = c^{-2}$ relationship. In Eq. the inductance L is found with respect to an infinite coaxial current return, which, in FDTD, leads to an expression depending on the space step and the wire radius. The maximum time-step for stability of the coupled scheme is proven to depend on L (Schmidt & Lazzi, 2004) and to be minored by the wire radius, which cannot be larger than half the space step. For wire bundles, this upper limit translates, roughly speaking, into that of a circumference enclosing the whole bundle. Hence, in practice the FDTD-MTLN Holland-Berenger method is often applied at cable-overbraid level, to find the currents flowing along the outermost sheath. Next, these currents are used to find the currents inside the bundle by using the transfer impedance (Feliziani & Maradei, 2002), found by using measurements, or analytical formulas (Vance, 1974; Yatsenko et al., 2000).

The limitation on the stability constraints imposed by the fully coupled MTLN-FDTD method also can be removed by using a pure field to transmission line (field-to-TL) model (Agrawal, et al., 1980). This may seem a simplification of the fully coupled technique in the sense that it just takes into account the coupling of the external field found by FDTD to the cables, without coupling back the currents flowing along the bundle wires to the FDTD domain (just the up-to-down coupling is kept in Eq. . However, field-to-TL is in no manner an approximation and becomes an exact application of Huygens's equivalence principle, when only TL modes are assumed to be present (Rachidi, 2012). However, concerns are raised when the current return is not perfect (for instance due to interruptions of the overbraid), and the structure (e.g. the fuselage in case of an aircraft) provides the current return instead. For these cases, the FDTD-MTLN Holland-Berenger formulation would provide a more convenient approach.

Modeling of Reverberation Chambers

RCs are highly-resonant overmoded metallic cavities which are large enough to accumulate enough statistically independent modes at low frequencies (Besnier & Demoulin, 2011). They typically have a mechanical stirrer that rotates to change the modal distribution inside the cavity for obtaining an average uniform field throughout on the volume occupied by the device under test. The international standards recommend at least 60 transmission modes for considering a reverberation environment (IEC 61000-4-21 Ed2, 2011 and IEEE 299.1, 2013). In the case of a cubic reverberation chamber, the number of modes can be written as function of frequency (Hill D.A., 2009):

$$N_S\left(f\right) = \frac{8\pi}{3} abc \frac{f^3}{v^3} - \left(a + b + c\right)\frac{f}{v} + \frac{1}{2} \tag{21}$$

where *a*,*b*,*d* are rectangular chamber dimensions, ν is the speed of light, and *f* is the chamber source frequency.

The numerical simulation of several stirrer positions inside an electrically large cavity is unaffordable for a time-domain solver. However, an equivalent model can be used by employing a superposition set of plane waves with a random uniform statistical distribution on their polarization, delays and direction of incidence (Hill D.A., 1998; Moglie & Pastore, 2006). Instead of a PEC cavity, the computational space can then be truncated by usual PML as absorbing boundary conditions and affordable simulations can be done. Each plane wave is inserted in FDTD by using the total field/scattered field formulation of (Taflove & Hagness, 2005) based in Huygens' equivalence principle.

CAD CLEANING, SIMPLIFICATION AND MESHING

In the previous section, we have described some of the basic desirable features that a numerical FDTD solver must include to be able to carry out EMC analysis of full aircrafts in a computationally affordable manner. To this end, some preliminary simplifications of the aircraft model need to be done to get the final numerical input for the solver. The starting digital mock-up provided by a CAD tool (*Figure 2*) is often too complex to be meshed and be directly used in the simulation. This is a crucial stage, since simulation results, and CPU times strongly rely on the quality of the numerical model fed to the solver: mesh, cables, sub-cell details, etc. In this section, we briefly discuss this process.

Typical CAD models include an unnecessary level of geometrical details from the EM point of view when they are provided by the mechanical engineering departments. Therefore, the simplification, cleaning, and meshing of complex CAD models is a very important part of the EMC simulation process in order to provide a mesh that can be fed to the EM solver. The main tasks of cleaning, simplification and defeaturing can be summarized as follows:

1. Solve typical problems in CAD models as lack of connectivity, bad or non-existing topology, bad definition of parametrical surfaces, residual curves from intersections.
2. Remove those pieces or details that are not involved in the simulation, either according to their electrical size (screws, rivets …) or to their composition, for instance, often lossless materials can be discarded.
3. Replace the complex parts with simpler ones able to represent with enough accuracy the reality from the point of view of the EM interactions.
4. Maintain the ohmic connections/disconnections: Sometimes the separation between objects must be increased to the cell size to avoid a false electrical connection in the mesh, some other times physically connected parts appear unconnected in the CAD file and must be fixed.
5. Dimension degeneration. One side of a 3D object can be reduced to a point when it is smaller than one cell size, implying a reduction on its dimensionality. For instance, a 3D panel thinner than the mesh size would be replaced by a 2D surface; in the same way cables and pipes which their section smaller than the cell size can be replaced by lines. This fact must be taken into account when the CAD model is treated because the dimensionality of an object restricts the models that can be used with it.

Obtaining a new geometry which is simpler than the original one but still able to capture the EM reality is a very time-consuming task. This phase requires experienced engineers with a good knowledge of the involved physics and with a deep understanding of the mesher and solver capabilities (Gil & Gutierrez, 2011, Gutierrez et al., 2014). We can see in *Figure 7*(b) the result of this process for the C-295 aircraft, which has led a reduction by a factor 500 in the size of the final CAD file, making it possible to be meshed and simulated.

For FDTD, the simulation mesh is typically obtained by a Cartesian mesher which converts a mesh of triangles/tetrahedrons into the voxels, surfels and linels which are handled by the FDTD method. This procedure automatically filters out sub-cell details, including those may be relevant from the EM point of view (thin panels, wires, slots, etc.). Special care must be taken to prevent this situation, and to assign explicitly appropriate material properties to model these details appropriately.

Numerical Models: Materials

Materials with low conductivities can be assumed to be equivalent to free space and removed from the model, *e.g.*, lossless air-like dielectrics, like plain fiberglass parts. On the other hand, materials with high conductivities, that when shaped as

Figure 7. a) Isometric view of the digital mock-up of an AIRBUS C-295 aircraft, with a detailed view of the cockpit. b) Simplified geometry with a detailed view of the cockpit (aerodynamic surfaces are depicted with a degree of transparency to allow the observation of the internal surfaces). (Reprinted from (Gutierrez et al., 2014) ©2014 IEEE, with permission from IEEE)

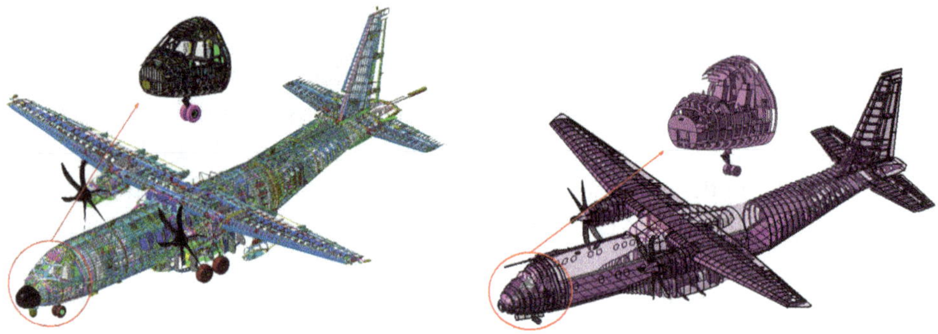

panels provide a SE over than 100 dB can be directly modeled as PEC surfaces, this includes most metals or composites which are reinforced with aluminum foils.

CFC panels are typically formed by multilayered plies of carbon fibers immersed in a matrix medium. They require a careful analysis to incorporate proper equivalent models of their macroscopic behavior. These models typically have a SE with a constant low-frequency profile which then increases with frequency due to skin-depth effects up to a certain maximum. After this maximum is reached, the SE may decay asymptotically due to the presence of inductive effects in the conductive fiber. Depending on the position of that maximum, either a plain homogeneous ohmic model, or a more complex dispersive one, must be employed. This analysis can typically be performed by looking at the microscopic geometrical properties of the panel: random, woven, non-woven, periodicity, number of layers, fiber density and orientation, conductivity of fibers, reinforcing meshes, etc. Several models are available to predict its shielding capabilities, or by measuring it with proper fixtures (Sarto et al., 2014).

Regarding apertures, they must be carefully handled, since they represent entry points of EM energy. Holes, gaps between different panels, thin slots, etc., can be treated as free space if their dimension is larger or comparable to the cell size. Otherwise, these must be managed by appropriate sub-cell models (Gkatzianas & Tsiboukis, 2003).

Finally, an important aspect in any numerical model is the effect of small variations or uncertainties in the different physical parameters, *e.g.*, position of cables, material properties, etc. This becomes apparent even at experimental level when data from different specimens of the same AV or from different test campaigns are compared and a disagreement is found for quantities that otherwise should coincide. A way to assess these uncertainties is to apply a statistical analysis by taking into account the variability of all the AV parameters. For instance, heuristic approaches based on an affordable set of simulations chosen as representative according to the engineer experience, can be used to analyze the impact of the inclusion of installations, the maintenance of electrical contacts, the assignation of materials, the selection of appropriate cables with their significant properties, etc. From the different simulations, the sensitivity of the simulation results to variations in the EM model can be derived. Next, theoretical guidelines can be drafted with quantitative notions of the error magnitudes introduced with these approximations. As a result, an increase in the reliability of the simulation results can be achieved in order to use them in a certification process.

VALIDATIONS METHODS

The workflow in *Figure 2* ends with the validation of the numerical results against experimental data. Validation methods are necessary to objectively evaluate the similarity between different datasets.

Although in many fields of research, a simple visual inspection is used as the validation method, numerous studies show that a direct comparison point-by-point is not feasible when large amounts of data are compared. Therefore, this method is not recommended to validate the results and much less to assign an absolute value of accuracy. This approach makes sense only for simple models, but in EMC simulations, the results are often very complex.

Feature Selective Validation

In the last years, several validation methods evaluating the similarity between measures, imitating the opinion of EMC experts have been developed: correlation, reliability factor, integrated error log frequency, etc. One of the most widely used, for its versatility and simplicity, is the FSV method based in IEEE 1597.1/2 standard (IEEE Std 1597.1, 2008; IEEE Std 1597.2, 2010). This method allows us the comparison to be quantified objectively, removing the element of subjectivity from the decision making process.

FSV has the advantage of analyzing the two major aspects that are widely considered paramount in any validation: the magnitude levels and the shape of the dataset graphs. It is based on the decomposition of the results into two groups; the first one discusses the difference in amplitude (ADM) and the second one the difference between the characteristic of the signals (FDM). The combination of these two indicators is a new indicator named GDM that measures the overall difference between both datasets (Duffy et al., 2006; Orlandi et al., 2006).

All these indicators (ADM, FDM and GDM) have the ability to be configured to perform a point-to-point analysis. The advantage of relying on a point-to-point data is to know which areas of the data sets have the major differences.

Another way to qualitatively analyze the FSV indicators is represented by a probability density function which is useful for a rapid and comprehensive analysis of the results. This indicator uses a histogram that can be divided into six categories: excellent, very good, good, fair, poor, and very poor, according to the IEEE standard (IEEE Std 1597.1, 2008) (*Figure 8*).

In the simulations described in the last part of this chapter, the FSV tool proposed in (Jauregui et al., 2010; Jauregui et al., 2013; Jauregui et al., 2014a) has been employed for validation[9].

With the aim of considering the most relevant data of the measurements and the simulations, several enhanced techniques can be used. The first one is a technique to provide equivalent uncertainty estimation in the FSV validation method (Jauregui

Figure 8. Confidence histogram rules [IEE-5971]

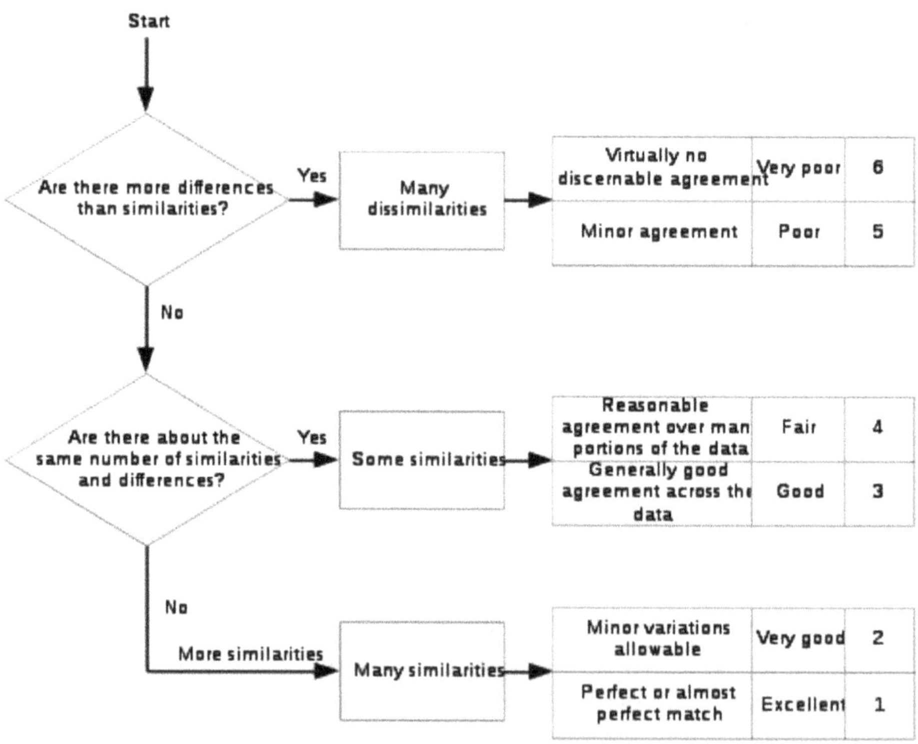

et al., 2013). This technique uses the level of uncertainty of one of the datasets, typically the measurements, to find an equivalent tolerance for the validation method. Another technique is to take into account a relative level in the comparison process. Because in many EMC cases the comparison analysis is focused on a certain level and ignores much of the data, this technique relies in the idea of weighting the different indicators of the traditional FSV method (Jauregui et al., 2014b). The weighting function may be changed to accomplish the application-specific needs of the case which is being validated, avoiding erroneous decisions on the final results.

APPLICATION CASE 1: C-295, NUMERICAL ASSESSMENT OF HIRF EFFECTS

In the remainder of this chapter the authors will describe the process carried out for the virtual testing of a series of complex AV EMC scenarios. The first of this series consists of a C-295 aircraft from AIRBUS under HIRF conditions, presented in (Gutierrez et al., 2014) by some of the authors of this chapter. The C-295 is a medium-weight transport aircraft. It employs standard aeronautic materials. The most relevant conductive ones, compliant with MIL-HDBK-5H (1998), are: aluminum alloys (2024, 6061, 7075, 7050), ferrous alloys (DAISI 4340, AISI 321, PH 13-8Mo.), and titanium alloys (Ti6Al4V).

Its fuselage is mostly metallic, which makes it robust against external electromagnetic threats. The engine nacelles, some parts of the central wing, the landing gear sponsons, small pieces in the stabilizers and wings, are all taken as PEC. Composite materials are also employed, mainly carbon fiber and fiberglass *e.g.* the radome is a sandwich-type structure with outer fiberglass laminates. Around 20% of the C-295 composition is based in CFCs. Leaks from EM fields and currents are mainly due to apertures around hatches, joints, hinges and the use of non-conductive materials in radomes, fairings, windows, etc.

The experimental setup consisted of a radiated LLSF test, carried out in the Airbus OATS in Getafe (Spain), consisting of a circular platform 90 m in diameter, made of concrete reinforced with an embedded metallic grid. The aircraft is radiated (far-field regime) with proper antennas between 100 MHz and 18 GHz. Care is taken to let at least 50 meters between the aircraft and any other source of EM noise, including any object able to create undesired reflections (see *Figure 9*). The tests were carried out according to (Eurocae ED-107A, 2010).

First of all, the calibration field is measured in free-space with the probes located at the six points that they will occupy in the LLSF test of the aircraft. Next, the same data is recorded at the same points within the aircraft, and normalized to the calibration levels. Two of these test-points are inside the cockpit, three in cargo

Figure 9. LLSF Tests at Airbus Military OATS Facility (left). Electric Field measurements during LLSF Tests. (right). (Reprinted from (Gutierrez et al., 2014) ©2014 IEEE, with permission from IEEE).

bay, and one in the engine. In order to get the worst case for the coupling through the aircraft slots and apertures, several angles and polarizations are employed to radiate each zone. For this, the maximum field level is found at each frequency and the envelope of all of them is found along the frequency band.

Modeling and Simulation Approaches

The FDTD-MTLN Holland-Berenger model of SEMBA-UGRFDTD is used to deal with cable bundles and/or cable overbraids, and the SGBC method to handle electrically thin composites. The mesh employs a uniform cell size of 20 mm which yields a problem with 734 Mcells (1228 x 1307 x 459). A time step 80% that of the CFL limit has been taken ($30 \cdot 10^{-12}$ s) to comply with cable-bundle stabilities mentioned earlier in this chapter. This discretization allows us to achieve a maximum of frequency of 10 GHz (instead of the 18 GHz achieved with measurements), since the resolution reaches the limit of PPW=15 cells/wavelength (in free space), to consider results accurate. Hence, a Gaussian modulated plane-wave, with a frequency range between 100 MHz and 10 GHz, is used to illuminate the aircraft, mimicking the antenna radiation. This will also have impact in the results, as will be seen later, in some mismatches between 100 and 200 MHz, where the far-field assumption for the antennas is in its validity limit. Finally, the model is truncated by means of PML boundary conditions.

For validation the FSV methodology has been employed. Additionally, results with a PFC used for validation under the HIRF-SE project will also be shown. The procedure to get the PFC results is summarized as follows.

1. A set of seven observation points occupying the volume of the receiving antenna is used to probe the EM field through an average operation. This process obtains a better representation of the receiving antenna used during the tests.
2. A minimum of 100 frequencies per decade above 100 kHz equally spaced on a log scale is measured.
3. The raw data is filtered by using an averaging bandwidth of 5% of the frequency of interest, as recommended in (Eurocae ED-107A, 2010, section 6.4.5).
4. The maximum value frequency-by-frequency considering all illumination angles and both polarizations, for each point, is calculated, and taken as the WC results.
5. The maximum values of the WC, within a sliding frequency window of 10% of the central frequency, are used to find the data envelopes. This permits to account for any shift in resonances between the aircraft installation and the modeled test setup.
6. Finally, the value of the difference between simulation and measurement in dB is calculated.
7. The data are assumed to pass the criterion if this difference falls below 6 dB.

Results

The transfer function for the electric field at 3 test points is shown in *Figure 11*, *Figure 12*, and *Figure 13*, comparing the worst-case numerical/experimental transfer function, as well as the envelopes resulting from the application of a sliding frequency

Figure 10. C-295 cockpit unstructured mesh (left). Cartesian mesh of the engine (right). (Reprinted from (Gutierrez et al., 2014) ©2014 IEEE, with permission from IEEE)

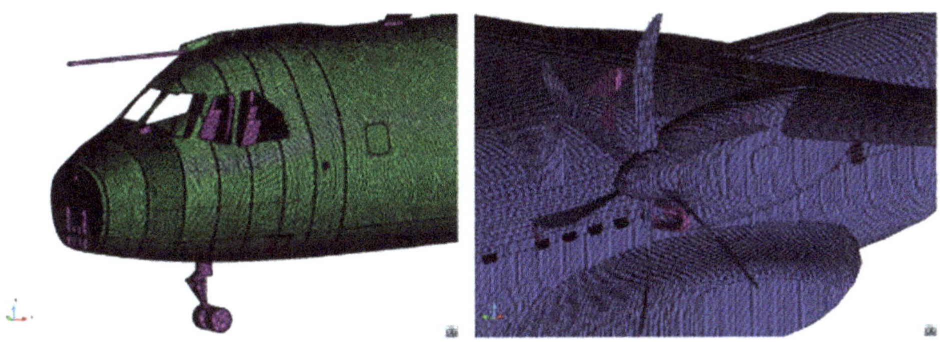

Figure 11. a) Comparison between measurements and simulations for the test point 1 located in the cockpit. b) Difference between simulation and measurement and 6 dB limit. (Reprinted from (Gutierrez et al., 2014) ©2014 IEEE, with permission from IEEE)

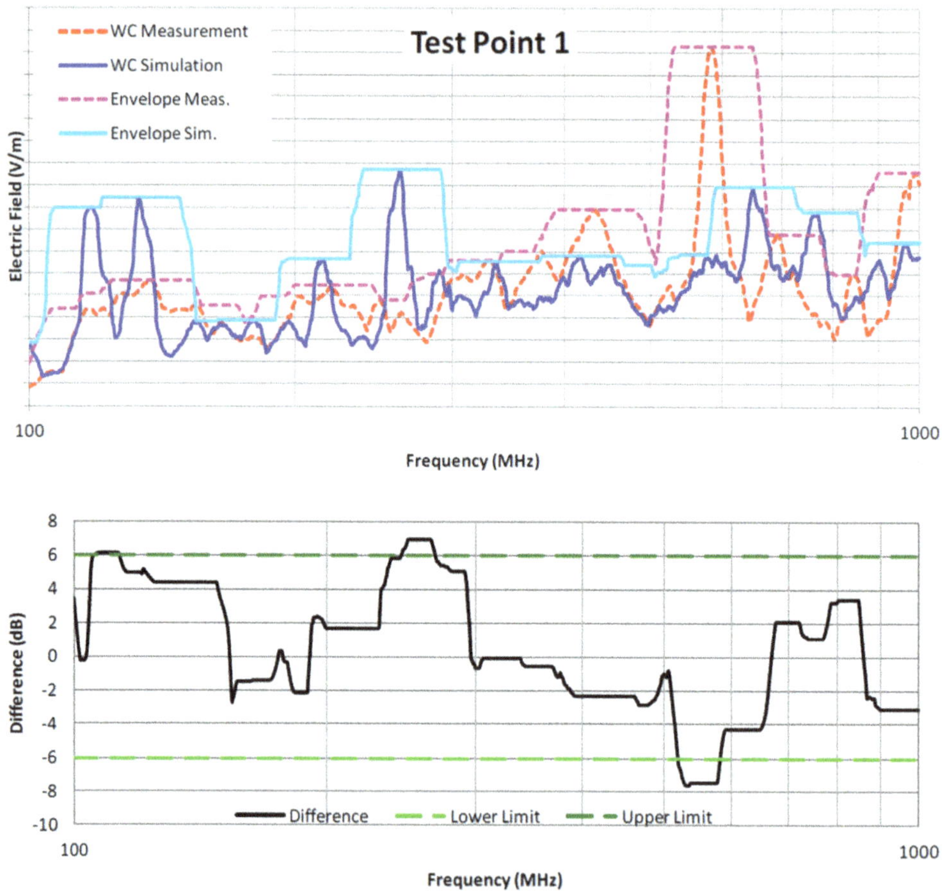

window. It bears noticing that numerical and experimental envelopes are within a ±6 dB range for most of the frequencies. *Table 1* shows the PFC results for all the six points. The FSV methodology has also been applied to the worst-case curves for comparison, finding fair agreements, in general (*Table 2*).

To become aware of the quality of experimental/numerical results matching in ±6 dB range, we have conducted another comparison with data found in two different test campaigns, with two different C-295 aircrafts, and different antennas and equipment. A point in the cockpit has just been taken and results are plot in *Figure 14*. It bears noticing that a similar ±6 dB matching pencil is also found in

Figure 12. a) Comparison between measurements and simulations for the test point 3 located in the cargo bay. b) Difference between simulation and measurement and 6 dB limit. (Reprinted from (Gutierrez et al., 2014) ©2014 IEEE, with permission from IEEE)

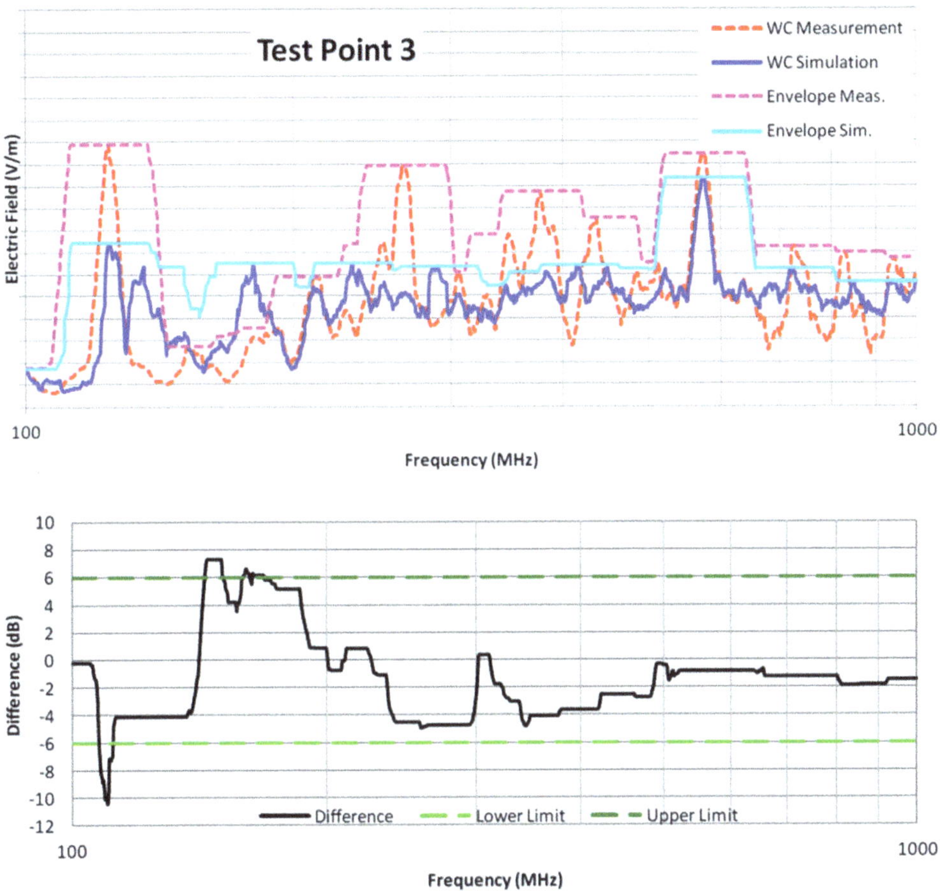

measurement/measurement comparisons. Fair agreement is also found by FSV, though not shown here, between experimental/experimental data. This provides a good idea of the real uncertainty of this kind of EMC problems, and points out the way for revisiting the FSV standard (IEEE Std 1597.2, 2010; Jauregui et al., 2014) to yield rates adapted for this specific topic.

Figure 13. a) Comparison between measurements and simulations for the test point 6 located in the engine. b) Difference between simulation and measurement and 6 dB limit. (Reprinted from (Gutierrez et al., 2014) ©2014 IEEE, with permission from IEEE)

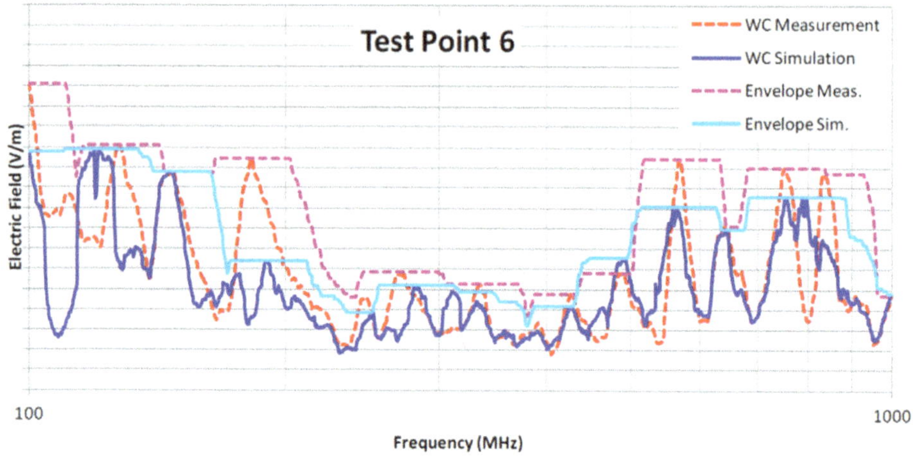

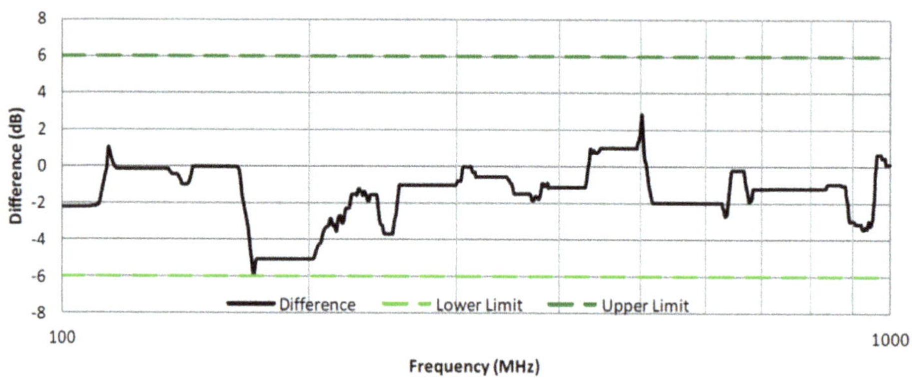

Table 1. PFC results for the UGRFDTD code. Percentage of points inside the limit, greatest difference between envelopes and mean difference between envelopes. (Reprinted from (Gutierrez et al., 2014) ©2014 IEEE, with permission from IEEE).

Pass/Fail Criterion	Points inside Limit	Greatest Difference	Mean Difference
Test Point 1	88%	-7.65	0.51
Test Point 2	89%	16.29	1.03
Test Point 3	94%	-10.44	-1.34
Test Point 4	81%	-14.41	-2.52
Test Point 5	76%	-12.20	-1.55
Test Point 6	100%	-5.95	-1.41

Table 2. FSV results for the C-295 test-case. (Reprinted from (Gutierrez et al., 2014) ©2014 IEEE, with permission from IEEE).

FSV	GDM	ADM	FDM
Test Point 1	0.9936	0.6123	0.6880
Test Point 2	0.7764	0.4588	0.5896
Test Point 3	0.7689	0.4640	0.5572
Test Point 4	0.7943	0.5702	0.5042
Test Point 5	0.8357	0.4876	0.5682
Test Point 6	0.6295	0.3418	0.4934

APPLICATION CASE 2: MILANO CENTRAL FUSELAGE IN A REVERBERATION CHAMBER, A TEST-SETUP TO ASSESS HIRF EFFECTS

The next test-case consists of the central fuselage of the MILANO UAV. This platform is a medium altitude long endurance remote piloted aircraft system developed by INTA. The MILANO UAV, made mostly in carbon-fiber composite, has a wingspan of 12.5 m, a length of 8.52 m and a height of 1.43 m. The UAV structure consists of independent composite modules with metallic fittings in order to assemble and disassemble the aircraft for transport. Only the central fuselage was used for the work described in this section, in green in Figure 15. Its top fairing that can be removed, exposing then three different bays or compartments. The big apertures of the central fuselage located at the bottom, left hand side and right hand side (see Figure 16) are covered with metallic plates.

The MILANO UAV is suited to hold equipment usually classified as Design Assurance Level A[10]. Hence, this equipment requires very high level of protection. The best solution is to home it within highly shielded enclosures. However, in practice, there is also a degradation of the shielding mainly due to apertures, seams and low-conductivity materials. Furthermore, for weight requirements, light composite materials are used, which worsens the situation, since the enclosure is more penetrable to RF.

The SE has been evaluated from 200 MHz up to 18 GHz at each of the three main bays of the fuselage, conducting a LLSF test inside a reverberation chamber, following the standard IEEE Std 299.1 (2014).

Figure 14. a) Comparison between two measurements on a point in the C-295 cockpit from different test campaigns. b) Difference between the two measurements and 6 dB limit. (Reprinted from (Gutierrez et al., 2014) ©2014 IEEE, with permission from IEEE)

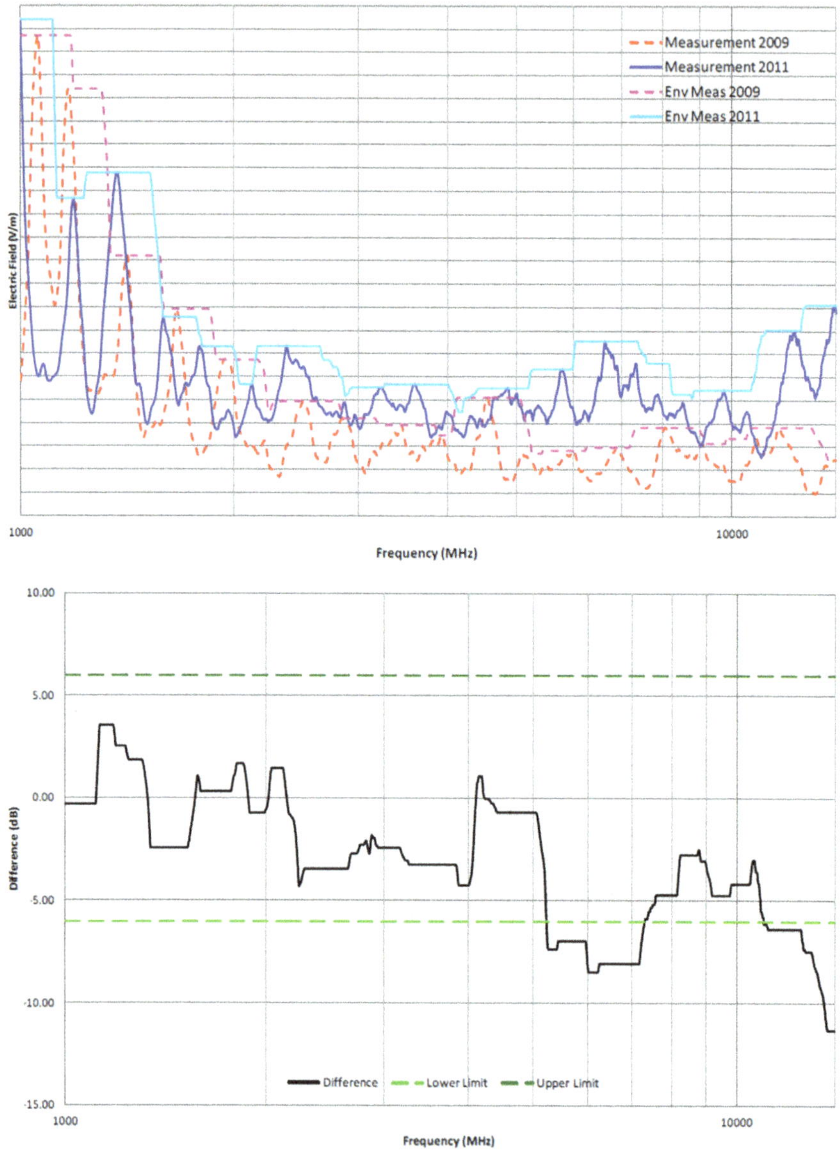

Figure 15. MILANO overview. Central fuselage in green

Figure 16. Layout of the central fuselage (dimensions in mm)

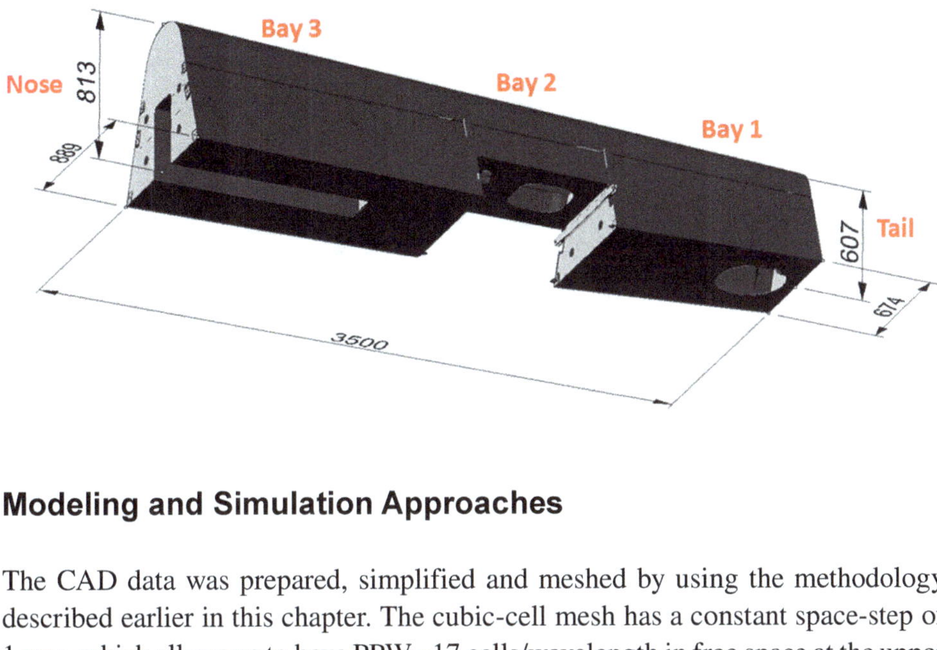

Modeling and Simulation Approaches

The CAD data was prepared, simplified and meshed by using the methodology described earlier in this chapter. The cubic-cell mesh has a constant space-step of 1 mm, which allows us to have PPW=17 cells/wavelength in free space at the upper frequency limit of 18 GHz. A problem with 5.2 GCells is yielded (3822 x 1205 x 1135). In our case, the fuselage incorporates just two kinds of materials: PEC and CFC. The CFC material has been taken as a homogeneous conductive thin-panel, 2.5 mm thick, and a conductivity of $10\ \mathrm{kS}/\mathrm{m}$. The Shielding effectiveness of a planar slab made on this CFC is illustrated in *Figure 19*. We have employed the SGBC method to simulate it with SEMBA-UGRFDTD. PML boundary conditions are used to truncate the model.

Figure 17. Overview of front part inside the RC

For the source, the RC model describer earlier in this chapter has been used. A Gaussian-modulated time-profile covering the frequency range 200 MHz – 10 GHz has been taken.

Results

The SE was evaluated at three observation positions, one at the center of each bay. Results with SEMBA-UGRFDTD solver and experimental data are shown in *Figure 20*, *Figure 21* and *Figure 22*. The results show a reasonable agreement at each probe position.

Figure 18. Example of a probe (at P2)

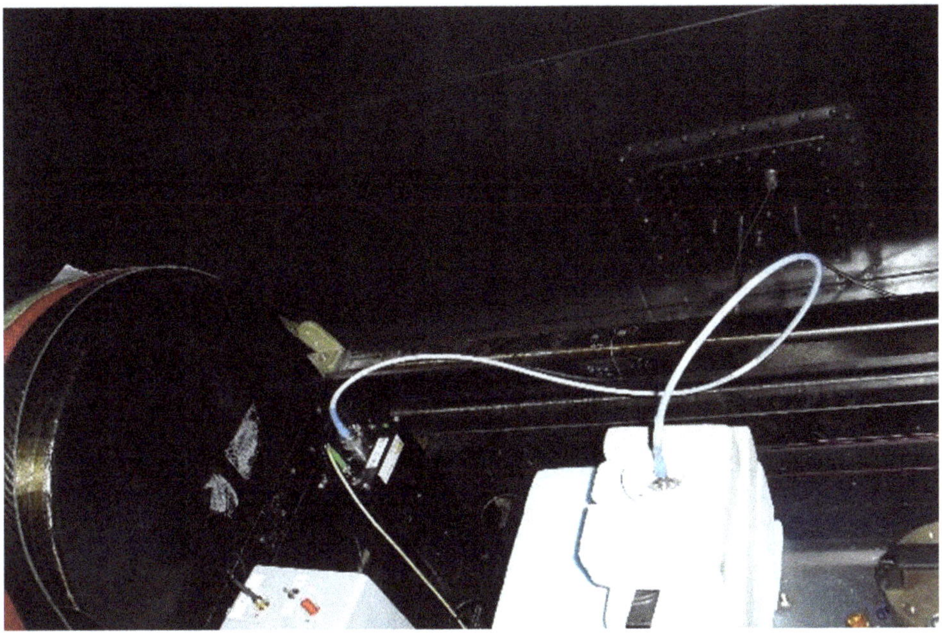

Figure 19. Shielding effectiveness of a planar slab of the Milano CFC

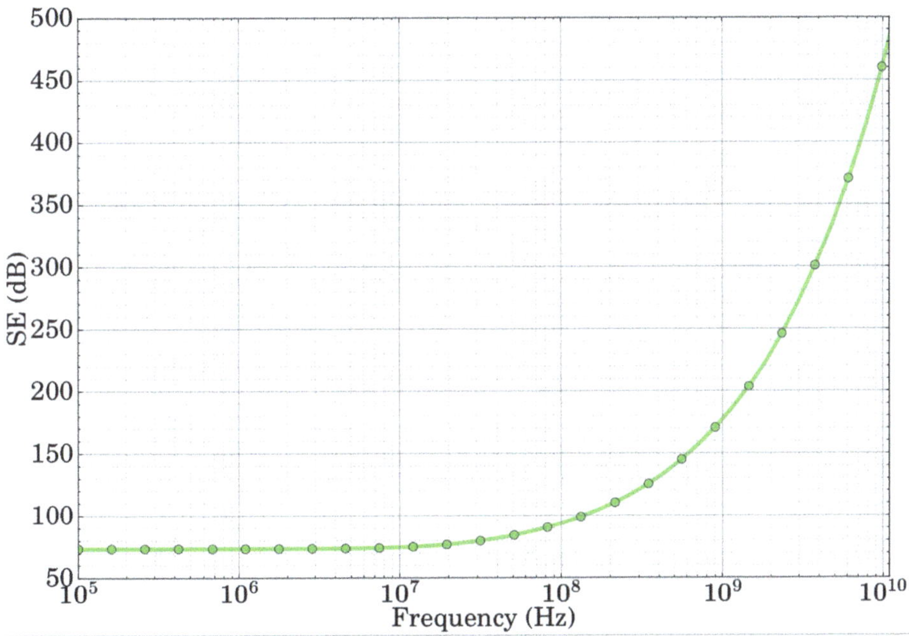

Figure 20. Shielding effectiveness at P1

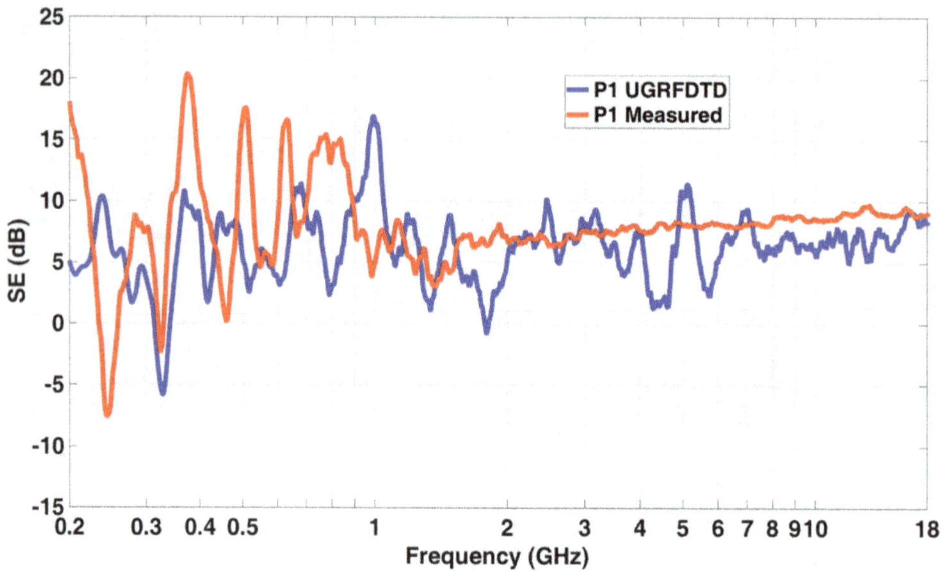

Figure 21. Shielding effectiveness at P2

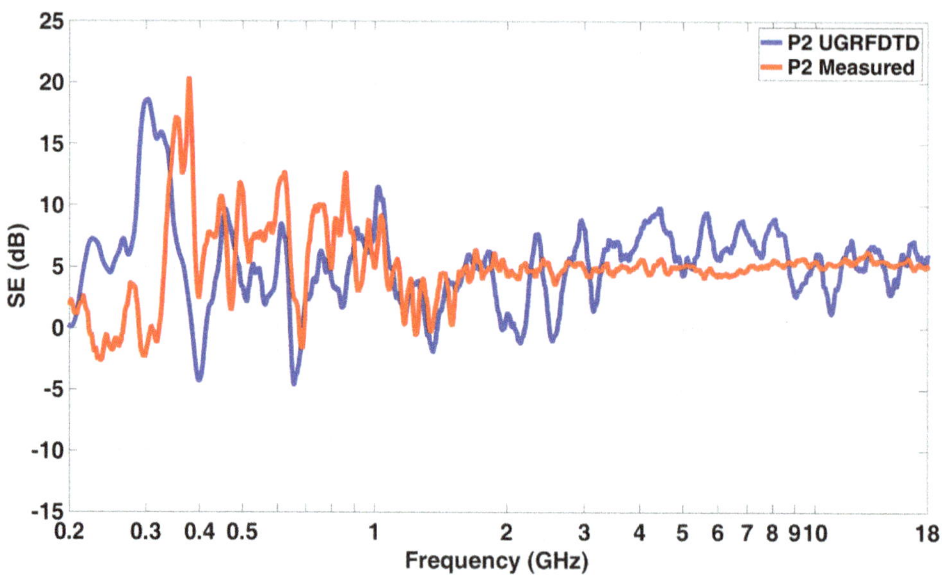

Figure 22. Shielding effectiveness at P3

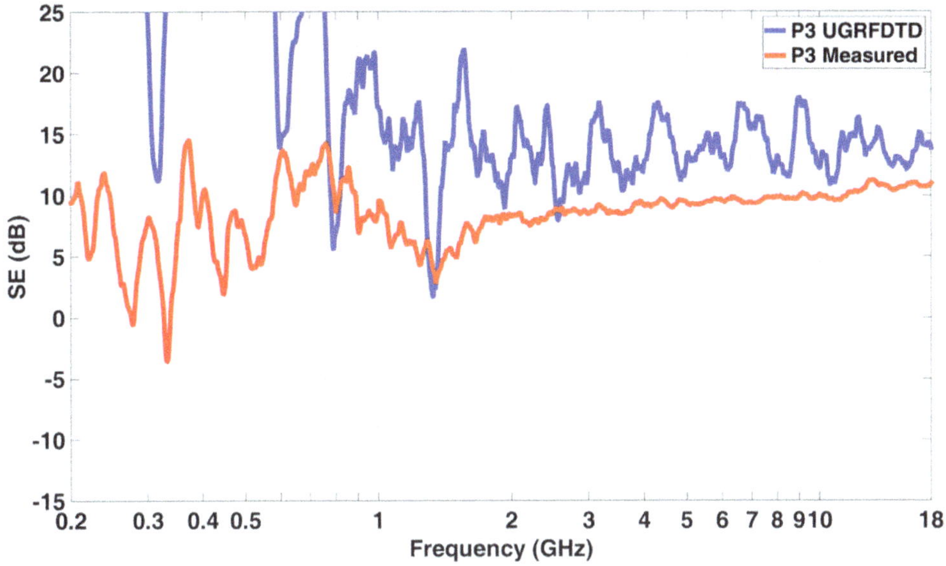

APPLICATION CASE 3: SIVA, A DCI TEST-CASE TO ASSESS LIE EFFECTS

The last case presents the experimental arrangement for the SIVA UAV, published in (Cabello et al., 2017c) by most of the authors of this chapter. This set-up has been tested with the DCI procedure in a LLDD setup (*Figure 23*). Results are shown

Figure 23. DCI test setup using coaxial return of SIVA-UAV in INTA's OATS (Reprinted from (Cabello et al., 2017c) ©2017 IEEE, with permission from IEEE)

mainly for the LIE frequency band and compared to experimental data conducted by INTA. The SIVA system consists of a ground-control station plus an UAV mostly made in CFC and fiberglass. It has a wingspan of 5.81 m, a length of 4.025 m, and a height of 1.63 m.

For this UAV, in the absence of large apertures, otherwise usual in typical aircrafts, the main mechanisms of EM energy entrance are due to apertures covered by fiberglass (radomes, fairings), hatches, joints, and hinges, and through the CFC materials themselves.

Experimental data for the DCI test setup of *Figure 23* measured in INTA's OATS have been used for validation. The spacing between the metallic wires of the cage return network is 30 cm. They are separated 63 cm from the UAV upper surface, and 37 cm for the lower surface. The RF power was injected through 2 wires connected to the propeller screws. The injected current was measured and recorded for normalization of the current induced on different cables installed inside the UAV, located in four positions (*Figure 24*). Most SIVA equipment was removed, keeping just four representative ones for our validation purposes: the FTCU, the PCU, the Airbag Bottle and the wings lights, together with the three cable bundles routed among them.

Figure 24. DCI probes: CP1, CP3, CP5, and CP7 (Reprinted from (Cabello et al., 2017c) ©2017 IEEE, with permission from IEEE)

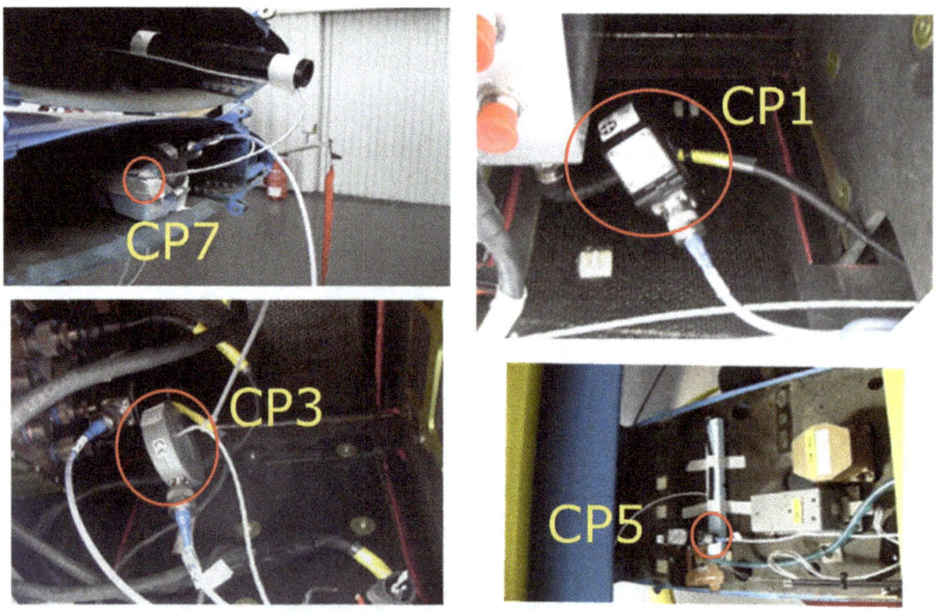

Modeling and Simulation Approaches

The original CAD model *Figure 25* has been cleaned up and defeatured, according to guidelines given earlier in this chapter, to yield the simplified model of *Figure 26* . It has been meshed with a uniform isotropic grid of 6 mm space-step for the UAV region, combined with a non-uniform one ending into a 20 mm space step at the PML boundaries (*Figure 27*), resulting into a very well resolved problem at the upper 200 MHz frequency: 75 cells/wavelength in free-space. A 160 Mcells (1168 x 498 x 277) model is finally yielded. The time step corresponds to a CFLN=0.75, imposed by the thin-wire hybrid FDTD solver stability condition.

Regarding the materials, just CFC ones were maintained, while fiberglass ones were removed. The equipment PCU, FTCU, airbag bottle (*Figure 25*) have been modeled as PEC. The UAV skin is a CFC, assumed to have a constant conductivity of 10 kS/m and an average thickness of 0.92 mm, treated by the SGBC method in SEMBA-UGRFDTD.

Cables and bundles inside the UAV employ the FDTD-MTLM Holland-Berenger method. Only representative cables have been kept: the one from the FTCU to igniter harness, and the one from the wing light to the PCU, including their resistive ending connectors.

Figure 25. Original CAD model including all geometrical details (Reprinted from (Cabello et al., 2017c) ©2017 IEEE, with permission from IEEE)

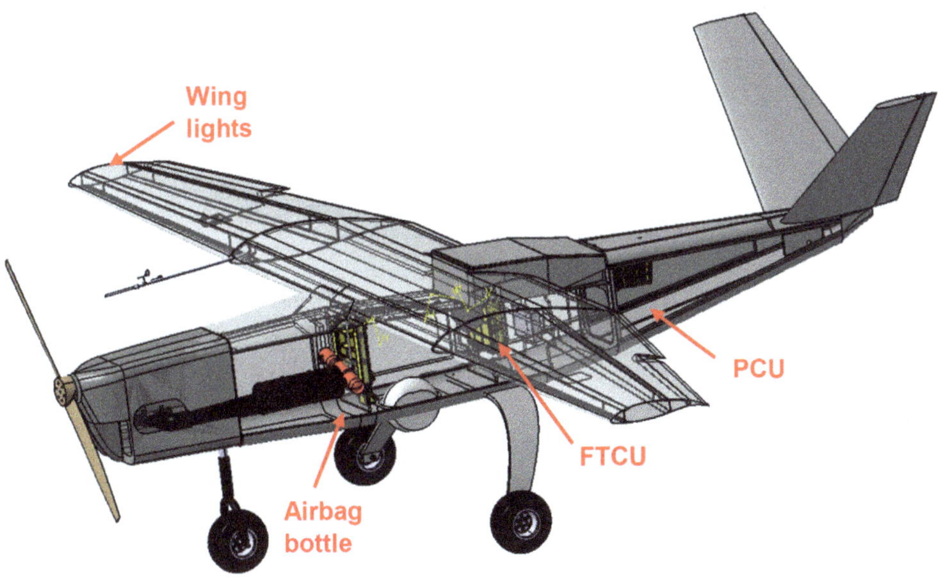

Figure 26. CAD model simplified after cleaning and including equipment and current probe locations (Reprinted from (Cabello et al., 2017c) ©2017 IEEE, with permission from IEEE)

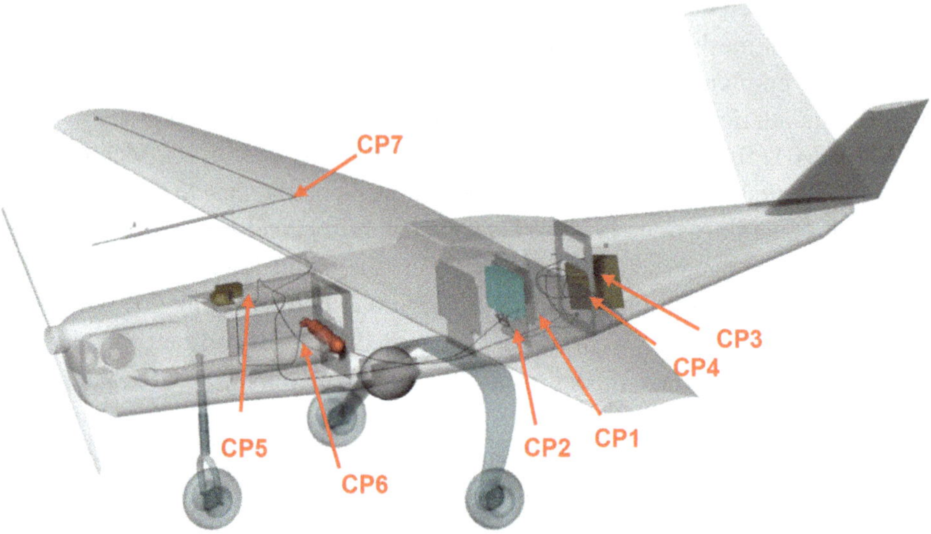

Figure 27. General and zoom views of final FDTD mesh. The upper right inset shows the DCI injection wires attached to the nose and tail of the UAV. In red the internal wire bundles. (Reprinted from (Cabello et al., 2017c) ©2017 IEEE, with permission from IEEE)

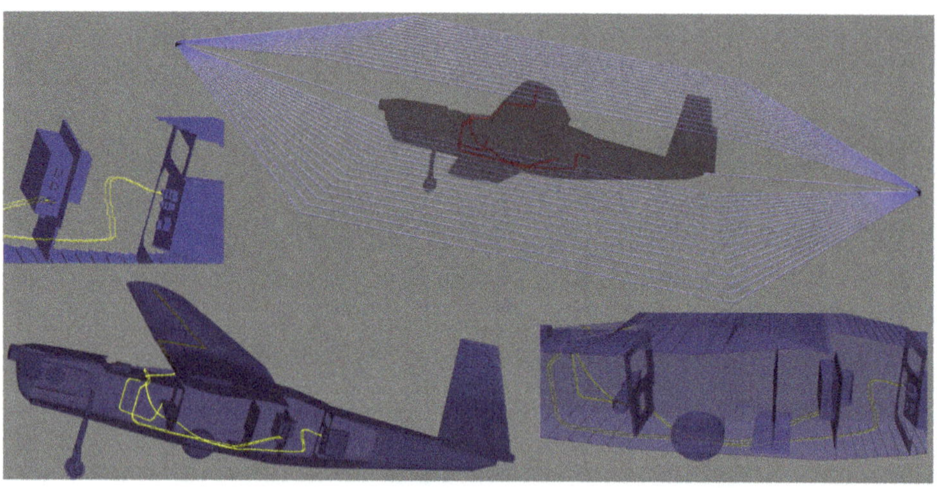

For the numerical simulations of the DCI test, the current is directly injected through the cable at the attachment point *Figure 23* . The cable at the exit point has a lumped resistance of 50 ©. A Gaussian signal with -3 dB decay at 400 MHz is fed as a voltage source in a thin wire located at the beginning of the coaxial return. The surfaces and lines of the coaxial return array have been modeled as PEC lines.

The OATS concrete ground is also included in the numerical model, though the results in this frequency range are insensitive to this, for being a DCI test-setup and field is assumed to be confined into the system formed by the fuselage and the coaxial return.

For validation the FSV methodology has been employed.

Results

The induced current at different probes (CP1, CP3, CP5, CP7 (See *Figure 24*, *Figure 26*)) are evaluated (*Figure 28*, *Figure 29*, *Figure 30*, *Figure 31*) and their values normalized in the frequency domain with respect to the injected current to find the transfer function as $T_{dB} = 20\,\log_{10}\frac{\left|I_{cable}\right|}{\left|I_{injected}\right|}$.

Figure 28. Current in probe CP1 (Reprinted from (Cabello et al., 2017c) ©2017 IEEE, with permission from IEEE)

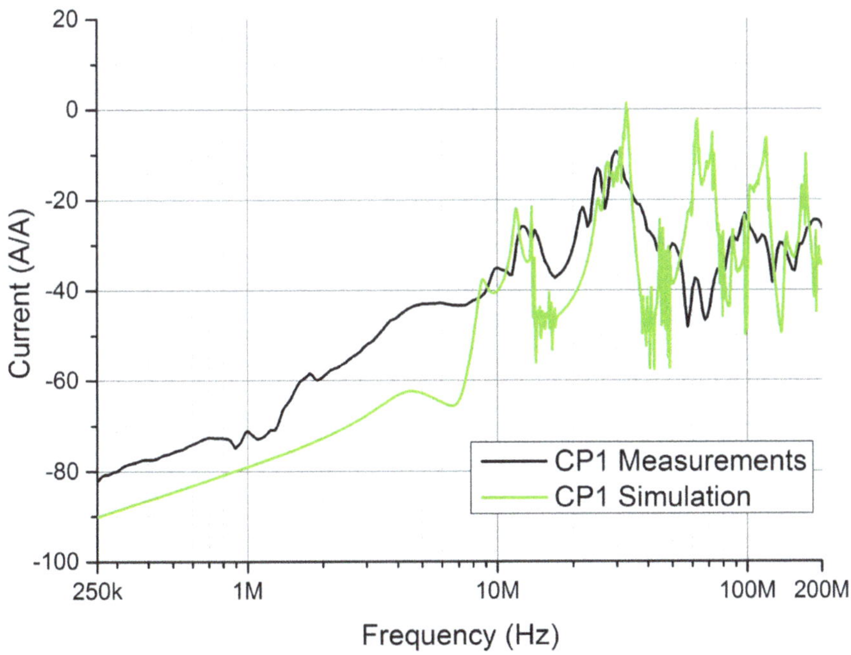

Figure 29. Current in probe CP3 (Reprinted from (Cabello et al., 2017c) ©2017 IEEE, with permission from IEEE)

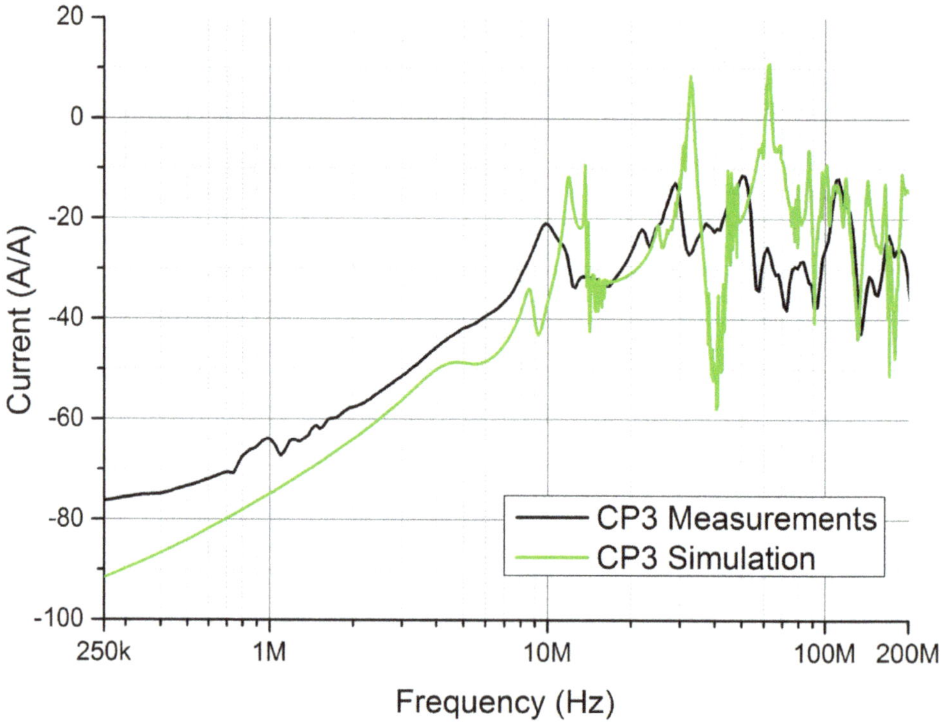

We can appreciate a low-frequency inductive trend at the points CP1, CP3, CP7, since these cables are being grounded through a resistive connector to the fuselage. However CP5 was directly grounded in the experimental setup, but the numerical model made use mistakenly of a grounding resistance. This explains the constant trend in the experimental transfer function at low frequency, compared to the linear one numerically found.

The FSV validation methodology (Jauregui et al., 2014) has been applied. *Table 3* summarizes the results. They yield a *good - very good* equivalent qualitative opinion of the experts, according to the threshold levels defined in the standard IEEE Std 1597.1 (2008), hence allowing us to state that the numerical model of the SIVA is representative of the real test setup. As a means of clarifying the implication of the GDM results presented in Table 3, a bar diagram of the FSV results comparing the measurements and simulations of the current probe CP3 is given in *Figure 32*. The bar diagram shows the experts opinion for the amplitude, the shape and the global result, when CP3 simulations and measurements are compared. From the diagram of the global indicator (GDM), it is shown that around the 30% of the experts will

Figure 30. Current in probe CP5 (Reprinted from (Cabello et al., 2017c) ©2017 IEEE, with permission from IEEE)

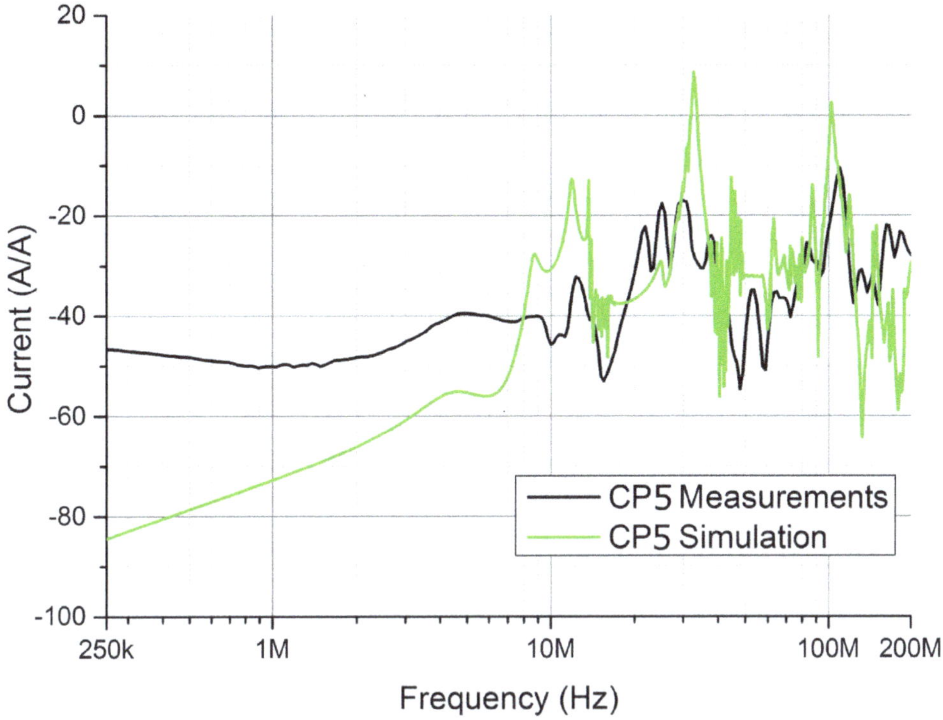

conclude that the similarity between the graphs is very good, another 30% will consider it a good agreement, around the 15% a fair agreement, around 12% a poor fitting, and finally less than 10% a very poor similarity. Elsewhere, the bar results for the amplitude indicator (ADM) and the shape indicator (FDM) are also shown, highlighting that the experts' opinion clearly assumes that the fitting is stronger in terms of shape than in agreement of amplitude.

Finally, it bears noticing that the global frequency trend observed by concatenating the SIVA and the MILANO results in frequency (e.g. *Figure 20* and *Figure 28*), agree to that given by the generic transfer function of *Figure 4*.

FUTURE RESEARCH DIRECTIONS

Accepting simulation for certification purposes is an ambitious goal. It requires counting with certification authorities, and with the synergy of test houses, airframers and software providers. In this sense, there is a long way to go in order to achieve a

Figure 31. Current in probe CP7 (Reprinted from (Cabello et al., 2017c) ©2017 IEEE, with permission from IEEE)

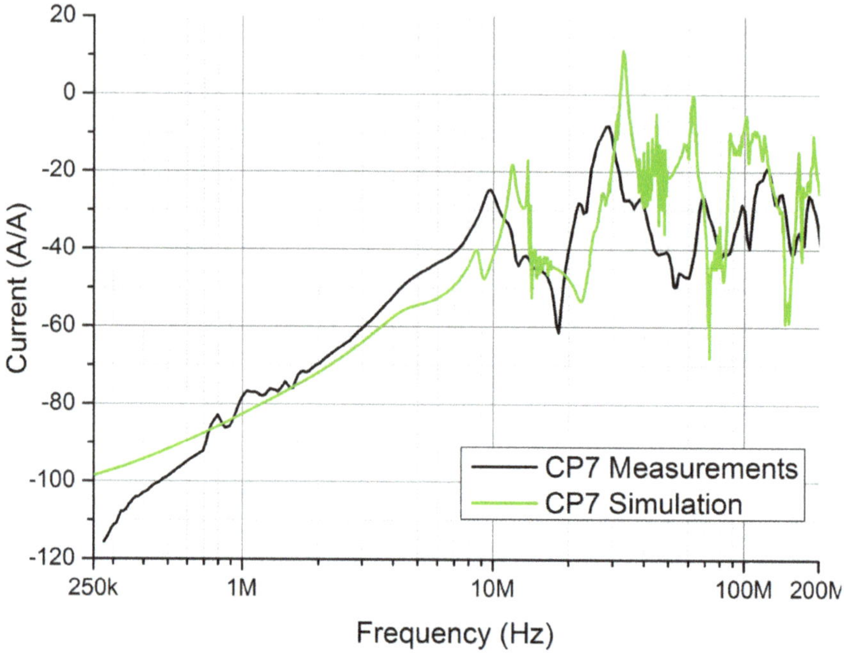

Table 3. FSV indicators.

Reference	ADM	FDM	GDM
CP1	0.20	0.53	0.62
CP3	0.35	0.55	0.76
CP5	0.27	0.61	0.74
CP7	0.26	0.53	0.67

good level of confidence in numerical solvers in a task which classically relies just in experimental testing. The validation of any numerical model of physical features (material, junctions, cables, etc.) with experimental data is the only way to advance in this line. Some open lines are:

1. The precise characterization of aeronautic materials, which is a hectic field of research. Methods for the measurement of their characteristics as well as macroscopic causal models are a must.
2. The role of nonlinearities in materials, often tiptoed over because of the complexity required to deal with them, also requires further works. These

Figure 32. FSV bar diagram (ADM (top), FDM (middle), GDM (bottom)) comparing measurements and simulation of current probe CP3. Ex: Excellent, VG: very good, G: good, F: fair, P: poor, VP: very poor (Reprinted from (Cabello et al., 2017c) ©2017 IEEE, with permission from IEEE)

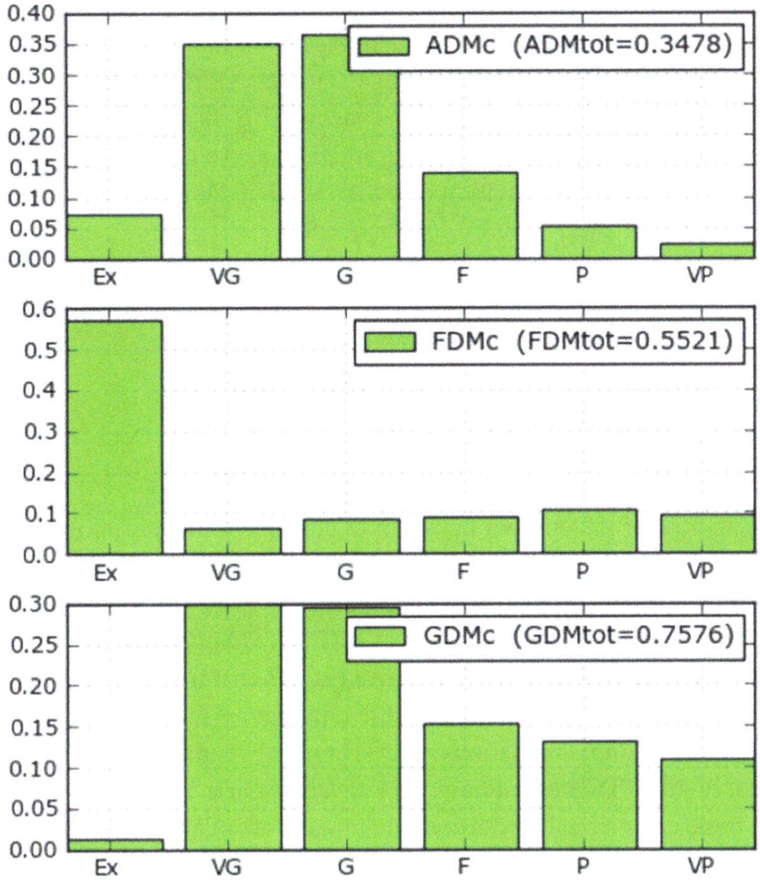

effects may be critical for instance in cases with extremely large field values like the study of lightning indirect effects.

3. The complexity of AVs is constantly increasing. Deterministic methods begin to fall short to manage the incertitude, including its life-cycle, in the physical parameters. This topic must be addressed in a smart manner: stochastic methods, reduced-order modeling…
4. Experimental EM measurement methods and procedures must be further developed to be able to measure with high confidence the output produced by the numerical simulation.

5. The evaluation of uncertainties cannot longer be accomplished by deterministic methods because of the increasing complexity. Statistical techniques must be investigated
6. A single numerical technique is not enough to address the whole frequency band. FDTD is currently limited in modern computers by the amount of memory and CPU time taken for convergence. In practice it is unfeasible to find results over 18 GHz for average-sized aircrafts. At the same time the low-frequency convergence is slow, especially for highly-resonant structures. Physical Optics, Power Balance, etc. are methods suitable to be used above 18 GHz, while mixed formulation of the Method of Moments in frequency may provide a good alternative for low frequency.
7. Finally, cables by themselves, together with their shielding and grounding, still deserve research, to be able to handle the full complexity of the wiring of an aircraft in a computer feasible manner.

CONCLUSION

The ultimate goal of the authors has been to advance in the definition of a whole methodology to address EMC certification scenarios, combining numerical techniques with experimental ones, to be actually acceptable by EMC certification authorities. This chapter has presented a whole roadmap for it, including: i) the main certification scenarios for EMC, with reference to current standards; ii) the desired features of a numerical FDTD solver to cope with the complexity found in EMC problems; iii) a novel specific implementation of FDTD to deal in an efficient way with lossy thin panels; iv) the FSV methodology for the validation of numerical results; v) the process to build a numerical model computationally affordable and representative of the full complexity. Three representative test cases of real AVs have been described to illustrate this roadmap. They compare experimental and numerical results of aircrafts, including their actual complexity (CFC materials, cable bundles, etc.), in three real certification scenarios.

ACKNOWLEDGMENT

This research was supported by the Spanish MINECO and EU FEDER [TEC2013-48414-C{1,2,3}-R, TEC2016-79214-C{1,2,3}-R, TEC2015-68766-REDC]; AIRBUS [Alhambra-UGRFDTD FEUGR 3713-00]; and the J. de Andalucía (Spain) [P12-TIC-1442]. This work has also received support from the EU PRACE project

SREDIT which funds simulation tasks under the Marconi KNL supercomputing cluster of CINECA (Italy).

REFERENCES

AC 20-158A. (2014). *The certification of aircraft electrical and electronic systems for operation in the high-intensity radiated fields (HIRF) environment.* Advisory U.S. Department of Transportation Circular Federal Aviation Administration.

Agrawal, A. K., Price, H. J., & Gurbaxani, S. H. (1980). Transient response of multiconductor transmission lines excited by a nonuniform electromagnetic field. *IEEE Transactions on Electromagnetic Compatibility*, *EMC-22*(2), 119–129. doi:10.1109/TEMC.1980.303824

Baum, C. E., Liu, T. K., & Tesche, F. M. (1978). On the analysis of general multiconductor transmission-line networks. *Interaction Note*, *350*(6), 467–547.

Berenger, J. P. (1994). A perfectly matched layer for the absorption of electromagnetic waves. *Journal of Computational Physics*, *114*(2), 185–200. doi:10.1006/jcph.1994.1159

Berenger, J. P. (2000). A multiwire formalism for the FDTD method. *IEEE Transactions on Electromagnetic Compatibility*, *42*(3), 257–264. doi:10.1109/15.865332

Besnier, P., & Demoulin, B. (2011). *Electromagnetic reverberation chambers*. John Wiley & Sons.

Cabello, M. (2017d). *Subcell FDTD techniques for electromagnetic compatibility assessment in aeronautics (PhD dissertation)*. University of Granada. Retrieved from http://www.sembahome.org/wp-content/uploads/2017/PhD_dissertation_MRCabello_definprint.pdf

Cabello, M. R., Angulo, L., Alvarez, J., Flintoft, I. D., Bourke, S., Dawson, J. F., & Gómez Martín, R. (2017b, May). A Hybrid Crank Nicolson FDTD Subgridding Boundary Condition for Lossy Thin-Layer Modeling. *IEEE Transactions on Microwave Theory and Techniques*, *65*(5), 1397–1406. doi:10.1109/TMTT.2016.2637348

Cabello, M. R., Angulo, L. D., Alvarez, J., Bretones, A. R., Gutierrez, G. G., & Garcia, S. G. (2015, August). A New Efficient and Stable 3D Conformal FDTD. *IEEE Microwave and Wireless Components Letters*, *26*(8), 553–555. doi:10.1109/LMWC.2016.2588579

Cabello, M. R., Angulo, L. D., Bretones, A. R., Martin, R. G., Garcia, S. G., & Alvarez, J. (2017a, May). A novel subgriding scheme for arbitrarily dispersive thin-layer modeling. In *Numerical Electromagnetic and Multiphysics Modeling and Optimization for RF, Microwave, and Terahertz Applications (NEMO), 2017 IEEE MTT-S International Conference on* (pp. 266-268). IEEE. doi:10.1109/NEMO.2017.7964255

Cabello, M. R., Fernández, S., Pous, M., Pascual-Gil, E., Angulo, L. D., López, P., & Garcia, S. G. et al. (2017c). SIVA UAV: A Case Study for the EMC Analysis of Composite Air Vehicles. *IEEE Transactions on Electromagnetic Compatibility, 59*(4), 1103–1113. doi:10.1109/TEMC.2017.2648507

Dey, S., & Mittra, R. (1997). A locally conformal finite-difference time-domain (FDTD) algorithm for modeling three-dimensional perfectly conducting objects. *IEEE Microwave and Guided Wave Letters, 7*(9), 273–275. doi:10.1109/75.622536

Duffy, A. P., Martin, A. J., Orlandi, A., Antonini, G., Benson, T. M., & Woolfson, M. S. (2006, August). Feature selective validation (FSV) for validation of computational electromagnetics (CEM). Part I-The FSV method. *IEEE Transactions on Electromagnetic Compatibility, 48*(3), 449–459. doi:10.1109/TEMC.2006.879358

Edelvik, F. (2002). *Hybrid solvers for the Maxwell equations in time-domain* (Doctoral dissertation). Acta Universitatis Upsaliensis.

Eurocae ED-105A. (2013). *Aircraft Lightning Test Methods. Revision A*. Author.

Eurocae ED-107A. (2010). *Guide to certification of aircraft in a high-intensity radiated field (HIRF) environment*. Author.

Eurocae ED-14G. (2011). *Environmental conditions and test procedures for airborne equipment*. Author.

Feliziani, M., & Maradei, F. (2002). Full-wave analysis of shielded cable configurations by the FDTD method. *IEEE Transactions on Magnetics, 38*(2), 761–764. doi:10.1109/20.996197

GarciaS. G. (2013). *SEMBA*. Retrieved from http://www.sembahome.org

Garcia, S. G., Perez, I. V., Martin, R. G., & Olmedo, B. G. (1999). BiPML: A PML to match waves in bianisotropic media. *Microwave and Optical Technology Letters, 20*(1), 44–48. doi:10.1002/(SICI)1098-2760(19990105)20:1<44::AID-MOP12>3.0.CO;2-W

Garcia, S. G., Rubio, R. G., Bretones, A. R., & Lopez, R. G. (2007, November). Revisiting the Stability of Crank-Nicolson and ADI-FDTD. *IEEE Transactions on Antennas and Propagation*, *55*(11), 3199–3203. doi:10.1109/TAP.2007.908794

Gil, E. P., & Gutierrez, G. G. (2011). Simplification and cleaning of complex CAD models for EMC simulations. *Int. Symp. Electromagnetic Compatibility (EMC Europe)*.

Gkatzianas, M., & Tsiboukis, T. (2003). Thin-slot/thin-layer subcell FDTD algorithms for EM penetration through apertures. *Electromagnetics*, *23*(2), 119–133. doi:10.1080/02726340390159469

Guiffaut, C., Reineix, A., & Pecqueux, B. (2012). New oblique thin wire formalism in the FDTD method with multiwire junctions. *IEEE Transactions on Antennas and Propagation*, *60*(3), 1458–1466. doi:10.1109/TAP.2011.2180304

Gustavsen, B. (2006, July). Improving the pole relocating properties of vector fitting. *IEEE Transactions on Power Delivery*, *21*(3), 1587–1592. doi:10.1109/TPWRD.2005.860281

Gutierrez, G. G., Romero, D. M., Cabello, M. R., Pascual-Gil, E., Angulo, L. D., Gomez, D. G., & Garcia, S. G. (2016). On the design of aircraft electrical structure networks. *IEEE Transactions on Electromagnetic Compatibility*, *58*(2), 401–408. doi:10.1109/TEMC.2016.2514379

Han, Dutton & Fan. (2006, March). Model dispersive media in finite-difference time-domain method with complex-conjugate pole-residue pairs. *IEEE Microwave and Wireless Components Letters*.

Hao, Y., & Mittra, R. (2009). *FDTD Modeling of Metamaterials Theory and Application*. Artech House.

Hill, D. A. (1998). Plane wave integral representation for fields in reverberation chambers. *IEEE Transactions on Electromagnetic Compatibility*, *40*(3), 209–217. doi:10.1109/15.709418

Hill, D. A. (2009). *Electromagnetic fields in cavities: deterministic and statistical theories* (Vol. 35). John Wiley & Sons. doi:10.1002/9780470495056

HIRFSE. (2008). *HIRF -SE project EU FP7, 2008-2013*. Retrieved from http://HIRFse.axessim.eu/

Holland, R., & Simpson, L. (1981). Finite-difference analysis of EMP coupling to thin struts and wires. *IEEE Transactions on Electromagnetic Compatibility*, *EMC-23*(2), 88–97. doi:10.1109/TEMC.1981.303899

IEC 61000-4-21. (2011). *Electromagnetic compatibility (EMC) - Part 4-21: Testing and measurement techniques - Reverberation chamber test method*. Author.

IEEE Std 1597.1. (2008). *IEEE Standard for Validation of Computational Electromagnetics Computer Modeling and Simulations, IEEE 1597.1-2008*. IEEE.

IEEE Std 1597.2. (2010). *IEEE Recommended Practice for Validation of Computational Electromagnetics Computer Modeling and Simulations, IEEE Standard 1597.2-2010*. IEEE.

IEEE Std 299.1. (2014). *IEEE Standard Method for Measuring the Shielding Effectiveness of Enclosures and Boxes Having all Dimensions between 0.1 m and 2 m. IEEE Std 299.1-2013*. IEEE.

Jauregui, R., Aragon, M., & Silva, F. (2013, February). The role of uncertainty in the feature selective validation (FSV) method. *IEEE Transactions on Electromagnetic Compatibility*, *55*(1), 217–220. doi:10.1109/TEMC.2012.2227262

Jauregui, R., Pous, M., & Silva, F. (2014b, September). Use of reference limits in the Feature Selective Validation (FSV) method. *IEEE 2014 International Symposium on Electromagnetic Compatibility*, 1031-1036.

Jauregui, R., Zhang, G., Rojas-Mora, J., Ventosa, O., Silva, F., Duffy, A. P., & Sasse, H. (2014a, August). Analyzing transient phenomena in the time domain using the feature selective validation (FSV) method. *IEEE Transactions on Electromagnetic Compatibility*, *56*(4), 825–834. doi:10.1109/TEMC.2013.2246167

Jauregui, T. R., Silva, F., Orlandi, A., Sasse, H., & Duffy, A. (2010, April). Factors influencing the successful validation of transient phenomenon modelling. *2010 IEEE Asia-Pacific Symposium on Electromagnetic Compatibility (APEMC)*, 338-341.

Kobidze, G. (2010). Implementation of collocated surface impedance boundary conditions in FDTD. *IEEE Transactions on Antennas and Propagation*, *58*(7), 2394–2403. doi:10.1109/TAP.2010.2048859

Kunz, K. S., Steich, D. J., & Luebbers, R. J. (1997, May). *Determining an Aircraft's Attenuation Characteristics of HIRF Testing*. SAE Technical Paper.

Kunz, S., Steich, D. J., & Luebbers, R. J. (1993, August). Low Frequency Shielding Effects of a Conducting Shell with an Aperture: Response of an Internal Wire. *IEEE Transactions on Electromagnetic Compatibility*, *34*(3), 370–373. doi:10.1109/15.155859

Ledfelt, G. (2001). *Hybrid time-domain methods and wire models for computational electromagnetics* (Doctoral dissertation). Numerisk analys och datalogi.

Meyer, M., Flourens, F., Rouquette, J. A., & Delnevo, A. (2008, September). Modeling of lightning indirect effects in CFRP aircraft. *2008 International Symposium on Electromagnetic Compatibility (EMC Europe)*, 1-5. doi:10.1109/EMCEUROPE.2008.4786842

MIL-HDBK-5H. (1998). Department of defense test method standard: Metallic Materials and Elements for Aerospace Vehicle Structures. *Section, 5*(1).

MIL-STD-202H. (2015, April). *Department of defense test method standard: Test method standard electronic and electrical component parts.*

MIL-STD-461G. (2015). *Department of defense test method standard: Requirements for the control of electromagnetic interference characteristics of subsystems and equipment.*

MIL-STD-464C. (2010). *Department of defense test method standard: electromagnetic environmental effects requirements for systems.*

Mittra, R. (2016). *Computational Electromagnetics*. Springer-Verlag.

Moglie, F., & Pastore, A. P. (2006). FDTD analysis of plane wave superposition to simulate susceptibility tests in reverberation chambers. *IEEE Transactions on Electromagnetic Compatibility*, *48*(1), 195–202. doi:10.1109/TEMC.2006.870793

Nayyeri, V., Soleimani, M., & Ramahi, O. M. (2013). Modeling graphene in the finite-difference time-domain method using a surface boundary condition. *IEEE Transactions on Antennas and Propagation*, *61*(8), 4176–4182. doi:10.1109/TAP.2013.2260517

Orlandi, A., Duffy, A. P., Archambeault, B., Antonini, G., Coleby, D. E., & Connor, S. (2006, May). Feature selective validation (FSV) for validation of computational electromagnetics (CEM). part II-assessment of FSV performance. *IEEE Transactions on Electromagnetic Compatibility*, *48*(3), 460–467. doi:10.1109/TEMC.2006.879360

Parmantier J., Issac F., Gobin V. (2012, December). Indirect Effects of Lightning on Aircraft and Rotorcraft. *Journal of Aerospace Lab., ONERA*, 1-27.

Rachidi, F. (2012, August). A Review of Field-to-Transmission Line Coupling Models With Special Emphasis to Lightning-Induced Voltages on Overhead Lines. *IEEE Transactions on Electromagnetic Compatibility*, *54*(4), 898–911. doi:10.1109/TEMC.2011.2181519

Radasky, W. (2014) Electromagnetic Warfare is Here. *IEEE Spectrum*. Retrieved from http://spectrum.ieee.org/aerospace/military/electromagnetic-warfare-is-here

Radasky, W. A., Baum, C. E., & Wik, M. W. (2004). Introduction to the special issue on high-power electromagnetics (HPEM) and intentional electromagnetic interference (IEMI). *IEEE Transactions on Electromagnetic Compatibility*, *46*(3), 314–321. doi:10.1109/TEMC.2004.831899

Rasek, G. A., & Loos, S. E. (2008, August). Correlation of direct current injection (DCI) and free-field illumination for HIRF certification. *IEEE Transactions on Electromagnetic Compatibility*, *50*(3), 499–503. doi:10.1109/TEMC.2008.926872

Rasek, S., Tobola, R., & Loos, T. (2015, February 3). HIRF transfer function observations: Notes on results versus requirements and certification approach. *IEEE Transactions on Electromagnetic Compatibility*, *57*(2), 195–202. doi:10.1109/TEMC.2014.2381095

Rasek & Gabrisak. (2011, April). Wire bundle currents for high intensity radiated fields (HIRF) and indirect effects of lightning (IEL) with focus on bulk current injection (BCI) test. *21st International Conference Radioelektronika.*

Rierson, L. K. (1997). *Determining an Aircraft's Attenuation Characteristics of HIRF Testing* (No. 971488). SAE Technical Paper.

Romero, S. F., Morales, A. L., Cancela, M. A., & Gutierrez, G. G. (2012, September). Validation procedure of low level coupling tests on real aircraft structure. In *Electromagnetic Compatibility (EMC EUROPE), 2012 International Symposium on* (pp. 1-6). IEEE. doi:10.1109/EMCEurope.2012.6396706

RTCA/DO-160 G. (2014). *Environmental Conditions and Test Procedures for Airborne Equipment.*

Sarto, M. S. (1999). A new model for the FDTD analysis of the shielding performances of thin composite structures. *IEEE Transactions on Electromagnetic Compatibility*, *41*(4), 298–306. doi:10.1109/15.809798

Sarto, M. S., Greco, S., & Tamburrano, A. (2014). Shielding effectiveness of protective metallic wire meshes: EM modeling and validation. *IEEE Transactions on Electromagnetic Compatibility*, *56*(3), 615–621. doi:10.1109/TEMC.2013.2292715

Schmidt, S., & Lazzi, G. (2004). Use of the FDTD thin-strut formalism for biomedical telemetry coil designs. *IEEE Transactions on Microwave Theory and Techniques*, *52*(8), 1952–1956. doi:10.1109/TMTT.2004.832019

Schneider. (2010). *Understanding the Finite-Difference Time-Domain Method.* Retrieved from http://www.eecs.wsu.edu/~schneidj/ufdtd/ufdtd.pdf

Schroder, Bruns, Kucera, & Schuster. (2014, February). Analysis of High Intensity Radiated Field Coupling into Aircraft Using the Method of Moments. *IEEE Trans. Electromagnetic Compatibility, 56*(1).

Schuster, C., Christ, A., & Fichtner, W. (2000). Review of FDTD time-stepping schemes for efficient simulation of electric conductive media. *Microwave and Optical Technology Letters, 25*(1), 16–21. doi:10.1002/(SICI)1098-2760(20000405)25:1<16::AID-MOP6>3.0.CO;2-O

STANAG 4370 Ed 5. (2014). NATO Standardization Agreements.

Taflove, A., Oskooi, A., & Johnson, S. G. (Eds.). (2013). *Advances in FDTD computational electrodynamics: photonics and nanotechnology*. Artech House.

Taflove & Hagness. (2005). *Computational Electrodynamics The Finite- Differences Time Domain Method*. Norwood, MA: Artech House.

Vance, E. F. (1975). Shielding effectiveness of braided-wire shields. *IEEE Transactions on Electromagnetic Compatibility, EMC-17*(2), 71–77. doi:10.1109/TEMC.1975.303389

Yang, Y., Chen, R. S., & Yung, E. K. (2006). The unconditionally stable Crank Nicolson FDTD method for three-dimensional Maxwell's equations. *Microwave and Optical Technology Letters, 48*(8), 1619–1622. doi:10.1002/mop.21684

Yatsenko, V. V., Tretyakov, S. A., Maslovski, S. I., & Sochava, A. A. (2000). Higher order impedance boundary conditions for sparse wire grids. *IEEE Transactions on Antennas and Propagation, 48*(5), 720–727. doi:10.1109/8.855490

Yee, K. (1966, May). Numerical solution of initial boundary value problems involving Maxwell's equations in isotropic media. *IEEE Transactions on Antennas and Propagation, 14*, 202–307.

Yee, K. (1966, May). Numerical solution of initial boundary value problems involving Maxwell's equations in isotropic media. *IEEE Transactions on Antennas and Propagation, 14*, 202–307.

Yu, Mittra, Su, Liu, & Yang (2006). *Parallel Finite-Difference Time-Domain Method*. Artech House, Inc.

Zivanovic, S. S., Yee, K. S., & Mei, K. K. (1991). A subgridding method for the time-domain finite-difference method to solve Maxwell's equations. *IEEE Transactions on Microwave Theory and Techniques, 39*(3), 471–479. doi:10.1109/22.75289

KEY TERMS AND DEFINITIONS

ADM: Amplitude difference measure.
ANSI: American National Standards Institute.
AV: Air vehicles.
CAD: Computer-aided design.
CAE: Computer-aided engineering.
CAM: Computer-aided manufacturing.
CEM: Computational electromagnetic.
CFC: Carbon fiber composite.
CFL: Courant-Friedrichs-Lewy.
CFRC: Carbon fiber reinforced composite.
CFRP: Composite fiber reinforced plastic.
CPU: Central processing unit.
DCI: Direct current injection.
DD: Direct drive.
E3: Electromagnetic environmental effects.
EASA: European Aviation Safety Agency.
EM: Electromagnetic.
EMC: Electromagnetic compatibility.
EMI: Electromagnetic interference.
EMP: Electromagnetic pulses.
ETD: Exponential time differencing.
EUROCAE: European Organization for Civil Aviation Equipment.
EWIS: Electrical wiring interconnect system.
FAA: Federal Aviation Administration.
FDM: Feature difference measure.
FDTD: Finite difference time domain.
FSV: Feature selective validation.
FTCU: Flight termination control unit.
GDM: Global difference measure.
HIRF: High intensity radiated fields.
HIRF-SE: HIRF-synthetic environment.
INTA: Spanish National Institute of Aerospace Techniques.
LIE: Lightning indirect effects.
LLDD: Low-level direct drive.
LLSC: Low-level swept current.
LLSC: Low-level swept coupling.
LLSF: Low-level swept fields.
MTLN: Multiconductor transmission line.

NIBC: Network impedance boundary conditions.
OATS: Open area test site.
PCU: Power control unit.
PEC: Perfect electric conductors.
PFC: Pass/fail criterion.
PML: Perfectly absorbing matching layer.
PoE: Point of entry.
PPW: Points per wavelength.
RC: Reverberation chamber.
RF: Radio frequency.
RTCA: Radio Technical Commission for Aeronautics (USA).
SAE: Society of Automotive Engineers (USA).
SE: Shielding effectiveness.
SEMBA: EMC broadband simulator.
SGBC: Sub-gridding boundary conditions.
UAV: Unmanned air vehicles.
UGR: University of Granada.
UPC: Polytechnic University of Catalonia.
UPV: Polytechnic University of Valencia.
WC: Worst Case.

ENDNOTES

1 Some projects on this line are: FULMEN, ILDAS, EM-HAZ, STRUCTURES, GENIAL, ARROW, SAFETEL, GEMCAR, SARITSU, FLAVIIR, MOVEA, etc.

2 See http://cordis.europa.eu/result/rcn/140447_en.html

3 The referenced EUROCAE documents used by European certification authorities have a direct correspondence with US ones. For instance, ED-105A is an expanded version of MIL-STD-1757, ED-107A is published by SAE under the ARP5583A reference, ED-14 and RTCA DO-160 are similarly worded…

4 In fact the ED-107A guide requires the individual equipment wire bundle currents to be normalized with respect to the aircraft skin current analysis, completed by calculation to relate aircraft skin current to external RF fields.

5 For a uniform discretization.

6 We will not address here computer requirements issues, and the reader is referred to the cited literature. However, let us simply state that the current Xeon-Phi CPU MPI/OpenMP implementation of SEMBA-UGRFDTD scales

at a speed of 2 Gcells/CPU. For instance, let us assume a 1 GCell problem (in the order of magnitude of those simulated in this chapter), meshed with space steps of 1 mm to reach 18 GHz. Using a typical CFLN=0.8, a physical time of requires roughly 3 hours of CPU in 32 nodes of a modern supercomputer cluster (like the Marconi KNL one that we currently employ).

7 Note that the condition (15) implies that good space resolutions automatically yield good resolutions in time.

8 It can easily be formulated also in a conformal manner to handle curved materials.

9 The GVT FSV tool of the UPC can be freely downloaded from http://www.upc.edu/web/gcem/.

10 DAL-A, Catastrophic: The failure of systems in that level would prevent the safe operation of the aircraft.

Chapter 6
Transient Electromagnetic Phenomena-Standardization and EMC Requirements-Mitigation Techniques

Eleni P. Nicolopoulou
National Technical University of Athens, Greece

Panagiotis K. Papastamatis
National Technical University of Athens, Greece

ABSTRACT

In this chapter, the spacecraft charging effects that lead to ESD events, as well as the associated immunity testing required, are presented in detail and compared to the human-induced discharges, mainly simulated by most commercial ESD generators. Additionally, design guidelines for mitigating the effects of these ESD occurrences, essential for every space application designer, are discussed. Finally, the measurement requirements, as well as the test setups, regarding the immunity testing to power leads transients are described thoroughly. This chapter aims at reviewing the current international standards and the recent bibliography, summarizing and commenting on the EMC requirements, regarding the transient phenomena, pinpointing the differentiations of these standards and their effects on the design procedure and testing creditability.

DOI: 10.4018/978-1-5225-5415-8.ch006

INTRODUCTION

Spacecraft in synchronous or lower orbits often experience a buildup of charge on the exposed and the internal materials as a result of geomagnetic storms. The resulting large potential differences can lead to insulator breakdown events and the generation of electrostatic discharges with peak values of a couple kilovolts. The generated electric current is coupled, in turn, to the electronic equipment ports disrupting sensitive digital and analog electronics with probably disastrous effects and potential system failure. Over the last decades, space engineers have introduced many ESD mitigation methods (selection of specific surface materials, grounding techniques etc.) on modern spacecrafts; nevertheless, the equipment-level immunity testing to ESD still remains the most reliable mechanism of failure prevention. International Standardization provides testing specifications which simulate as accurately as possible the flight ESD phenomena. In this chapter, the different test methods are described in detail.

Later, this chapter describes the phenomena of power bus transients induced by the switching on/off of loads on the bus presenting and commenting on the suggested measurements and immunity test methods by the associated Standards. Some of the Standards mentioned herein are the Military Standard MIL-STD-461, and its most recent version (G), the ISO 14302 Standard, specifying the Electromagnetic Compatibility requirements for Space systems, as well as the ECSS-E-ST-20-07C European Standard on Electromagnetic Compatibility for Space engineering by the European Cooperation for Space Standardization.

Summarizing, the scope of this chapter is to present the types and causes of transient electromagnetic phenomena encountered in the space environment, review and comment on the recent standardization efforts by the various associated bodies and suggest sufficient protection measures for a robust and electromagnetically compatible space system design.

More specifically, in the “BACKGROUND” section the electromagnetic phenomena expected to occur in the space environment and affect the aerospace systems, as well as their causes are described. The “INTERNATIONAL STANDARDS: TESTING AND MEASUREMENT PROCEDURES” section presents earlier and current testing methods according to different International Standards trying to establish the most accurate simulation and testing of the aforementioned phenomena. Differentiation points as well as the authors’ comments on the preferable methods are presented herein. Finally, the “SOLUTIONS AND RECOMMENDATIONS” section includes design guidelines and mitigation techniques against those types of electromagnetic disturbances crucial during the design process of every aerospace system.

BACKGROUND

Charging Effects in Space

Charging effects in space applications are attributed to the interaction during the flight between the surrounding space plasma environment and the materials and electronic systems of the spacecraft. The primary charge source comes usually from the Sun in the form of solar wind, magnetic storms and ion clouds. The various regions of a spacecraft with different materials and configurations bombarded by space particles act as a set of electrically floating capacitors charged at different rates and different potentials with respect to the space plasma potential. These surface potentials are developed according to the physical concept of current balance at equilibrium, for which the sum of all currents adds up to zero.

The ambient space plasma is a partially ionized gas consisting of electrons, protons and other ions of variable energies. Protons are usually not considered for the procedure of spacecraft charging due to their greater mass which does not allow them to penetrate into matter. On the contrary, electrons and ions, depending on their energy content and on the composition of the target material will penetrate matter up to some depth leading to the following categorization of charging mechanisms and the resulting discharges:

1. **Surface Charging:** It is caused by the deposition of charge on the outer surface of the spacecraft and specifically by lower energy charges penetrating just under the surface of the exposed structures. The resulting surface discharges have an impact primarily on victims such as external wiring and antenna feeds and less on electric systems in the interior of the spacecraft as the various coupling mechanisms cause attenuation to their transferred energy. This surface charge deposition can result to unstable conditions in which charge can be lost from the spacecraft to space. Spacecraft-to-space breakdowns are generally similar to dielectric surface breakdowns but involve only small discharges.
2. **Internal Charging:** it is caused by penetrating energetic particles which are accumulated either on dielectric surfaces or on ungrounded internal conductors. The time scale of this charging procedure is longer compared to surface charging. It is also known by the terms buried charging or deep dielectric charging although the use of the word dielectric can be misleading, since ungrounded internal conductors can discharge with a higher peak current and a higher rate of change of current than a dielectric.

In both types of charging, discharge is initiated when the electric field -which is intensified by the high concentration of charges combined with possible geometry factors, such the arrangement and thickness of the material- exceeds the dielectric strength of the material or when the potential difference between neighboring surfaces exceeds a critical value. During a surface discharge process, the stored charge is ejected to space, thus producing an electric field pulse of several tens of kilovolts per meter in the immediate vicinity of the site. This blow-off of the electrons induces a replacement current pulse which flows in the spacecraft structure and the electronic systems and generates transient parasitic voltage and current pulses at the equipment inputs through coupling on the wires.

Local breakdowns in dielectrics in cases where the internal electric field exceeds $2 \cdot 10^7$V/m (Garrett & Whittlesey, 2011) cause the formation of rapidly expanding ionization channels which allow current conduction and heat, leading to damages of instruments. Apart from these direct effects, transient currents induced through electromagnetic coupling, permanent degradation of material properties and leakage currents are some indirect effects (European Cooperation for Space Standardization [ECSS], 2008). Spacecraft charging may cause due to transient electrostatic discharge effects (ESD) damages or even total destruction of subsystems (such as power, navigation, communications or instrumentation). Furthermore, charge deposition is responsible for the accumulation of contaminants which affect the thermal properties of surface materials thus degrading the performance of optical instruments and solar arrays. Contaminants can also distort plasma measuring instruments as they can change particle trajectories (Garrett & Whittlesey, 2011).

Orbits

The level of exposure to the abovementioned space charging phenomena, their severity and the extent of their impact highly depends on the orbit of the examined spacecraft. In order to estimate the expected ESD threat for a given satellite orbit, it is necessary to know the electron charging environment along that orbit. Satellites move around the Earth on Medium Earth orbits (MEO), Low Earth orbits (LEO), Geosynchronous Earth orbits (GEO) and Polar Earth orbits (PEO). Potentials attributed to charging phenomena at altitudes above 400km reach 400-500V, values which may cause discharges. Surface charging is in most cases negative because of the higher relative velocity of electrons, although in periods without intense magnetic storms the photoemission current emitted from the surface of the spacecraft in the presence of sunlight might charge a spacecraft slightly positively (Lai, 2011). Geosynchronous orbits are of high importance concerning surface charging, as the majority of telecommunication satellites is gathered in this region. Incident electrons in GEO orbits have energies in the range 2eV up to 2keV (ECSS, 2008). The surface charging levels at the GEO latitudes may reach hundreds of volts or even several kV and they are mainly developed when the spacecraft is not sunlit, taking into consideration that energetic ions and electrons enter these regions at about midnight. As expected, spacecrafts experience much higher charging levels (typically hundreds of kV) at auroral latitudes (60° to 70°) occurring even at lower altitudes. Events such as solar coronal mass ejection clouds or passage from the so-called "radiation belts" where very high energy electrons and ions prevail, are responsible for deep dielectric charging due to charge penetration into the interior (Lai, 2011).

The different charging mechanisms described above, as well as the abovementioned characteristics of the selected spacecraft orbit, are the main parameters for the identification and characterization of the resulting current waveform, presented in the next section and used as reference for the associated immunity testing procedures.

Conventional ESD Transient Current Characterization

The amplitude and duration of the resulting pulse are dependent on the charge deposited. Additionally, charge accumulation is not only related to the incident charge flux but also to the ability of retaining the charge after deposition which is highly affected by material conductivity and grounding schemes. Materials with resistivity in the order of $10^{12}\Omega\cdot$cm, if grounded, will effectively dissipate away GEO fluxes which are anticipated to be in the order of 10^{12}A. Dielectric materials such as Teflon® and Kapton®, flame retardant 4 (FR4) circuit boards and conformal coatings widely used in modern spacecraft applications often have resistivities that may cause problems. Meanwhile, metallic surfaces, although conductive, might pose a threat if they are electrically isolated by more than $10^{12}\Omega$ (Garrett & Whittlesey, 2011).

Based on recordings of ESD events from telemetry measurements, it is concluded that the signature of a transient current depends on the location of the ESD and the characteristics of the ESD pulse are governed by the following parameters: the rise time by the plasma cloud expansion rate, the pulse width and the peak current by the ESD initiation point and the size of the affected area, whereas the fall time at the tail of the pulse by the residual charged area at the end of the phenomenon. Typical values of the ESD waveform parameters for various arc initiation points as calculated from incidents mainly on the Galileo and Voyager satellites are shown in Table 1 (National Aeronautics and Space Administration [NASA], 2013b):

Extra material on the presented spacecraft charging phenomena and the respective mitigation techniques can be found in (Dunn, 2016), (Pisacane, 2016) and (NASA, 2007). In addition to the environmentally induced transient phenomena, such as the electrostatic discharges, the equipment installed in a spacecraft can also suffer from system level transient disturbances, termed as power bus switching transients, that are introduced below.

Table 1. Typical values of the ESD waveform parameters for various arc initiation points as calculated from incidents mainly on the Galileo and Voyager satellites (NASA, 2013b)

ESD generation point	Capacitance, C [nF]	Breakdown voltage, V_B [kV]	Energy, E [mJ]	Peak current, I_{PK} [A]	Discharge current rise time, t_r [ns]	Discharge current pulse width, t_D [ns]
Dielectric plate to conductive substrate	20	1	10	2	3	10
Exposed connector dielectric	0.150	5	1.9	36	10	15
Paint on high-gain antenna	550	1	150	150	5	2400
Conversion coating on metal plate	4.5	1	2.25	16	20	285
Paint on optics hood	550	0.360	35000	18	5	600

Power Bus Load-Induced Switching Transients

System level switching transients have always been an important consideration point for EMC engineers, especially in the last decades where the electronic systems, reduced in size, are collocated and interacting with a variety of other systems, systems that are consuming high currents leading to the generation of switching transient sources affecting every system. Sensitive electronics, powered from the same source, may be affected by these disturbances resulting in undesired state changes, resets, false readings or even total system failure. Standardization Organizations are constantly working on identifying these phenomena, proposing measurement and testing methods, thus ensuring the EMC robustness of electrical and electronic systems.

The mechanism of the switching transient generation is based on the switching on/off of a heavy load attached to the power bus or on a potential short circuit, occurring before the central protection units can react. These transients are divided into two categories 1) The turn-on or negative going transients and 2) the turn-off or positive going transients.

Turn-on transients occur, while the EUT is switched on, when a switch is closed somewhere on the power bus and current attempts to flow through the load. This change of the current is opposed by the LISN inductance by dropping the source voltage across itself until the current is depleted. The transient time characteristics depend on the LISN inductance and the RC constant of the load.

Turn-off transients are the result of the LISN 50μH inductance reaction to the switching off of a heavy load somewhere on the power bus. This inductance raises its output voltage relative to its input voltage. This voltage peak is limited by the line to line capacitance and the stray capacitance while the transient time characteristics are independent of the switched off load impedance (NASA, 1999).

In the case of current limited buses or fuse-protected loads, in addition to these types of transients that require measurement and control, there are requirements regarding the peak inrush current during the switching on of the EUT.

Space agencies and Standardization Organizations have made great efforts over the last decades to establish robust and reproducible test methods either for the measurement of those transients or for the immunity testing to these phenomena.

INTERNATIONAL STANDARDS: TESTING AND MEASUREMENT PROCEDURES

Aerospace systems and especially systems that are designed to be installed in unmanned vehicles have always suffered from the difficulty of "ad hoc" maintenance and troubleshooting, while the associated failure events are dramatically more difficult to handle in comparison to ground systems. Therefore, pre-flight testing and compliance with the international emissions and immunity standards is broadly considered the safest and the most cost-efficient precaution available. Space systems, in particular, operating in an electromagnetic environment which is still relatively unpredictable (magnetic storms, charging effects, different orbits etc.) require additional consideration regarding their electromagnetic compatibility robustness.

Electrostatic Discharges

As mentioned earlier, the most reliable mitigation technique against ESD events is the pre-flight compliance testing, where designers and manufacturers can evaluate in practice the performance of their selected protection measures, some of which will be described in a later section. ESD testing in space applications is divided, based on the source of the discharges, into two major categories:

1. Immunity testing to personnel borne electrostatic discharges. Simulation of the discharge source is based on the human body capacitive charging mechanism and the immunity requirement is applicable to electrical/electronic equipment and systems with a man-machine interface.
2. Immunity testing to machine-induced electrostatic discharges, simulating flight ESD events on an unmanned spacecraft.

For a space equipment/system, designed for a manned spacecraft, to be labeled as EMC compliant and to be, in turn, installed in the final spacecraft system both of the immunity testing procedures should be carried out.

Personnel Borne Electrostatic Discharges

The standing human body, insulated from ground by footwear and/or floor covering, can be considered an insulated conductor with a known capacitance value, termed as "Human Body Capacitance" (HBC), which has been experimentally measured to be in the range of 100-150pF. The charging process of the HBC, based on friction generated charge, is especially favored by the combination of synthetic fabrics and dry atmosphere.

Immunity testing to this phenomenon, for space applications, is carried out according to the Military Standard, MIL-Std-461(Department of Defense: United States of America, 2015) and the testing method described in the section CS118. This method, newly added in the latest version (G) of the Standard, addressing effects of electrostatic discharging on electrical and electronic equipment caused by human ESD, is based entirely on the IEC Standard 61000-4-2 (International Electrotechnical Commission [IEC], 2009), the test reference Standard for ESD immunity testing used in the majority of EMC tests and certifications and required by most Product Family Standards.

MIL-STD-461G Section CS118

As it is stated in the aforementioned Standard, the Equipment Under Test (EUT) shall exhibit no malfunction or degradation of performance below a specific performance limit or tolerance values specified by the manufacturer. The evaluation is performed according to the system or subsystem specifications. The EUT is subjected to the appropriate test level, presented in Table 2 (Department of Defense: United States of America, 2015) while discharging from a 150pF capacitor through a 330Ω resistor with a circuit inductance not exceeding 5μH. Contact discharges are applied to

Table 2. ESD Test Level according to MIL-STD-461G - CS118

Level	Test Voltage	Method
1	±2kV	Air
2	±4kV	Air
3	±8kV	Contact/Air
4	±15kV	Air

Table 3. ESD Test Level according to IEC 61000-4-2

Level	Contact Discharge	Air Discharge
1	±2kV	±2kV
2	±4kV	±4kV
3	±6kV	±8kV
4	±8kV	±15kV

conducting surfaces only, while air discharges are applied to surfaces where contact discharge testing is not possible. This is also the first point of differentiation between MIL-STD-461G and IEC 61000-4-2. Comparing the test levels of Table 2 with the test levels of IEC 61000-4-2, presented in Table 3 (IEC, 2009), it can be easily identified, that contact discharges at levels below 8kV are omitted in the Military Standard making the equipment possibly susceptible to lower level ESD events.

The waveform of the resulting current is presented in Figure 1 (Department of Defense: United States of America, 2015), while in Table 4 and Figure 2 the current parameters for the ±8kV test level and a simplified diagram of an ESD generator are presented accordingly.

The general test set-up, as presented in Figure 3, shows significant differentiations compared to the IEC 61000-4-2 test up of Figure 4, such as the LISN requirement,

Figure 1. Ideal contact discharge current waveform at 8 kV (Department of Defense: United States of America, 2015)

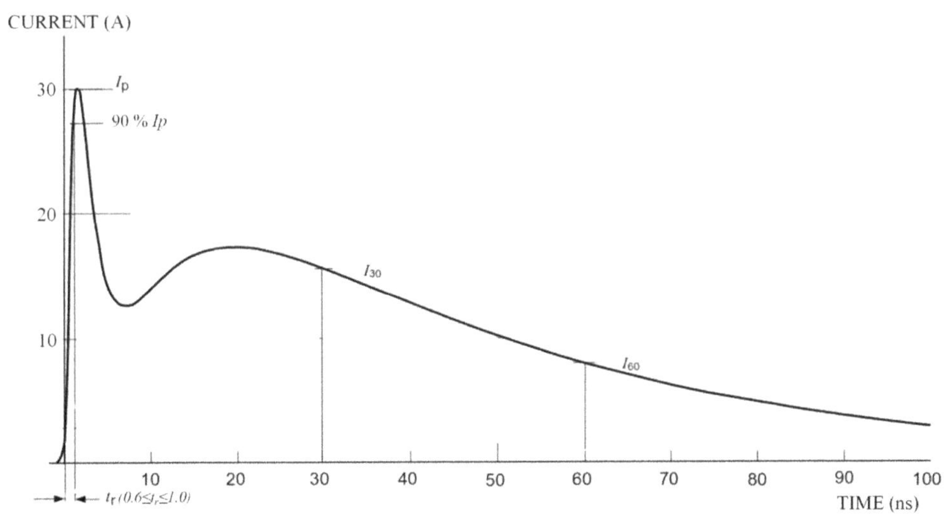

Figure 2. Simplified ESD generator diagram (IEC, 2009)

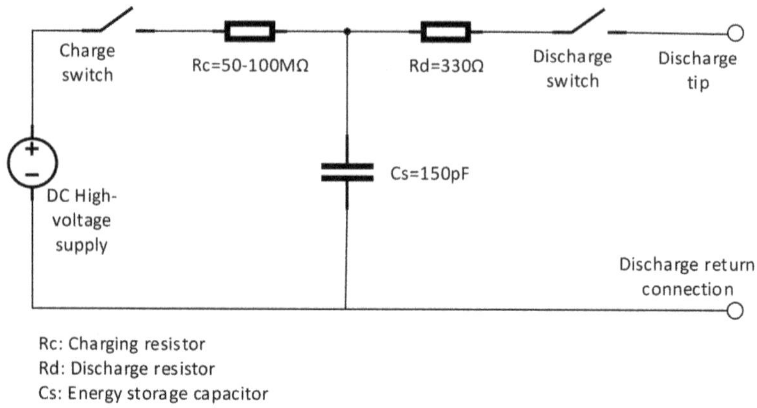

Table 4. Contact discharge current parameters for 8kV test level (Department of Defense: United States of America, 2015)

Displayed Voltage(kV)	First Current Peak, ±15% (A)	Rise time (ns)	Current at t=30ns, ±30% (A)	Current at t=60ns, ±30% (A)
±8	30	$0.6 \leq t_r \leq 1.0$	16	8

Figure 3. MIL-STD-461G ESD Test set-up (Department of Defense: United States of America, 2015)

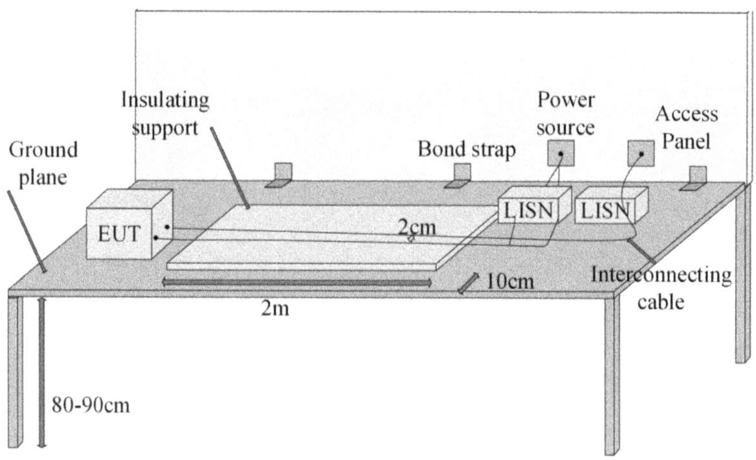

Figure 4. IEC 61000-4-2 ESD Test set-up (IEC, 2009)

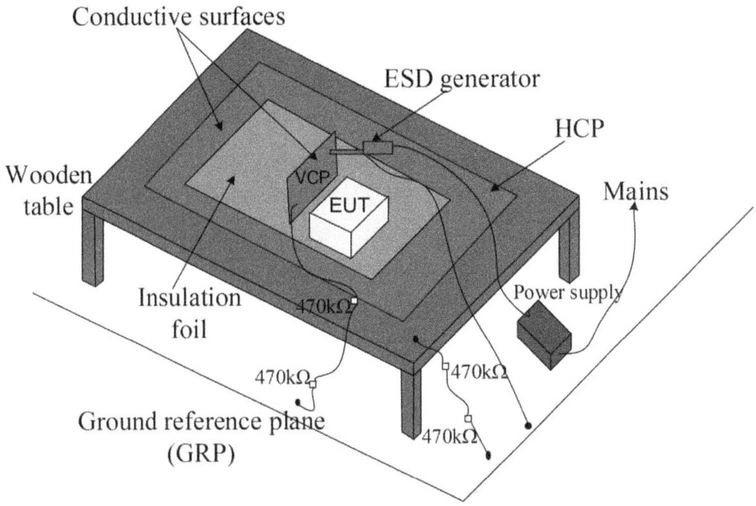

the omission of indirect discharges to the horizontal and vertical coupling planes and the strict requirements regarding EUT cabling.

The test point selection is based on the accessibility of conducting/non-conducting points or surfaces to the operator and personnel. Points of increased susceptibility risk are: metallic screws, seams and apertures, LCD displays, connector shells and exposed connector pins. For each selected test point MIL-STD-461G requires 5 positive and 5 negative discharges to be applied. For the case of air discharge testing, where no conductive ground path is available, the residual charge shall be removed through a 1 MΩ resistor, an ionizer or by waiting for the charge to dissipate. It should be noted that humidity conditions affect the charge dissipation severely, therefore, it is advised (in the IEC 61000-4-2 Standard this is a requirement) that the tests are performed in conditions with relative humidity values in the range of 30-60%.

Concluding this section, the authors would like to note that, as far as human-induced ESD immunity testing is concerned, the abovementioned test method, mainly due to the fact that it is based on a widely used testing standard (IEC 61000-4-2) with proven repeatability and reproducibility, presents a well-established testing procedure for unit-level ESD testing. On the contrary, the standards applicable for non-human induced electrostatic discharges, described in the next section, exhibit significantly less coherent and reproducible characteristics.

Non-Human Induced Electrostatic Discharges

As it has been mentioned before, the reproduction of ESD events in space, contrary to the personnel borne electrostatic discharges described above, is based on a variety of ESD test-setups used in space industry. The methods of ESD application depend mainly on the expected ESD environment and the types of discharge current each of the parts of the vehicle are subject to. The three most common methods are 1) Radiated Field ESD Testing 2) Single-Point discharges and 3) Structure Current Testing (Garrett & Whittlesey, 2011). More specifically:

- **Radiated Field Testing:** During this type of ESD testing, the susceptibility to RF disturbances, produced by electrostatic discharges, of the spacecraft communications and sensor systems is evaluated. Due to the very small rise time of the ESD pulses, the spectral content of these disturbances may include frequencies up to several hundreds of MHz, resulting into false readings or even resets of sensitive system components.
- **Single-Point Discharges:** Local discharge currents are simulated by discharging of the arc current on a metallic spacecraft surface with the return current wire in close proximity.
- **Structure Current Testing:** The definition of the actual replacement current pulses flowing though the spacecraft structure and coupled at the input wires of any equipment, due to the blow-off of charges from its surface, is almost always an impossible task for any space safety assessment. Therefore the testing against those phenomena requires specific analysis identifying the test currents and the test locations. Figure 5 presents some possible test paths for structure current testing.

Figure 5. Suggested test paths for structure current testing

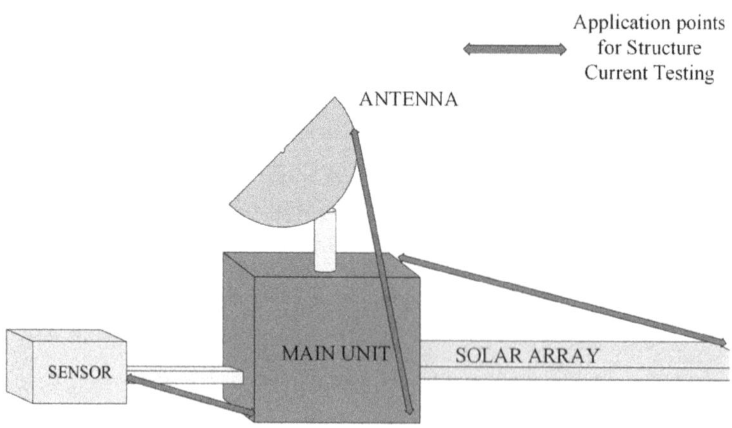

Non-human induced ESD immunity testing has been addressed in the last decades by a variety of international standards, such as MIL-STD-1541A, ISO 14302 and ECSS-E-ST-20-07C, that will be extensively discussed in the next section.

MIL-STD-1541A

For decades, the main Standard for ESD immunity testing was the Military Standard MIL-STD-1541A (Department of the Air Force: United States of America, 1987), introducing an indirect, radiated field based, ESD simulation, as well as direct discharges to the enclosure of equipment exposed directly to space environment. This Standard distinguishes two cases of testing required, based on the orbit (synchronous or geosynchronous). In the first case, a 10kV pulse discharge at a pulse rate of 1pulse/sec for 30 seconds is established, while the discharging electrodes are held at a 30cm distance from each exposed face of the equipment under test (EUT). This procedure is repeated either by applying direct discharges to each of the top corners of the EUT or by driving the discharge current to the mounting surfaces of the EUT if, according to its specifications, it is to be installed within a shielded vehicle. For geosynchronous orbits the selected test level is determined according to the Inter- and Intrasystem analysis and hazard evaluation. MIL-STD-1541A suggests a configuration for the arc discharge source, as illustrated in Figure 6 (Department of the Air Force: United States of America, 1987).

As it is presented in Table 5 (Department of the Air Force: United States of America, 1987), the only adjustable parameter of this test method is the spark gap spacing, therefore the only adjustable ESD parameter is the breakdown voltage. This limitation to the test waveform parameters such as the pulse rise time, fall time or width was one of the reasons this specific test method became obsolete and is

Figure 6. Arc discharge source according to MIL-STD-1541A (Department of the Air Force: United States of America, 1987)

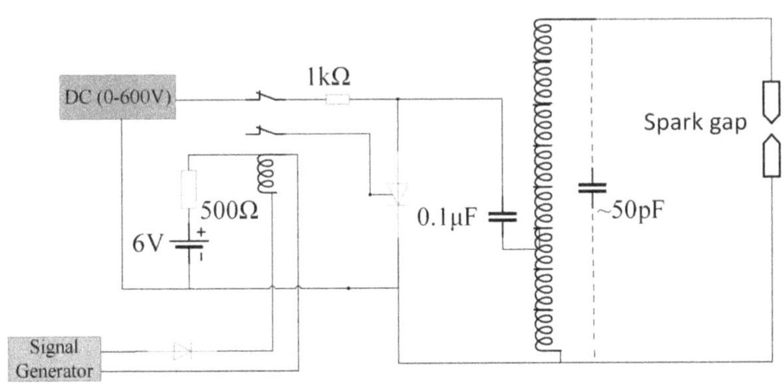

Table 5. Gap-spacing and Breakdown levels according to MIL-STD-1541A (Department of the Air Force: United States of America, 1987)

Gap (mm)	V_b(kV)	Approximate Energy Dissipated (μJ)
1	1.5	56.5
2.5	3.5	305
5	6	900
7.5	9	2000

not referenced by any recent Military Standard version or newer space applications Standards.

The aforementioned ESD test methods have been replaced by a magnetic-coupling ESD unit-level test method, introduced in the ISO 14302 Standard (International Standardization Organization [ISO], 2002) and adapted by the European Cooperation for Space Standardization (ECSS) in ECSS-E-ST-20-07C (European Cooperation for Space Standardization [ECSS], 2012). According to the guidelines of ECSS-E-ST-20-07C this test should be performed on equipment designed to be installed under the following conditions:

- Outside the main structure of the vehicle.
- Connected to units located outside of the vehicle, such as sensors, actuators etc.
- Comprising of a high voltage power source.

A critical point of ESD testing in space applications is the constraint regarding the testing on final, ready to become flight models, versions of the equipment. Due to its potentially destructive effects and the possibility of latent failures, in combination with the limited availability of products to be tested, ESD testing is performed only in a representative manner, in equipment configurations that may differ from the actual flight configuration. Despite the efforts of simulating the conditions (interconnecting cables, auxiliary equipment etc.) as closely as possible, the magnetic coupling nature of the ISO 14302 test method makes the performance evaluation, under laboratory conditions, of the equipment immunity less credible.

ISO 14302

According to ISO 14302 the suggested test configuration is illustrated in Figure 7 (International Standardization Organization [ISO], 2002). The ESD current is coupled in common mode with a parallel wire to the system harnesses. The harness under test may be either the harness of the interconnecting cables of two EUTs or the harness connecting the EUT to its ground support equipment.

The set-up of Figure 7 consists of the following components:

1. The length of the floating discharge circuit tightly coupled to the harness under test is 20cm.
2. This resistor, termed as damping resistor, is used in order to minimize the damping and therefore the oscillations of the discharge circuit. In order for critical damping to be achieved the value of the damping resistor should be selected according to the circuit capacitance and self-inductance values. The typical value of this resistor is 47Ω.
3. The spark gap with a typical value of 6kV. ISO 14302 forbids the use of air gap, because the atmospheric conditions, and mainly the pressure values, could affect its discharge characteristics. Additionally, the selected spark gap should exhibit high *dI/dt* capabilities to adequately handle the fast ESD pulse (Pelissou, Daout, & Wolf, 2015).
4. Choking resistors with a minimum value of 10Ω are used in order to block the high-frequency component of the discharge current from flowing in undesired directions.
5. Ground plane.
6. For the high-voltage source the Standard suggests two alternatives. 1) A high-voltage DC supply in combination with a choke resistor with a resistance value of more than 10MΩ or 2) An ESD generator, as described in the IEC 61000-4-2 Standard (see Section "Personnel borne discharges"), in air discharge mode, where the two discharge tips are connected to the choke resistors of Figure 9 accordingly.

The typical value for the high-voltage capacitor is 100pF, while the resulting transient current has a peak value of 30A and duration of 30ns (interval between

Figure 7. ESD test set-up according to ISO 14302 (ISO, 2002)

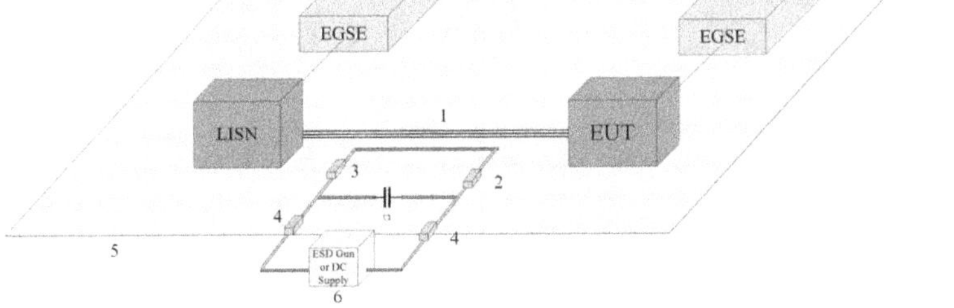

the two crossings at the 50% of the peak value). Before the adaptation of this test method by ECSS, many parameters of the testing procedure remained unclear. The separation distance between the coupling wire and the harness under test, a crucial parameter when a tight electromagnetic coupling is desired, the number and the repetition rate of the discharges, as well as the discharge circuit loop size, affecting the transient pulse duration, are some examples of undefined parameters. Combining these undefined parameters with the limited testing availability, described above, the repeatability and reproducibility levels of ESD testing according to ISO 14302 was significantly restricted until its adaptation by ECSS-E-ST-20-07C, described below.

ECSS-E-ST-20-07C Rev. 1

As mentioned above, the ESD test procedure of ISO 14302 was adapted and improved, largely, by ECSS, increasing its reproducibility, leaving nevertheless room for improvement, as it will be discussed later in this section. The test setup suggested by this EMC Standard is illustrated in Figure 8 (ECSS, 2012).

The EUT is mounted via the space vehicle mount on a conducting ground plane. The main and most crucial point of differentiation between the two setups of Figures 7 and 8 is the requirement in ECSS-E-ST-20-07C for an insulating support. The unclear term of the "floating discharge circuit" in ISO 14302, and the resulting inconsistencies regarding its stray capacitance, is replaced in ECSS-E-ST-20-07C with the reproducible usage of a 5cm non-conductive support with appropriate high-voltage insulating characteristics. Another setup requirement, presented in the ECSS Standard is the 10cm distance of the harness under test, closest to the front boundary, from the edge of the ground plane. Finally, contrary to ISO 14302, where

Figure 8. ESD Test setup according to ECSS-E-ST-20-07c Rev.1 (ECSS, 2012)

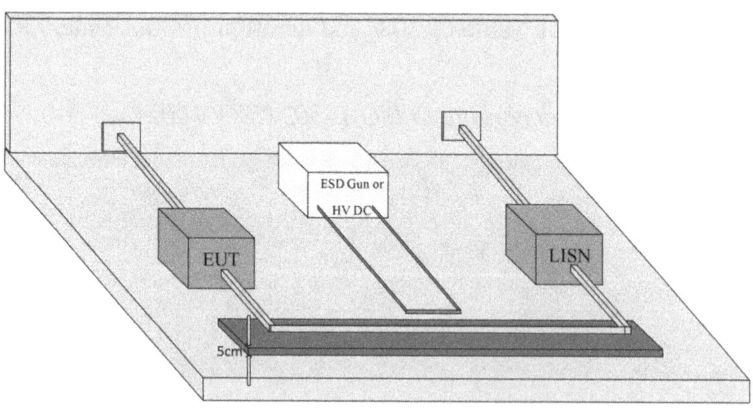

Figure 9. ESD calibration setup according to ECSS-E-ST-20-07C

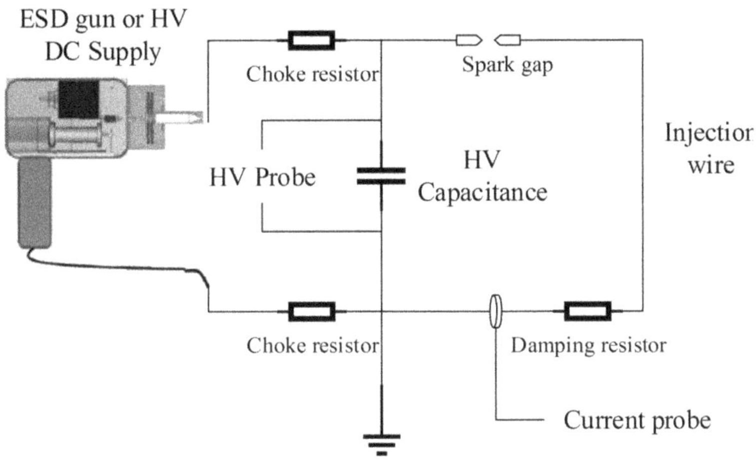

the ESD discharges should be applied in air discharge mode, ECSS-E-ST-20-07C requires the discharges to be applied in contact discharge mode.

Regarding the discharge circuit, the differences include only different values for some components, such as the case of the high voltage capacitor, which in ECSS-E-ST-20-07C has a typical value of 50pF with a less than 20nH inductance and the case of the choke resistors that have a typical value of 10kΩ. The spark gap is allowed to be an air spark gap, taking, however, its reduced stability and greater rise times, into consideration. As it has been mentioned before, due to the dependence of the air spark gap on atmospheric conditions, its usage reduces the reproducibility of ESD testing.

Contrary to ISO 14302, ECSS-E-ST-20-07C describes the calibration process of the testing configuration as well as the characteristics of the monitoring devices necessary. The calibration setup is illustrated in Figure 9, while in Figure 10 the ESD test configuration is presented.

According to Figure 10, for the calibration of the ESD pulse a high-voltage probe, grounded at the damping resistor side, with a 10kV range, 1MHz bandwidth and minimal input impedance as well as a current monitor, near the damping resistor at the capacitor side, with 100A peak capability and more than 100MHz bandwidth, are required. The desired waveform shall have a minimal rise time, a 30A peak value and 30ns "time to half value" duration. The rise time of the pulse can be reduced by adjustment of the damping resistor, minimization of the discharge loop size and by placing the choke resistors as close to the capacitor as possible. Unfortunately, ECSS-E-ST-20-07C does not set any tolerance values for the resulting pulse parameters, so the pulse suitability must be evaluated "ad hoc" by the test engineers and the EUT

Figure 10. ESD test setup according to ECSS-E-ST-20-07C

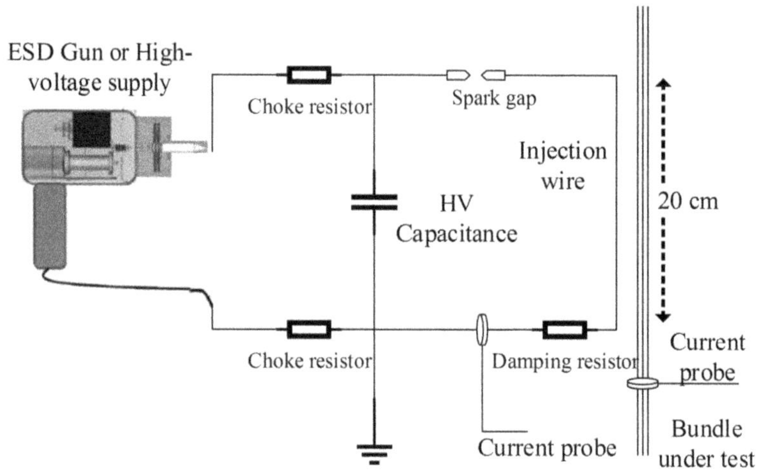

manufacturer according to the prior hazard analysis and the expected levels of the actual ESD events. Reports from different EMC laboratories show differentiations of the ESD duration that range from 28ns to 52ns and of the ESD rise time from 5.7ns to 14ns (Pelissou et al., 2015). Even though these differentiations may seem trivial and unimportant, it should be noted that the energy injected into the system from a pulse with almost double the "at mid height" duration is also almost twice that large. Additionally, the different rise time values result into different disturbance frequency spectra. A 5,7ns rise time pulse has significant spectrum content up to a 62MHz frequency, while a 14ns rise time pulse up to a 25MHz frequency. This variation could be potentially crucial when testing components and circuitry that operate at these frequency bands.

For the testing configuration, the high-voltage probe is removed and a second current probe is used for the measurement and monitoring of the coupled pulse on the bundle under test. This probe is placed at a 5cm distance from the EUT connector. The length of the parallel to the harness under test wire of the discharge circuit should not be larger than 20cm. ECSS-E-ST-20-07C requires the application of at least 15 positive and 15 negative pulses at the rate of 1Hz. The requirement for testing at lower levels than the worst-case level, as it was estimated in the hazard analysis, is not discussed in this Standard.

Points for Future Improvements of the Reproducibility of ESD Testing for Space Applications

The magnetic coupling nature of ESD testing suggested by ISO 14302 and ECSS-E-ST-20-07C presents significant sensitivity and uncertainty concerns regarding the adequate simulation of real ESD events and the reproducibility of the test method among different EMC laboratories. Some points that require further investigation are the following:

- **Discharge Circuit Component Placement:** The arbitrary placement of circuit parameters can result into different total impedance values of the discharge circuit and therefore different magnetic coupling factors. Strictly defining the component placement can be a difficult task especially due to the size limitation of the loop in order for the transient pulse duration objective to be achieved.
- **Coupling Wire Test Position And Separation Distance:** ECSS-E-ST-20-07C suggests a 1cm maximum distance between the injection wire and the bundle under test so a tight electromagnetic coupling is ensured. This suggestion can result into differentiations of the injection wire location and therefore the coupling factor.
- **Spark Gap Technology:** Probably the most important parameter that can affect the reproducibility of this ESD testing method is the selection of the proper spark gap technology. Excluding the air sparks, that are a poor option due to their dependency on the atmospheric conditions, the most attractive solution is the use of hermetically sealed, made of ceramic-metal, overvoltage spark gaps. With breakdown current rise times ranging between 1ns and 5ns and a current capability of up to hundreds of kA, they present a good alternative spark gap type.

Summarizing the ESD part of this section, it should be noted that the inductive characteristics of the aforementioned test methods present difficulties associated with the adequate simulation of the expected ESD events in space. These difficulties are not so evident during the second part of the transient analysis, the power bus switching transients, discussed in the following section.

Power Bus Switching Transients

As it has been briefly described in the "Background" Section of this chapter, the performance evaluation of an equipment or system regarding power bus load-induced switching transients includes the three following procedures:

- The measurement of the load-induced switching transients.
- The measurement of the peak inrush current.
- The immunity testing against power-line switching transients.

Load-Induced Switching Transients

Cycling loads on/off on a power bus can result in the generation of voltage transient on it, due to the power bus wiring impedance, as described above. The measurement method of this voltage is presented in ISO 14302 as well as in the NASA Handbook for switching transients (NASA, 1999). This LISN based method suggested is desirable since the parameter of interest is the reaction of the bus wiring to current rapid changes. In both cases the selection of the right LISN type, time domain or frequency domain, depends on the measured transient duration. Frequency domain LISNs, used mainly for conducted emissions measurements do not adequately control the impedance below 150 kHz, therefore the use of a time domain LISN, as illustrated in Figure 11 (ISO, 2002) is suggested. Additionally, Figure 12 (ISO, 2002) illustrates the suggested measurement setup.

The measuring apparatus can either be an oscilloscope with a minimum 10MHz bandwidth or an appropriate EMI meter. In order for worst-case emission measurements to be achieved, special consideration should be put on the selection of the switch. The maximum inrush current *dI/dt* is achieved when the EUT is cycled near the peak of the AC waveform.

Measurement of the Peak Inrush Current

As mentioned before in the case of current limited buses the peak inrush current of the EUT, which can result in voltage sags, shall remain within tolerance limits. These limits are derived from the following equation and are presented as a sample limit line in Figure 13 (ISO, 2002).

Figure 11. Time domain LISN for the measurement of Load-induced switching transients (ISO, 2002)

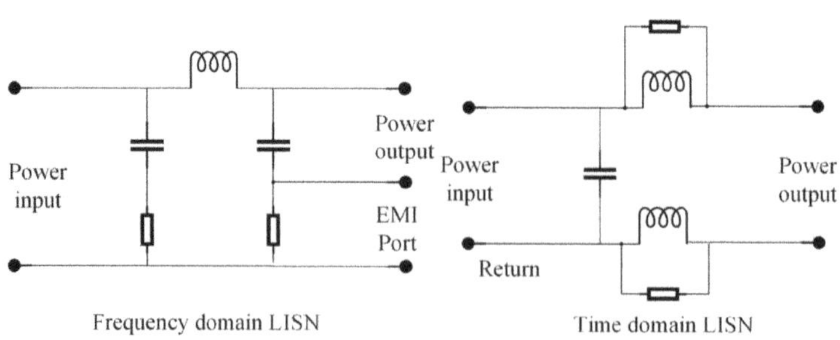

Figure 12. Measurement setup according to ISO 14302 (ISO, 2002)

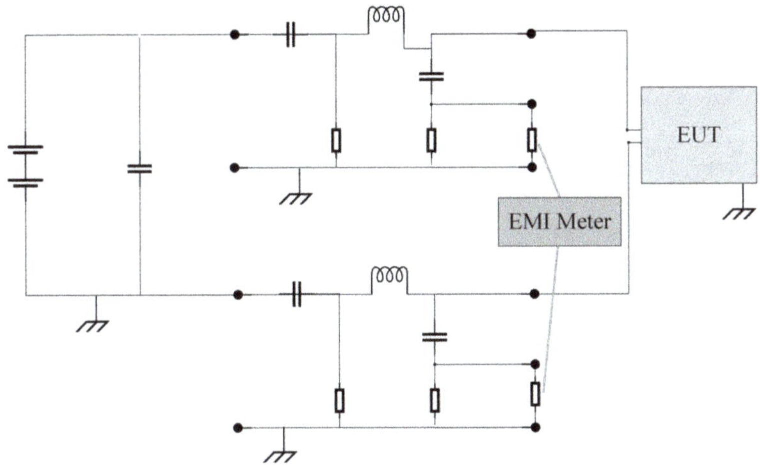

Figure 13. Transient inrush current sample limit (ISO, 2002)

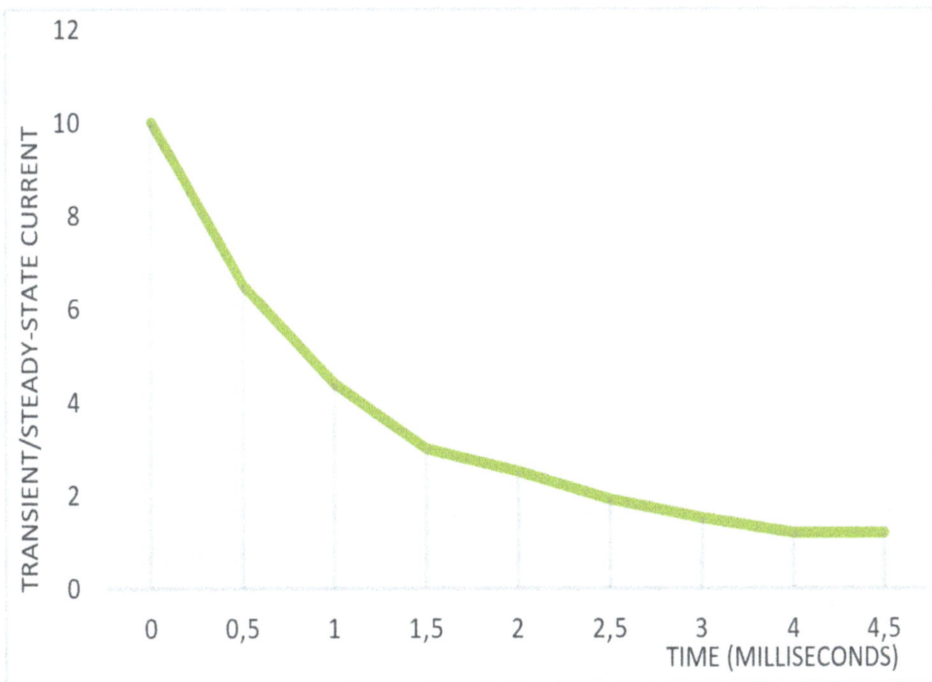

$$\Delta V_{trans}\left(V_{nom}-V_{min}\right)\leq\sqrt{\frac{I_{ss}}{I_{bus}}}$$

Where: ΔV_{trans}: maximum allowable voltage sag
V_{nom}: nominal bus voltage
V_{min}: minimum bus voltage
I_{ss}: EUT steady-state current draw
I_{bus}: power bus maximum steady-state current draw

The measurement setup suggested by ECSS in ECSS-E-ST-20-07C is presented in Figure 14 (ECSS, 2012). The inrush current is measured directly with a current probe, while the cycling of the EUT is established either with a fast bounce-free power switch or an actual power-controller in the case that the On/Off command is not implemented in the EUT. This method differentiates from the ISO 14302 test method, where the measured parameter is the voltage sag across a series resistor.

In order for the inrush current of an equipment to be limited, minimizing this way the resulting voltage sag on the power line, ECSS-E-ST-20-07C suggests the following limit values:

- The current rate of change for any nominal change of operation should be lower than 5×10^4A/s.
- During the cycling-On of the EUT this rate of change should be limited to 2×10^4A/s. In the case of pulsed radar or plasma thrusters these values may differ.

Figure 14. ECSS-E-ST-20-07C inrush current measurement setup (ECSS, 2012)

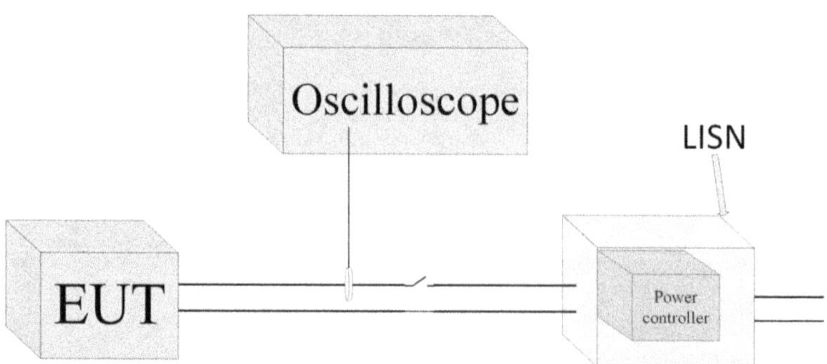

- Parameters, such as the duration of the inrush current (ms), the total transferred charge (mC) and the inrush current slope (A/μs), may also be limited according to the spacecraft power system specifications.

Immunity to Power-Line Switching Transients

The susceptibility of the EUT to positive or negative spikes superimposed on its input leads is evaluated by the series or parallel injection of spikes with a predefined waveform to the EUT and its operational monitoring during the test. The suggested setups across all the aforementioned Standards involve the use of a spike generator in series or parallel to the EUT inputs, a LISN as the power source, an oscilloscope for the recording of the applied spikes and a monitoring system for the performance evaluation of the EUT. A typical setup is illustrated in Figure 15 and Figure 16 (ECSS, 2012).

The parameters of the applied transient waveform shall be defined according to the hazard assessment, performed prior to the testing procedure, evaluating the expected characteristics of potential transient phenomena. Parameters such as the expected bus switched currents, the power bus LISN model, the type of loads attached to the bus as well as the hold-up capacitance need to be taken into consideration during the selection of the appropriate transient generator and the bandwidth of the oscilloscope used for the recording of the pulses.

An important point regarding immunity testing to power-line switching transient is the different parameters of the turn-on and turn-off transients. As illustrated in

Figure 15. Typical setups for immunity testing to power-line switching transients: Series configuration (ECSS, 2012)

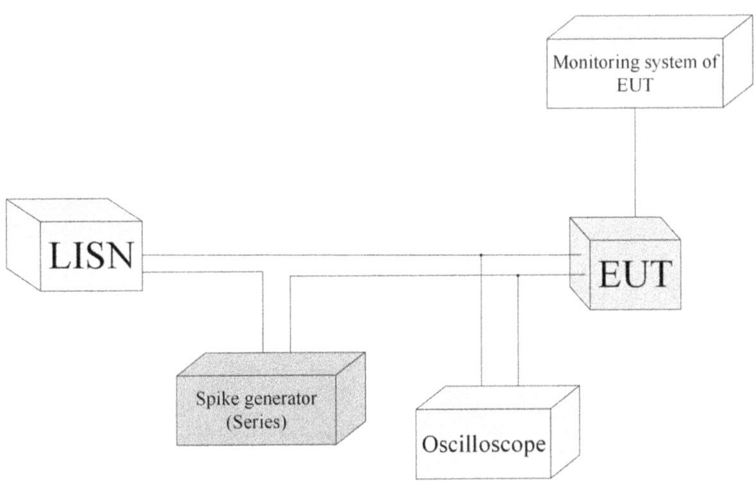

Figure 16. Typical setups for immunity testing to power-line switching transients: Parallel configuration (ECSS, 2012)

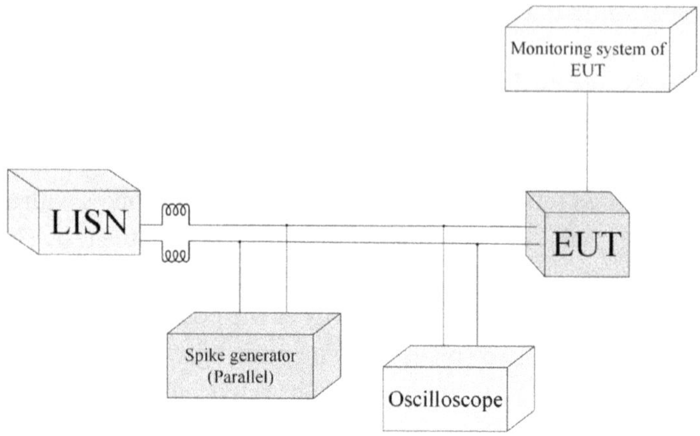

Figure 17 (ISO, 2002), for transient pulses with a rise time smaller than 1μs, the pulse amplitude and duration vary significantly. Therefore, the selection of different pulse width and amplitude values is strongly advised. ECSS-E-ST-20-07C suggests

Figure 17. Switching transient envelopes on a 28Vdc bus (ISO, 2002)

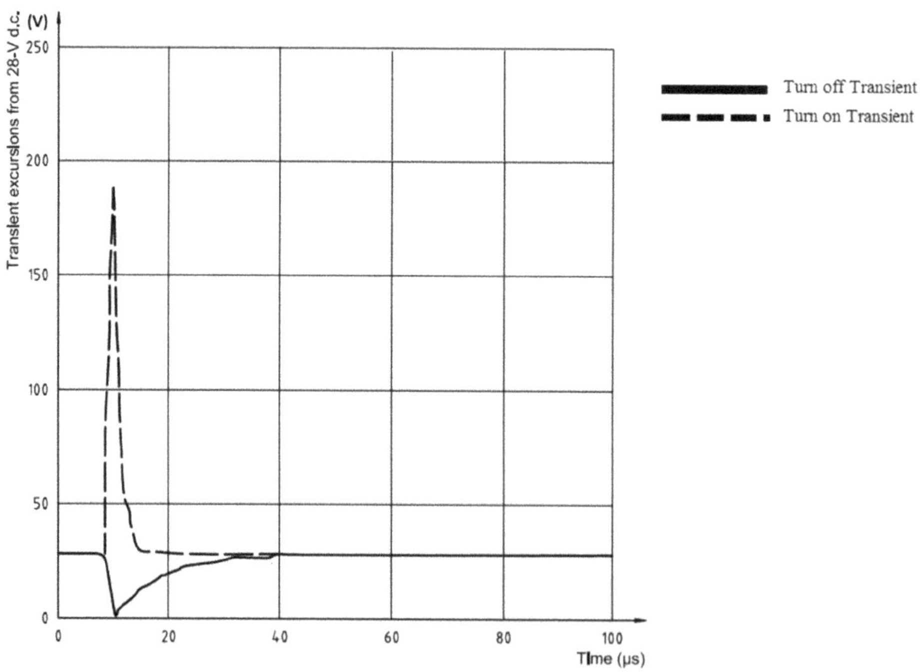

the application of pulses with a width value of 10μs and 0.15μs, with a 10 pulses/ second repetition rate.

Furthermore, it should be noted that the aforementioned test method addresses the control of power-bus switching transients and does not address externally induced transients, like field-to-wire coupled transients. The test methods for these types of phenomena are discussed and simulated using the bulk injection method in the CS115 and CS116 sections of MIL-STD-461G.

After presenting the various test standards, current or withdrawn/superseded, the authors find it necessary to comment on the preferable test method or combination of test methods, according to the current equipment-level International Standardization regarding the transient electromagnetic phenomena in space applications. Based on the criteria of the best-achieved accuracy of the phenomenon simulation, the repeatability and the reproducibility of the test method, the most reliable approach is the application of the MIL-STD-461G Section CS118 test procedure for the human-induced ESD immunity testing, as well as the ECSS-E-ST-20-07C test procedure for the non-human ESD immunity testing and the power-line switching transients immunity testing. Both standards, with their relatively recent revisions, offer the most precise and analytical specifications for the necessary immunity testing procedures contributing to the desired repeatability and reproducibility levels. Regarding the required emissions measurement methods the authors suggest, for the same reasons, the application of ISO 14302 for the measurement of the load-induced switching transients and the ECSS-E-ST-20-07C measurement method for the peak inrush current.

SOLUTIONS AND RECOMMENDATIONS

Failure Evaluation

One of the major concerns for EMC designers and engineers has always been identifying the reasons why an ESD immunity test failed. ESD failure occurrences are divided in two categories: 1) Failures involving physical damage to semiconductor components, termed as "Hard failures", due to a thermal or electric breakdown of the semiconductor materials, mainly attributed to the lack, false selection or aging of the ESD protection circuits 2) Failures involving momentary disruptions to the normal operation of internal circuitry, termed as "Soft Failures" due to the high amplitude transient electromagnetic field, which in turn is coupled as transient voltages on the traces and the wiring of circuits and due to the current generated from the ESD discharge leading to voltage reading fluctuations. These fluctuations can possibly be falsely interpreted by specific components, such as sensors or analog-to-digital converters, as a change-of-state input resulting into undesired indications and changes to the operation mode of the system. The failure evaluation process is significantly important, especially for the selection of the appropriate mitigation technique, as presented in the next section.

Mitigation Techniques

During the design phase of the mission numerous practical guidelines should be followed regarding the selection of materials, the geometrical arrangement of cable harness and circuit components and the protection of electronic systems in order to prevent the disastrous effects of surface and internal charging. These guidelines are categorized in the following main pillars:

1. **Orbit Selection:** Since the level of charging depends on the electromagnetic environment of the orbit, orbits and altitudes with intense particle background should be avoided where possible, although this measure cannot be implemented in most cases.
2. **Shielding:** The primary method for the mitigation of ESD threats is to create a physically and electrically continuous enclosure comprised of the main spacecraft structure, equipment enclosures and cable shields that will act as a Faraday cage shielding all vulnerable electronic elements. Radiated electromagnetic fields responsible for charging phenomena should be attenuated at least by 40dB through shielding measures involving layers of aluminum or magnesium foil in the form of films, blankets, face sheets, tight enclosures or even honeycomb structures. The effectiveness of the Faraday cage is highly dependent on the absence of openings, apertures and slits which interrupt the continuity of the shielding structure. Cable penetration points to the spacecraft structure should be carefully shielded along with externally exposed parts of the cabling (Garrett & Whittlesey, 2011). Usually, the first line of defense against ESD is the selection of the appropriate enclosure material for the device under test. Depending on the manufacturer's design specifications, the device enclosure can be metallic or plastic. Metallic enclosures, mainly selected for their shielding effectiveness against radiated fields and their emissions suppression capabilities can often become a liability regarding ESD immunity, especially when the grounding path for the enclosure runs in close proximity to sensitive nodes of the circuit boards. Nevertheless, a careful selection of this path can be a sufficient protection measure against ESD, providing a safe ground path for the discharge current. On the other hand, plastic enclosures, considerably less effective against radiated fields/emissions, can be an affordable solution, taking always the possible seams and apertures into consideration. The same

approach should be adopted for the screws and fixtures of the enclosure. The use of insulating plastic caps for the metallic screws has shown positive results during ESD immunity testing (Duvvury & Gossner, 2015).

3. **Bonding:** The abovementioned shielding procedure is based to a great extent on the method of bonding all conductive structural elements both on the surface of the spacecraft and in the interior, all mechanical parts, enclosures of electronic circuits etc. to each other and to a common ground reference. For internal conductive parts a maximum $10^{12}\Omega$ bonding impedance to chassis is required. Either through direct connection or through a resistor for charge bleed-off, a low-impedance path for the dissipation of ESD current pulses should be established. Especially for the bonding of surface materials which is not structural but only oriented for the mitigation of ESD threats, a less stringent $10^{9}\Omega$ surface-to-structure impedance is considered adequate. Signal and power grounds should be connected to the metallic structure as directly as possible, avoiding the formation of loops.

   An extremely important factor affecting bonding requirements is the selection of the proper material. In general, the extended use of dielectrics with high insulating properties raises issues of protection from internal charging, whereas metals require only the provision of an adequate charge-leakage path against surface charging. Conductivities in the region 10^{4}-10^{11} Ω•cm are considered sufficient for internally installed dielectrics. Exterior surfaces should be made conductive and if not possible -in regions where the use of dielectrics is necessary such as areas in the vicinity of antennas/receivers, sensitive detectors, areas where material purity and temperature are controlled- at least partially conductive and bonded to the spacecraft structure.

   Indicative materials suitable for this purpose are:

   a. Combinations of deposited metals (aluminum, gold, silver) on dielectric substrates (Inconel® on Kapton®, Teflon®, Mylar® and fused silica) in the form of sheets, strips, or tiles
   b. Thin, conductive front-surface coatings: ITO (250nm) is the most common material
   c. Conductive paints (white Zinc orthotitanate is the most conductive paint). They are required also for conductive substrates.
   d. Carbon-filled Teflon®, or carbon-filled polyester on Kapton®
   e. Conductive adhesives, necessary for the bonding of a conductor to the ground reference
   f. Exposed conductive facesheet materials (graphite/epoxy or metal).
   g. Etched metal grids not simply stretched over dielectric surfaces but effectively bonded. Anodizing procedures, fiberglass materials, uncoated Mylar®, Teflon® (even metalized Teflon® is extremely

vulnerable to ESD threats) and Kapton®, silica cloth, quartz and glass surfaces should be avoided.

4. **Circuit Filtering:** Low pass filters and CMOS circuits are some of the methods of electrical filtering in order to protect circuits from charge-generated ESD.
5. **Isolation of Transformer Primary Windings From Secondary Windings.**
6. **Optimum Cable Harness Layout:** Cable harnesses should be routed away from openings whereas cables coming from the exterior should be filtered and separated from internal cables to different bundles.

The abovementioned mitigation techniques, in combination with some more specific and quantitative surface/internal guidelines, described in the following sections, can contribute significantly to the general transient immunity of the space system, either at the spacecraft level or at the equipment level.

Quantitative Surface ESD Guidelines

Some qualitative design guidelines based on empirical equations for the achievement of the desired surface conductivity are the following (Garrett & Whittlesey, 2011; NASA, 1984):

1. 1. For the grounding of conductive components to the structure the grounding resistance R [Ω] must be:

$$R < 10^9 \;/\; A$$

Where: A = exposed surface area of the conductor [cm^2]

2. 2. For the use of partially conductive materials over a grounded conductive surface (e.g. conductive paint coating) the resistivity-thickness product must adhere to:

$$rt < 2 \cdot 10^9$$

Where: r = material resistivity [Ω•cm]

t = material thickness [cm]

3. 3. For the use of partially conductive materials over a dielectric grounded at the edges the resulting material resistivity must be:

$$rh^2 / t < 4 \cdot 10^9$$

Where: r = material resistivity [Ω•cm]

t = material thickness [cm]

h = maximum distance of the conductive surface to a ground point [cm]

The above rules lead to different limitations depending on the specific geometry and application. Some primary guidelines are: For the grounding of isolated conductors a resistance less than 10^6Ω is required as recommended in ECSS-E-ST-20-06C (ECSS, 2008). Conducting substrates require coating materials with bulk resistivities less than 10^{11}Ω•cm whereas dielectric materials require application of surface materials with resistivity less than 10^9Ω•cm^2.

Quantitative Internal ESD Guidelines

Similarly, some quantitative measures for the protection of the interior of the spacecraft from ESD due to charge penetration can also be applied (Garrett & Whittlesey, 2011; NASA, 1984):

- Conductive components which are not parts of circuit boards such as unused cables and circuit traces with surface area greater than 3cm^2 or length greater than 25cm should be grounded (the respective critical area for conductive regions on circuit boards is 0.3cm^2). More stringent requirements allowing a maximum 1cm^2 ungrounded metallic areas are proposed in ECSS-E-ST-20-06C (ECSS, 2008).
- Regarding the shielding measures to prevent electron fluxes from entering into the spacecraft, 110mils equivalent aluminum shielding is considered adequate for GEO environments. A general rule is that for a computed maximum flux at the examined orbit less than 0.1pA/cm^2 no additional shielding is required, for fluxes between 0.1pA/cm^2 and 0.3pA/cm^2 shielding to a maximum level of 0.1pA/cm^2 is necessary for Class 1 ESD-sensitive circuitry (according to MIL-STD-88G3) (Department of Defense: United States of America, 1996), for fluxes between 0.3pA/cm^2 and 1pA/cm^2 shielding to a maximum level of 0.3pA/cm^2 is necessary for Class 2 ESD-sensitive circuitry (according to MIL-STD-88G3) (Department of Defense: United States of America, 1996), whereas for fluxes exceeding 1pA/cm^2 internal ESD phenomena are anticipated.

- Apart from testing the immunity of components against ESD on an equipment level, a procedure for classifying the embedded microcircuits according to their susceptibility to ESD in order to specify appropriate packaging and handling requirements and define the abovementioned critical levels of maximum permissible flux is described in MIL-STD-88G3 (Department of Defense: United States of America, 1996). A current waveform with rise time below 10ns, delay time 150±20ns for charging voltage levels 500V, 1kV, 2kV and 4kV is applied to various combinations of circuit pins. Equipment withstanding voltages above 4kV is characterized as Class 3, voltages between 2kV and 4kV as Class 2 and equipment with threshold up to 2kV as Class 1.
- Filter circuits (usually RC circuits or diodes) should be dimensionalised so as to withstand a pulse with 20ns width, approximately 5kV, 250μJ stress resulting from the charging of a 20pF capacitance with 100nC (ECSS, 2008). Immunity against a higher ESD voltage level of 20kV is recommended.
- All methods should limit the electric field stress developed within dielectrics below $4 \cdot 10^6$ V/m.
- Circuit boards should be coated with partially conductive dielectrics providing a leakage resistance of about 10^9-$10^{13}\Omega$.
- A so far non-validated method is the filling of circuit board unused regions between dielectrics with grounded metal in order to reduce the volume of dielectrics possibly accumulating ESD charge.
- According to NASA practice (1993a) ESD threats near sensitive circuits and radio frequency (RF) receivers should be eliminated, whereas for other regions the energy of an ESD pulse depending on charge accumulation, geometry and material conductivity should be minimized to 3mJ.

The final step, supplementing the aforementioned guidelines, is the selection of the proper protection devices, presented in the next section, in order to eliminate, minimize or divert away from sensitive circuit the ESD energy.

Protection Devices

Mitigation techniques against "Hard ESD Failures" mainly include the placement of protection devices near the entry points of sensitive circuitry, ensuring that voltages do not exceed the overvoltage margins, specified by the components manufacturers. These devices work in conjunction with the on-chip ESD protection reducing the injected ESD energy coupled to IC entry points to levels within the operating range of the on-chip protection circuitry and are divided in two categories, current limiting and voltage limiting devices. Current limiting devices, placed in series with the sensitive component, as illustrated in Figure 18, have a very low resistance at low currents and a high resistance at higher currents. Two important disadvantages of this type of protection devices are their relatively long reaction time, inconsistent with the fast rise time pulses of ESD and the energy requirement for the reaction, which is considerably bigger than the injected ESD energy. Nevertheless, these devices can contribute to the general ESD protection of the equipment when used in combination with voltage limiting devices, described below and illustrated in Figure 19.

Voltage limiting devices, or Transient Voltage Suppressors (TVSs), are placed in parallel with the circuit to be protected, normally between a line and ground, and have a high impedance at low voltages and low impedance at higher voltages. At the low impedance mode, the injected ESD current is diverted away from the sensitive circuit inputs, ideally to ground. The most commonly used TVSs are listed and described briefly below (Duvvury & Gossner, 2015):

- **Silicon Based TVS Devices:** Examples of these TVS devices are diodes and more specifically Zener diodes. Despite being a cheap and easy to use measure, the Zener diode presents a big capacitance load and a high clamping voltage. The capacitance values of several pF, presented by most commercial Zener diodes, may be sufficiently low for low or medium speed data lines, but they may be too high for high or ultra-high speed lines, which require a maximum capacitance value of < 1 pF (Jiang, 2012). The relatively high clamping voltage (>5V) of a discrete Zener diode makes it also an inappropriate solution for low-voltage logic circuitry, circuitry that is starting to become a trend in modern applications.

Figure 18. Current limiting devices used as protection against ESD

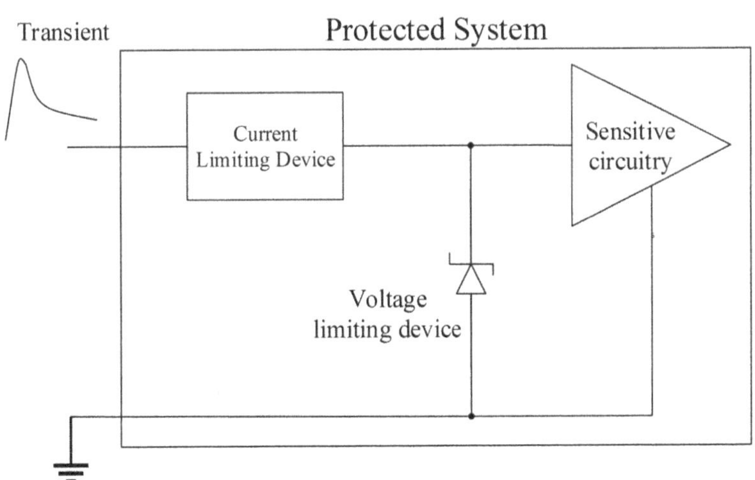

Figure 19. Voltage limiting devices used as protection against ESD

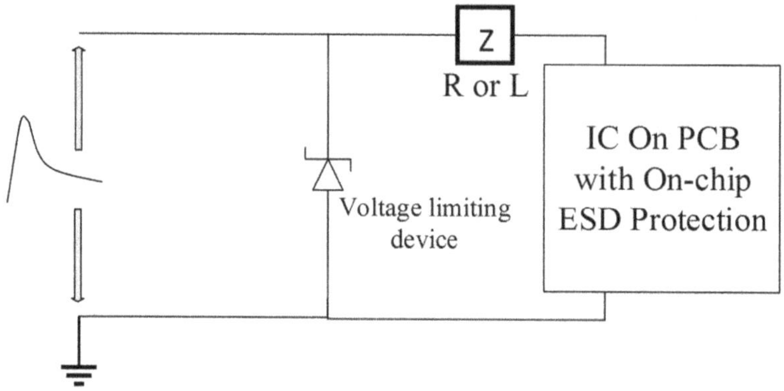

- **Metal Oxide Varistors (MOVs) and Multilayer Varistors (MLVs):** Varistors are another type of clamping devices, mainly fabricated by Zinc Oxide (ZnO) complimented with other oxides such as Bismuth, Cobalt and Manganese. They are often selected as a board level protection against most types of transient phenomena, mainly due to their ability to withstand high currents. Their high capacitance load presents the same limitations as in the case of the Zener diode, therefore they are a cost effective solution for power supply lines that can tolerate higher capacitance values, and are a good alternative for high wattage TVS diodes. Multi-line array configurations, the so called MLVs, are also commercially available providing additional high frequency noise filtering (Colby, 2011). Varistors are not considered the best choice for ESD protection due to their slow in-circuit response time (>35ns) and their high ESD clamping voltage. Additionally, due to their nonlinear resistance characteristic their voltage drop increases dramatically as the current increases.
- **PCB Spark Gaps:** PCB level spark gaps, despite being one of the simplest forms of ESD protection, they can be a very cost efficient solution, requiring only a small PCB area. A spark gap is actually a bidirectional arrangement of two conducting electrodes, of different shapes, separated by air or a controlled gas (Gas Discharge Tubes, GDTs). The high breakdown voltage of spark gaps (above 1.5kV) requires a high resistance (of the kΩ or even MΩ order) to be placed in series with the spark gap to limit the breakdown current, increasing the required PCB area even more. Another major disadvantage of this type of protection is its operational dependency on the environmental conditions. Breakdown voltage values increase at higher pressure levels. Dust, dirt

and other undesired traces on the electrodes can also negatively affect the insulating performance of spark gaps. These disadvantages result into limitations regarding the application of the protection devices with respect to the environmental conditions of the location the equipment is designed to be installed at.

All the aforementioned TVSs have been sufficient ESD protection measures over the last decades, but are starting to become obsolete in modern applications, mainly due to the emerge of high-speed data transmission and low-voltage IC logic. Each of them still finds applicability today but they are mostly used in conjunction with the two on-chip types of TVSs described below, the polymeric ESD Suppressors and the TVS Diode Arrays:

- **Polymeric ESD Suppressors:** For high-speed data, signal and control lines, where the major transient threat are the electrostatic discharges, as it is in the case of space applications, Polymeric ESD Suppressors present ultra-low capacitance values, in the range of 0.05-0.1pF, fast response times, in the range of 0.5-1ns, ideal against ESD pulses, and a very low leakage current reducing power consumption. The only disadvantage these devices present, compared to the TVS Diode arrays described below, is their relative high clamping voltage values (~ 30V).
- **TVS Diode Arrays:** A TVS Diode Array is an integrated multi-channel ESD protection device, very small in size, reducing the required board area and therefore the total system cost, providing very low, but sufficient, peak values, as well as the lowest clamping voltage values available due to their low dynamic resistance. In combination with a central TVS Diode, to which the ESD current will be directed, these arrays can safely absorb up to 30kV ESD transients allowing multi-channel protection and a low capacitance load.

FUTURE RESEARCH DIRECTIONS

Throughout this chapter it has been mentioned that International Standardization bodies are continuously investigating and trying to optimize the simulation of the encountered electromagnetic phenomena in the space environment. The identification of potential hazardous transient events, externally induced (ESD etc.) or internally produced (power bus switching transients), is gradually leading to even more adequate EMC testing prior to the installation of the equipment. The recently revised

Military Standard MIL-STD-461G, which was revised 8 years after the last version was published, has majorly contributed towards this effort. Nevertheless, there are aspects that require future research and experimentation. The best example for such an aspect is the reproducibility of the ESD testing methods and the solid definition of the testing parameters. As mentioned in this chapter, the active Standards leave significant freedom for differentiation regarding the ESD test method, reducing the reproducibility of the results across the various EMC test laboratories. Modern software packages can give engineers and researchers the necessary tools for simulating these phenomena and for experimenting with alternative setups and approaches (e.g. the coupling method for the ESD susceptibility test). Another aspect that would substantially contribute to the aforementioned effort is the collection of real flight data or simulation results regarding the power bus anomalies resulting into various switching transient on the power bus. Data from complete flight installation setups would benefit the identification of the potential transient waveforms and therefore the associated immunity testing parameters.

CONCLUSION

During the design process of every space system, the behavior of the system during transient events, either externally induced (space ESD events) or system induced (power bus transients), has been a point of considerable concern. As it has been extensively described in this chapter, the most reliable approach against these phenomena includes the identification of the transient sources, the selection of the proper protection measures against them and the evaluation of these measures through a reproducible testing procedure. Every aspect of this approach is presented with limitations and uncertainty factors, mentioned by the authors herein, establishing ground for future research, experimentation and optimization. The testing procedures, in their current form, still remain relatively vague and not accurately reproducible, a fact which in combination with the difficulties presented in the effective shielding procedure against these phenomena, can result in vulnerabilities of the equipment or system in its normal operating environment. It should be noted that the shielding procedure is a global process, ranging from the selection of the material for the outermost layer of the spacecraft surfaces to the selection of the proper ESD on-chip protection of the various microcircuits, contributing therefore to the necessity of a solid and systematic testing plan.

REFERENCES

Colby, J., (2011). *Application Note: ESD and Surge Circuit Protection.* Chicago: Littelfuse, Inc.

Department of Defense, United States of America. (1996). *MIL-STD-88G3, Test Method Standard: Microcircuits.* Author.

Department of Defense, United States of America. (2015). *MIL-STD-461G: Requirements for the Control of Electromagnetic Interference Characteristics of Subsystems and Equipment.* Author.

Department of the Air Force, United States of America. (1987). *MIL-STD-1541A (USAF), Electromagnetic compatibility, requirements for space systems.* Author.

Dunn, B. D. (2016). *Materials and processes - for Spacecraft and High Reliability Applications.* New York: Springer.

Duvvury, C., & Gossner, H. (2015). *System level ESD CO-Design.* Wiley, IEEE Press. doi:10.1002/9781118861899

European Cooperation for Space Standardization (ECSS). (2008). *ECSS-E-ST-20-06C: Space Engineering Spacecraft charging.* Author.

European Cooperation for Space Standardization (ECSS). (2012). *ECSS-E-ST-20-07C Rev. 1: Space engineering, Electromagnetic compatibility.* Author.

Garrett, H. B., & Whittlesey, A. C. (2011). Guide to mitigating spacecraft charging effects. Jet Propulsion Laboratory, California Institute of Technology, JPL Space Science and Technology Series.

International Electrotechnical Commission (IEC). (2009). *IEC 61000-4-2:2009 - Electromagnetic compatibility (EMC). Testing and Measurement Techniques. Electrostatic Discharge Immunity Test.* Author.

International Standardization Organization (ISO). (2002). *ISO 14302:2002 Space systems -- Electromagnetic compatibility requirements.* ISO.

Jiang, R. H-C. (2012). *System Level ESD/EMI Issues and Protection Design.* Academic Press.

Lai, S. T. (2011). *Spacecraft and human electrostatic discharge: a comparison of the two phenomena, Fundamentals of Spacecraft Charging: Spacecraft Interactions with Space Plasmas.* Princeton, NJ: Princeton University Press.

National Aeronautics and Space Administration (NASA). (1984). *NASA Technical Paper 2361: Design Guidelines for Assessing and Controlling Spacecraft Charging Effects*. NASA.

National Aeronautics and Space Administration (NASA). (1993a). *NASA Preferred Reliability Practices, Practice No. PD-AP-1301: Surface Charging/ESD Analysis.* NASA.

National Aeronautics and Space Administration (NASA). (1993b). *NASA Preferred Reliability Practices, Practice No. PT-TE-1414: Electrostatic Discharge (ESD).* NASA.

National Aeronautics and Space Administration (NASA). (1999). *NASA Handbook on electrical transients Specification, Measurement, and Control of Electrical Switching Transients, NASA/CR-1999-209574.* NASA.

National Aeronautics and Space Administration (NASA). (2007). *Low Earth Orbit Spacecraft Charging Design Handbook, NASA-HDBK-4006.* NASA.

Pelissou, P., Daout, B., & Wolf, K.-F. J. (2015). Improved ESD test method for testing spacecraft equipment. *Asia Pacific Symposium on Electromagnetic Compatibility (APEMC 2015).* DOI: doi:10.1109/APEMC.2015.7175316

Pisacane, V. L. (2016). *The Space Environment and its Effects on Space Systems.* AIAA Education Series. doi:10.2514/4.103537

KEY TERMS AND DEFINITIONS

Electromagnetic Immunity (EMI): The ability of a system/apparatus to maintain its proper functionality under exposure to various sources of electromagnetic disturbance.

Electrostatic Discharge (ESD): A transient current pulse occurring during breakdown of a dielectric due to charge accumulation and different potential levels between conductive and/or non-conductive surfaces.

Geosynchronous Earth Orbits: An orbit around the Earth at 35786km altitude with a 24-hour period.

Power Line Switching Transients: Positive or negative spikes superimposed on the input terminals of equipment due to on/off switching of loads on a power bus.

Shielding: A method for the mitigation of electromagnetic disturbances which is based on the concept of creating an electrically continuous enclosure around the under protection structure, mainly by the use of conductive materials.

Spacecraft Charging Effects: The interaction of a spacecraft structure with the ambient space plasma translated either as charge deposition on the exterior surfaces of the spacecraft or as charge penetration into the interior.

Transient Protection Devices: Devices installed near the entry points of sensitive circuitry in order to limit overvoltages or overcurrents.

Chapter 7
Magnetic Sensors for Space Applications:
Development and Magnetic Cleanliness Considerations

Neoclis Hadjigergiou
National Technical University of Athens, Greece

Marios Sophocleous
University of Cyprus, Cyprus

Evangelos Hristoforou
National Technical University of Athens, Greece

Paul Peter Sotiriadis
National Technical University of Athens, Greece

ABSTRACT

This chapter is composed of three parts. The first is an introductory part, providing general information about magnetism and related phenomena. Magnetic materials are also discussed and presented. Afterwards, the magnetic field and various measurement techniques are discussed. In the second part, different magnetic sensors used in a laboratory or space are presented. Magnetic sensors that are discussed include anisotropic magneto-resistance (AMR), giant magneto-resistance (GMR), giant magneto-impedance (GMI), flux-gate and superconducting quantum interference device (SQUID). Although some of them may be outdated and well known, they are widespread and they still pose an excellent choice for certain applications. Magnetic cleanliness is an important factor both in calibration and in normal operation of a system; in the third part, current techniques to isolate a system from the external magnetic field providing cleanliness are discussed.

DOI: 10.4018/978-1-5225-5415-8.ch007

INTRODUCTION

Discovery of Magnetism

Nowadays, magnetism has become a well-known principle. However, it is not really known who discovered the phenomenon of magnetism. The first scripts and references that are available today, suggest that magnetism was discovered in Greece prior to 600 BC. Diogenes Laërtius, Hieronymus of Rhodes, made a reference about the use of magnetic stones by Thales of Miletus. Aristotle, also talked about the use of lodestones ($Fe_3O_4 - Fe_2O_3$) by Thales of Miletus (Lucretius, 2015). The mineral lodestone, was in abundance in Magnesia of Greece. According to the previous historians, magnetism received its name from Magnesia of Greece (Daniel, 1988). Nevertheless, there is another reference about the origin of magnetic stones (lodestone) by Pliny. Pliny assumed that magnetism was named after the shepherd Magnes, "the nails of whose shoes and the tip of whose staff stuck fast in a magnetic field while he pastured his flocks." (William, 1991). However, this reference came much later and is relatively controversial. Another important milestone in the area of magnetism is the invention of the compass, which is dated between 300 and 200 BC during the Han dynasty in China. Later on, the Song Dynasty used the compass for navigation purposes (Lowrie, (2007); Merrill et al., (1983)). Nowadays, more complex and more precise systems are available for navigation and magnetic field sensing.

Nowadays

During the last few decades, the continuously increasing demand for accurate and reliable AC/DC magnetic field measurements, has paved the way for the development of various types of magnetic sensing systems as well as different measurement techniques. Sensor's sensitivity and linearity, signal-to-noise ratio, measurement range, cross-talk between sensors in multi-sensor applications are only some of the aspects that have been examined in the past.

Magnetic sensors are categorized primarily based on their sensing principle, which impacts directly the performance characteristics like measurement range, resolution, frequency response, working temperature and manufacturing cost. For example, Hall Effect and "pick up coils" based sensors have been proven to be practical and cost-effective solutions in many common applications. Nevertheless, "pick up coils" are not appropriate for measuring low frequency magnetic fields, whilst Hall-effect devices cannot provide the resolution levels needed in demanding applications. On the other hand, Anisotropic Magnetoresistance (AMR), Giant

Magnetoresistance (GMR), Giant Magneto-Impedance (GMI), Magneto-strictive and Flux-Gate sensors, although available at an elevated cost, are frequently employed in cases where higher resolution is required. Moreover, Superconducting Quantum Interference Devices (SQUID) offer the highest resolution and lower noise levels but requires cryogenic refrigeration.

Many industrial applications nowadays use magnetic sensors as measuring systems. These sensors have several advantages such as the fact that they can easily operate under harsh environmental conditions and elevated temperatures, they are very reliable due to the absence of moving parts and they can be embedded inside building materials. Magnetic Sensors are commonly used for laboratory measurements and for satellite magnetic measurements. Understanding the various types of magnetometers and their limitations is crucial for the interpretation of the satellite magnetometer results.

For space use, magnetic sensors must have some unique characteristics. For example, SQUID sensors may not be suitable for this type of applications due to the need for continuous cooling in cryogenic temperatures. In space, magnetometers can observe the fluctuations of a planet's magnetic field, from the planet's core or from the solar wind. On the other hand, in some cases there is a need to protect certain instruments from electromagnetic radiation. That is achieved by using a magnetic sensor to measure this disturbance and a closed loop technique to cancel it. Most common types of magnetic sensors for space, are the fluxgate and search coils magnetometers.

In laboratory, usually there is the flexibility of using more complex and delicate equipment. In accurate laboratory measurements, there is the need for an anechoic chamber to provide a magnetically clean environment. This chamber can be used for numerous applications such as:

- Calibration of magnetic sensors
- Noise measurements for magnetic sensor
- Identification of the magnetic signature of the Equipment Under Test (EUT)

There are passive and active techniques to create a magnetically clean room and they depend on the shielding specifications/requirements.

This chapter, will explore the different types of magnetic sensors and their applications, as well as some techniques for electromagnetic shielding at low frequencies.

BACKGROUND

A good knowledge of magnetic fields and magnetization is paramount in order to understand how the operating principle of the magnetometers. This information is also important because it will help identify the best sensor type for each specific application.

Magnetic Field

The definition of the magnetic field is most probably the most fundamental concept in magnetism. The magnetic field, which is generated by a magnetic source, covers a certain (infinite but reciprocal dampened) area around the source. In the vicinity of the magnetic field, there is an energy gradient, which can lead to the generation of a force in the presence of another magnetic field. For example, the needle of a compass is a magnetic dipole, which has its own magnetic field, which in turn interacts with the earth's magnetic field. Therefore, a torque is applied to the needle in order to come in rest, at the point with the least energy. (Daniel, 1988; Fraden, 2010; Jiles, 2015).

Magnetization and Magnetic Moment

The two quantities that represent the response of magnetic materials in a magnetic field, are magnetic moment and magnetization. According to Ampere's theorem, a current in an electrical circuit generates a magnetic field. The simplest circuit that can generate a magnetic field, is a loop of a conductor carrying an electric current. If the cross-sectional area of the loop is ***A*** (m^2) and the loop carries a current ***I*** (*A*), the magnetic dipole moment ***m*** (*Am*) is $m = IA$. In a magnetic field ***B*** (*Tesla*), the torque ***A*** (*Nm*) applied on a magnetic dipole moment ***m*** (*Am*) is:

$$\tau = m \times B \tag{1}$$

And in free space (vacuum):

$$\tau = \mu_0 \left(m \times B\right) \tag{2}$$

The magnetization ***M*** (*A/m*) is the vector field that expresses the density of permanent or induced magnetic dipole moments in a magnetic material. The magnetization is generated by the magnetic moments in the material that can be either microscopic electric currents resulting from the motion of electrons in atoms, or the spin of the electrons or the nuclei. The material responds to an external magnetic field by rotating its unbalanced magnetic dipole to reduce the energy of the magnetic field (e.g. in ferromagnets). Magnetization is not uniform within a solid due to imperfections such us dislocations and solute atoms or even vacancies. Magnetization also describes how a material responds to an applied magnetic field as well as the way the material changes that magnetic field (Jiles, 2015; Ripka 2001).

Magnetization can be defined as the magnetic moment per unit volume of a solid:

$$M = \frac{m}{V} \tag{3}$$

where $\boldsymbol{V}$ is the volume of the solid (m^3). Magnetization defines the auxiliary magnetic field $\boldsymbol{H}$ (*A/m*) as:

$$\boldsymbol{B} = \mu_0 \left(\boldsymbol{H} + \boldsymbol{M} \right) \tag{4}$$

According to the Jiles-Atheron model for the magnetic hysteresis that appears in a material, the magnetization is composed of two components: one hysteretic and one anhysteretic. The hysteretic magnetization is the cause of a hysteresis in a B-H loop. The anhysteretic magnetization is composed by the isotropic magnetization and the anisotropic magnetization. (Jiles, 1984).

MEASURING MAGNETISM

As with any other natural phenomenon, what puzzles physicists is how they can measure magnetism. There are numerous methods for measuring the magnetic field, different methods have limitations in the parameters of the measurement such as frequency range, measurement range, resolution, sensitivity, and noise. There are two major categories in which magnetometers, otherwise known as magnetic sensors, can be classified. The first category consists of the sensors that depend on magnetic induction using coils. The second category includes those which depend on measuring changes in various properties of materials due to the presence of the magnetic field. (Tumanski, 2011; Ripka, 2001).

Induction Methods for Measuring the Magnetic Field.

All induction methods are based on the same principle; they are measuring the magnetic flux on the basis of electromagnetic induction - Faraday's Law. The electromagnetic force (E.M.F) or voltage $\boldsymbol{V}$ (*V*) that is induced in a circuit, is equal to the time derivative of the magnetic flux $\boldsymbol{\Phi}$ (*Wb*) in the magnetic circuit.

$$V = -\frac{d\Phi}{dt} \tag{5}$$

If the coil has N turns and a cross-sectional area A (m^2), the magnetic induction is $B = \Phi / A$. Therefore, equation 5 becomes:

$$V = -NA\frac{dB}{dt} \tag{6}$$

Method With a Stationary-Coil

A stationary-coil method can only measure time-variant magnetic induction by measuring the induced voltage. This type of sensors measures the magnetic induction according to the mathematical expression:

$$B = -\frac{1}{NA}\int V dt \tag{7}$$

The main purpose of this type of sensor is to measure the magnetic properties (e.g. hysteresis graph) for soft magnetic materials. Nevertheless, this type of sensors needs to be adjusted to avoid integration errors, which may mislead to a linear dependency of the magnetic induction. This problem can be ignored in cases where the magnetic flux level is smaller than 10^{-10} Wb (0.1 maxwell).

Method With a Moving-Coil

In this case, instead of a time-varying magnetic flux passing through a coil, a coil moves inside a magnetic field. The integral of the voltage provides the difference of the magnetic fields as described in the following equation, where B_i is the initial magnetic field and B_f is the final magnetic field.

$$B_f - B_i = -\frac{1}{NA}\int V dt \tag{8}$$

An application of this moving-coil is the diversion of a galvanometer in which, the deflection is proportional to the change of the magnetic field.

Method With a Rotating-Coil

In this method, the calculation of the magnetic induction becomes possible from the use of various moving-coil sensors. The simplest type of these sensors is the rotating coil. A coil is rotating at a fixed angular velocity ω (*rads/s*). The magnetic field is calculated by:

$$B(t) = B_o(t)\cos(\omega t) \tag{9}$$

where t is the time (*s*). The generated voltage is then:

$$V(t) = -NA\sin(\omega t)\frac{dB_o(t)}{dt} \tag{10}$$

Usually, this type of sensors can measure magnetic fields in the range of 10 down to 10^{-7} Tesla. The precision of these sensors is 100 ppm. The major source of error in dealing with small voltages is the electrical connection to the rotating coil, such as slip rings.

Method of Vibrating-Coil

The same principle of the electromagnetic induction (Faraday's Law) is applied also in vibration-coils. This type of coils is used for determining magnetization and consequently the primary and minor B-H loops of a material. When a coil is around a magnetic material, the magnetic field is:

$$\boldsymbol{B}_m = \mu_0(\boldsymbol{H} + \boldsymbol{M}) \tag{11}$$

Whenever the coil moves away from the sample, then the magnetic induction is:

$$\boldsymbol{B}_0 = \mu_0\boldsymbol{H} \tag{12}$$

The change in the magnetic flux is then:

$$\Delta \boldsymbol{B} = \mu_0 \boldsymbol{M} \tag{13}$$

This technique to sense magnetization is sensitive to noise caused by the variation of the auxiliary magnetic field. It is preferable to vibrate the sample instead of the coil, if this is possible.

Method of Vibrating the Sample

A sensor that employs the method of vibrating the sample, is known as a Vibration Sample Magnetometer (VSM). It follows the same principle as the vibration-coil magnetometer with the exception that the moving component is the sample instead of the coil. The VSM was first described by Foner (Foner, 1959) and it is widely used for determining the saturation magnetization $\boldsymbol{M}_s$ of the sample.

The VSM has an accuracy better than 2% and has the capability to sense a magnetic moment as small as $50 kAm^2$. However, it should be noted that in cases of small samples, this method is not suitable for determining the magnetization curve or the hysteresis loop.

Fluxgate-Magnetometers

The last sensing technique that utilizes the electromagnetic induction effect is the Saturable-core magnetometry or fluxgate magnetometers. They were first developed in the 1930s for measurements of the Earth's magnetic field. They were used in magnetic mines for ships and later in space applications to examine the planets' magnetic field.

Fluxgates can only measure the component of the magnetic field vector that is parallel to the coils. They are capable of measuring magnetic induction in the range of 10^{-11} up to 10^{-2} Tesla. Their accuracy is better than 4‰. This type of sensor will be thoroughly analyzed later in this chapter.

MEASURING THE MAGNETIC FIELD USING MATERIAL PROPERTIES

The magnetic field does not affect only ferromagnetic materials or coils. It can affect the trajectory of an electron. As it was presented earlier, magnetic field is produced by the movement of an electric current, but a magnetic field can also affect or produce an electric current (Jiles, 2015; Fraden 2010; Ripka 2007; Prudenziati 1999).

Magneto-Optic Magnetometers

Magneto-optic magnetometers are based on three physical effects, the Faraday effect, the Kerr effect, and the Cotton-Mouton effect.

Faraday effect or Faraday rotation, refers to the polarized light that is transmitted through a medium and is subjected to a magnetic field along the direction of propagation. It was discovered in 1845 by Faraday and Maxwell completed the theoretical analysis in 1870. The implementation of the Faraday effect is straightforward in measuring the magnetic field strength, since the polarized light rotates as it passes through a transparent paramagnetic material, such as $MgCe(PO_4)_2$ and this can give a measurement of the magnetic field. The angle of rotation of the polarization plane $\boldsymbol{\theta}$ is given by:

$$\theta = V\boldsymbol{H}l \tag{14}$$

Where V is the Verdet constant (0.001 to 0.1 *minute/A*), $\boldsymbol{H}$ is the magnetic field strength (*A/m*) and $\boldsymbol{l}$ is the length of the path that the light travels in the material (*m*). In ferromagnetic or ferrimagnetic materials, the angle of rotation $\boldsymbol{\theta}$ can also be related to magnetization by:

$$\theta = K\boldsymbol{M}l \tag{15}$$

Where $\boldsymbol{K}$ is the Kundt's constant (typically up to $350 \cdot 10^{6\circ} / Tm$). Both, Faraday effect and Cotton-Mouton effect can be used for domain observation.

The Kerr effect, also called the Surface Magneto-Optic Kerr Effect (SMOKE), is the change of rotation of a linearly polarized light on the surface of the sample. The light beam should not be normal to the surface of the ferrimagnet or ferromagnet, since there must be a vector component of the magnetization on the surface of the sample parallel to the direction of the light. It should be noted that the surface domains are oriented with their magnetization to the plane of the surface. When a light beam is reflected in a magnetic domain, its polarization plane rotates by an angle $\boldsymbol{\theta}$, which is related to the magnetization of the domain by:

$$\theta = K_r\boldsymbol{M} \tag{16}$$

Where K_r is the Kerr constant (mV^{-2}).

A typical rotation is 54° for saturated nickel and 120° for saturated cobalt and iron. However, this technique can be used if and only if, the magnetization can be directly related to the magnetic field.

Method of Magnetic Resonance

Sensors of magnetic resonance refer to all resonance techniques used to measure the magnetic field, including electron spin resonance, nuclear magnetic resonance and proton precession. They are extremely precise and sensitive and have a resolution of up to 10^{-14} Tesla. An important note about this type of magnetic sensors is that they are measuring the total magnetic field and not the vector components.

The electron spin resonance (ESR) was first discovered by Zeeman (Ripka et al., 2007). When a magnetic field is applied to electrons, they can be exited to one higher state and depending on the spin polarity and quantum number, the rotation frequency can be increased or decreased. The same theory is also applied to nuclear magnetic resonance (NMR) but instead of electron spin resonance, the magnetic field resonates the nucleus.

Hall Magnetometers

The Hall effect was discovered by Dr. Edwin Hall in 1879. The Hall effect refers to the phenomenon that a voltage is been generated perpendicular to the current direction when the conductive medium is being placed in a magnetic field.

Today Hall effect sensors are the most versatile and widely used magnetic sensors. They are being used for measuring the linear or angular velocity in closed loop systems. (Honewell, 2011).

Thin-Film Magnetometers

Thin-film magnetic sensors are constructed by a thin layer of 200-5000 ångström in thickness. The material of the layer must come from a non-magnetostrictive alloy such as $Ni_{20\%}$-Fe. During manufacturing, a magnetic field is applied parallel to the easy magnetization axis. The sensing direction is the hard magnetization axis. This type of sensors has uniaxial anisotropy and can be used as a variable inductor in circuits. The sensing range is from 10^{-7} to 10^{-3} Tesla.

Magneto-Resistor Magnetometers

Magnetoresistance is the effect of change in the electrical resistivity of the material when subjected to an external magnetic field. The change in resistivity is about 2% at saturation in nickel and 0.3% in iron. Other materials can perform better than 100% change in resistivity but they cannot sense low magnetic fields. This type of sensors, like Hall sensors, can be easily fabricated in a chip. They can easily achieve a resolution of 10^{-5} A/m.

Magneto-Strictive Magnetometers

Magneto-strictive sensors change their shape when an external magnetic field is applied. The disadvantage of magnetostriction is that it is nonlinear and has hysteresis. The phenomenon of magnetostriction is known as the fraction of length change over the initial length ($\Delta l / l$) of the sample.

Both magnetostriction and magnetoresistance are interconnected. They are being generated by the spin orbit coupling so that the electron distribution at each ionic site is rotated. Magnetoresistance is a phenomenon in which the conductive electrons scatter by the change in electron distribution. Magnetostriction phenomenon is the change in ionic spacing when the moments of the electrons are rotated (Hristoforou et al., 1997; Hristoforou et al., 1991).

SQUID Magnetometers

The SQUID sensor (Superconducting Quantum Interference Device) is the ultimate magnetometer. This type of magnetometer is extremely sensitive and it is used to measure extremely low magnetic fields, as low as 4 fTesla (Drung, 2007). The SQUID magnetometers are based on the Josephson effect and on Josephson junction. The Josephson effect is the phenomenon of a supercurrent that flows indefinitely without any voltage applied in a superconductive ring. The Josephson junction is where two superconductors are coupled by a very thin layer of a non-superconductive metal (Josephson, 1962).

The SQUID is a split superconductive ring and is connected with a weak join. When there is no magnetic field, the current I splits into the two branches equally. If an external magnetic field is applied, a small current will appear in the ring to counteract the external magnetic field. If the magnetic field is more than half of the magnetic flux quantum, then the current changes direction and tries to increase the magnetic flux in order to be an integer number of the magnetic flux quanta.

This increase of the current and subsequently of the magnetic flux, is more energetically preferred than trying to reduce it. The current-voltage characteristic of the SQUID has a hysterical behavior and several techniques are used to reduce this effect.

Noise in Sensors and Circuits

Noise is defined as the unwanted electric signal that is arising within the circuit (inherent noise) or is picked up from outside the circuit (interference noise). The noise is present everywhere in electronics, it appears in the sensing element, in the transition lines and in the electric circuit itself. The main source of noise occurred when the electron travels in the conductor and is scattered by the movements of phono lattice. The noise can only be prevented from accumulating in a circuit and cannot be eliminated afterwards.

Even the best sensor design is susceptible to noise, which means it is unable to make an identical representation of the impute physical stimulus. The noise is probably the most important factor determining the accuracy of a sensor. There are many parameters that introduce errors in a sensor reading, which have to do with the drift of the calibration parameters of the sensors. This part will only emphasize the noise sources, classifications and some techniques to prevent the noise entering the system. As mentioned earlier, there are two major noise classifications, the inherent noise and the interference noise.

Inherent Noise

The sensing element, passive or active, rarely produce a voltage or current signal large enough to be picked up by an analog to digital converter or a meter. This small signal needs to be amplified by a series of operational amplifiers (OPA) however, each new electronic component introduced in a circuit, generates its own noise. The noise of the sensor elements and electronics can be classified according to the source and the frequency range. The main inherent noise sources are, thermal noise, shot noise, flicker noise and burst noise.

Thermal Noise

Thermal noise or Johnson-Nyquist noise is generated by the random thermal motion of the electric charge carriers (e.g. electrons), inside an electrical conductor due to scattering from the vibration of the photon lattice.

Thermal noise is unavoidable and is approximately white noise, which means that is nearly equal and independent of the frequency. The noise amplitude has a probability density function close to Gaussian. The amplitude of the noise can be modeled by a voltage source (or current source) representing the noise of the non-ideal resistor in series (or parallel) with an ideal noise free resistor. The one-sided voltage power spectral density, or voltage variance (mean square value) per hertz of bandwidth $\overline{v_n^2}$ (*V/Hz*), is given by (Fraden, 2010):

$$\overline{v_n^2} = 4k_B TR \tag{17}$$

where $\boldsymbol{k_B}$ is Boltzmann constant (*J/K*), $\boldsymbol{T}$ is the absolute temperature (*K*) and $\boldsymbol{R}$ is the total resistance (*Ω*). The current power spectral density, or current variance (mean square value) per hertz of bandwidth $\overline{\iota_n^2}$ (*A/Hz*), is given by:

$$\overline{\iota_n^2} = \frac{4k_B T}{R} \tag{18}$$

Some sensitive to noise electronics are kept in cryogenic conditions to suppress the effect of the thermal noise. This type of electronics can be found in very sensitive laboratory pre-amplifiers or in radio telescopes.

Flicker Noise

Flicker noise; also referred to as 1/f noise or pink noise, is a type of electronic noise that has 1/f behavior in its power spectral density (Fraden, 2010). The origin of this noise is unknown and there are only hypothetical mathematic models to explain it. Flicker noise is very important in electronic devices especially at low frequencies. It is also a factor that produces phase noise in oscillators, since it can be mixed up to frequencies close to the carrier. In most cases, the flicker noise has a Gaussian distribution. It has a linear behavior in resistors of FETs and a non-linear behavior in BJTs and diodes. The flicker noise has stronger influence in carbon-based and thick-film resistors as opposed to the wire-wound resistors. It is characterized by the corner frequency (f_c), which is the threshold value when it becomes smaller than the thermal noise.

In some cases, the flicker noise can be minimized by converting the signal at a higher frequency (chopping) where the electronic circuit is less influenced by this type of noise. After the signal conditioning, the signal can be down-converted to the baseband.

Shot Noise

Shot noise originates from the discrete nature of electric charges. Shot noise is less significant compared to the flicker and thermal noise while it is temperature and frequency independent. This noise is introduced mainly in p-n junction (diodes, transistors). This noise can be modeled as a Poisson process and calculated as the mean square value of current fluctuation σ_i (*A*) (Fraden, 2010):

$$\sigma_i = \sqrt{2qI\Delta f} \tag{19}$$

Where q is the charge of an electron (*C*), Δf is the bandwidth in (*Hz*) over which is considered, and I is the DC current flowing through the conductor. The voltage created from the shot noise at the end of a resistor is

$$\sigma_v = \sigma_i R \tag{20}$$

Burst Noise

Burst noise or popcorn noise, is a sudden step in the base current of a BJT or in the threshold voltage of a FET. This name was given due to the sound produced when played over a speaker; it sounds like a popcorn popping. It usually occurs at low frequencies less than 1kHz. It has a non-Gaussian distribution. The repeatability of a burst is varying and it may happen several times in a second or it may take several minutes to occur.

The most probable causes of the popcorn noise are assumed to be the charge traps or microscopic defects in the semiconductor material. Heavy metal atoms or heavy ion implantation or even the unintentional side-effects such as surface contamination can produce such defects.

Interference Noise

The interference noise or transmitted noise, is an external signal that it is picked up by the system. It can be produced by numerous sources such as power supplies, magnetic fields, electrostatic fields, electromagnetic emissions, thermal variations, mechanical vibrations, humidity, ionizing radiation, chemical agents, parasitic capacitances or inductive coupling, bad wire coupling and non-ideal metal connections (Fraden, 2010).

These types of noise can be either additive or multiplicative. An additive noise e_n is added to the useful signal V_s and mixed up with an independent voltage (or current).

$$V_{out} = V_s + e_n \tag{21}$$

A multiplicative noise affects the transfer function of the nonlinear component of the circuit in a way that the signal V_s is altered or modulated by the noise source.

$$V_{out} = \left[1 + e_n\right] V_S \tag{22}$$

Although the interference noise can cause severe problems to the circuitry and to the stability of a system, it can be eliminated or suppressed. Twisted pair or coaxial cables, the used of shielding, and the avoidance of the use of metals with different Fermi energies (thermoelectric effect) are some of the ways to suppress this type of noise.

TYPES OF MAGNETIC SENSORS

Magnetic sensors (magnetometers) are a key element for space applications. Geostationary satellites can be affected by the earth's magnetic field. However, sometimes the earth's magnetic field is destroyed by solar wind (flare) and magnetic sensors can detect this phenomenon. A solar wind can destroy electronic circuits due to the fact that it transfers electronic charges and therefore, circuits should be shut down in these circumstances. However, magnetic sensors are also used to measure the magnetic fingerprint of rock in other planets. Sometimes, there is a need to determine the magnetic field that is produced by the electronics of the space craft.

Through the years, scientists have developed a series of methods to measure magnetism. In this section, only the following sensor types will be explored:

- Anisotropic Magnetoresistance (AMR)
- Giant Magnetoresistance (GMR)
- Giant Magneto-Impedance (GMI)
- Flux-Gate sensors
- Superconducting Quantum Interference Device (SQUID)

Each sensor type has certain properties and each application favors the use of a specific sensor type. Sometimes, there is also the need to use more than one sensor type due to the application requirements. Some of the application requirements to be identified to evaluate the suitability of the sensor types are:

- Noise levels and resolution
- Offset and loge-term stability
- Sensitivity, nonlinearity and range of measured values
- Frequency bandwidth
- Temperature stability
- Immunity to orthogonal excitations
- Hysteresis (mostly due to magnetic material)
- Cost, availability of materials and production equipment
- Available installation space
- Sensor maintenance
- Power consumption

It is well known that the highest resolution possible is achieved by the SQUID magnetometer. However, the SQUID magnetometers need cryogenic temperatures and the cost to maintain that state is extremely high. For unmanned space missions, there is the need for the components to be working perfectly for decades and while the sensor element of a SQUID magnetometer can last, the cooling system may not be able to work seamlessly.

Nevertheless, one of the most important parameters is the noise level that leads to the resolution of the sensor. The noise is measured through the power spectral density (PSD) which is the power of the noise concentrated around the frequency bandwidth. It is customary to refer to it as the power spectrum even when there is no physical power involved. This PSD is the sum of all noise sources, but the most important noise source is the internal noise of the sensing element. PSD, $S(f)$, is defined as:

$$S(f) = \frac{V_n^2}{\Delta f} = \left(\frac{V_n^2}{\sqrt{\Delta f}} \right)^2 \tag{23}$$

where V_n is the voltage integral (V) and f is the frequency range (*Hz*). In the following table, the average PSD values that are usually found in each sensor type, are presented.

Anisotropic Magnetoresistance (AMR)

Anisotropic Magnetoresistance (AMR) was discovered in 1897 by William Thompson (Lord Kelvin). The evolution continued through the years and in 1960, it was fully modeled. Materials that exhibit AMR were used in the first magnetic random-access memory (MRAM) that was developed by IBM. Nowadays, AMR magnetometers are used in high current detection, position sensing, tachometry, low-magnetic-field anomaly detection and multi-axis compasses. Philips Electronics and Honeywell are the main manufacturers of this type of sensors. In order to test this type of magnetometers and calibrate them, a 3D Helmholtz coil is used. This is necessary due to the fact that AMR sensors have a non-linear behavior and there is no mathematical expression to calculate the AMR behavior in advance. The most common material used in AMR magnetometers is NiFe permaloy.

Physical Model

Anisotropic Magnetoresistance can be divided into two types, ordinary and anisotropic. Usually non-magnetic metals and semiconductors appear to be have ordinary magnetoresistance. This effect is due to the scattering of the electric carriers from the magnetic field (Hall effect) which increases the path length of the carriers that subsequently increases the resistance. The equation for the ordinary magnetoresistance is:

$$\frac{\Delta\rho}{\rho_0} = C\mu B^2 \tag{24}$$

Where $\boldsymbol{C}$ is a constant ($sT^2m^{-2}V^{-1}$), $\boldsymbol{\mu}$ is the carriers' mobility (m^2Vs^{-1}), $\boldsymbol{A}$ is the electric resistivity of the material (Ωm) and $\boldsymbol{B}$ is the magnetic field (*Tesla*). An AMR magnetometer reaches a saturation level when a 25-30 Oersted (*Oe*) magnetization field strength or 25-30 Gauss (*Gs*) for magnetic field density is applied. For an AMR sensor, this behavior can be simplified according to the following equation:

$$\frac{\Delta R}{R_0} = \Delta R_{max}\ cos^2\left(\theta\right) \tag{25}$$

The following graph presents that behavior.

Table 1. List of magnetic sensors PSD noise (Tumanski, 2011).

Sensor Type	S(f)
SQUID	$5fT / \sqrt{Hz}$
Hight T_C SQUID	$100fT / \sqrt{Hz}$
AMR	$15nT / \sqrt{Hz}$
GMR	$100pT / \sqrt{Hz}$
GMI	$3pT / \sqrt{Hz}$
Flux-Gate	$2pT / \sqrt{Hz}$

Figure 1. AMR element behavour in an apply magnetic field

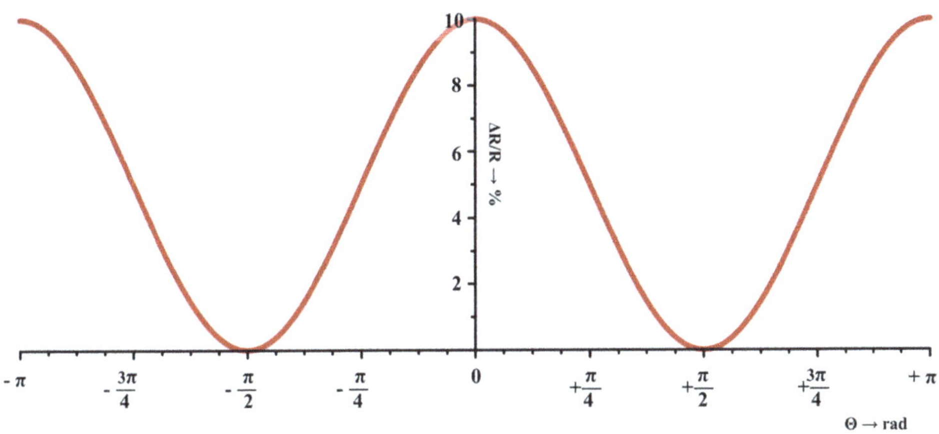

The Construction of an AMR Barber-Pole

As stated previously, an AMR element has a highly non-linear behavior. Moreover, the AMR element does not sense the difference in the direction of the magnetic field due to the square law in Equation (25). To overcome this problem, the sensing element should be biased and the simplest way to do that is to make the current travel at a shallower angle in to the magnetic field instead of 90°. The best angle to have the element biased is in the angle of 45°. In order to achieve this bias, as shown in figure 1, a series of alternating straps of aluminum and magnetic material is placed as a thin film in a non-contractive substrate like a barber-pole (Ripka et al., 2007).

The linear region is moved to the 45^0. The magnetic domain of the magnetic material is not necessarily polarized to the direction of the current. The linear region is moved to the 45^0 due to the orientation of the aluminum stripes as shown in Figure 2. The 45^0 angle is very convenient because it turns the linear region in the middle of the rotating angle. All the magnetic domain must be polarized in the same direction in order to achieve maximum resolution. To achieve this, a series of set/reset pulses must be applied to a coil around the barber-pole in the easy magnetization axis (long axis). A set pulse polarizes the magnetic domain in the direction of the current flow and a reset pulse polarizes it in the opposite direction. Figure 3 presents the change in the magnetic domain and the two operating regions of the magnetic material (Grosz et al., 2017).

Construction of a AMR Sensor Bridge

An AMR sensor is composed by the four AMR barber-pole sensing element, which is placed in a Wheatstone bridge configuration (Figure 4). For a complied magnetometer, there is the need for two extra coils in different orientations. The set/reset coil ($L_{S/R}$) is oriented in such a way that the magnetic field is in the direction of the easy magnetization axis of the elements. The offset strap (L_{offset}) is a coil that produces a magnetic field parallel to the sensitive axis. The use of this coil is to bias the bridge or to cancel the external magnetic field for closed-loop applications.

The AMR Wheatstone bridge is now a fundamental part in all AMR sensors. The construction of the sensing element is very important, especially for low noise

Figure 2. AMR (Barber-Pole) sensor element, aluminum shorting straps for redirecting the current (Ripka et al., 2007)

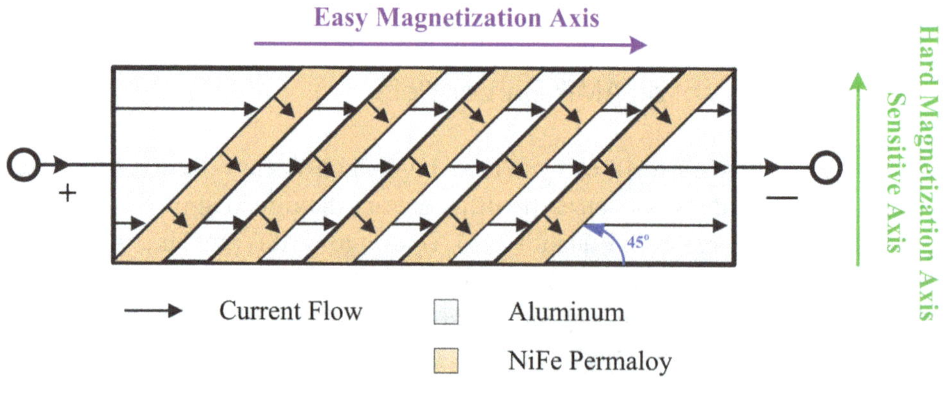

Figure 3. Resistivity behavior of a barber-pole with two directions of polarization

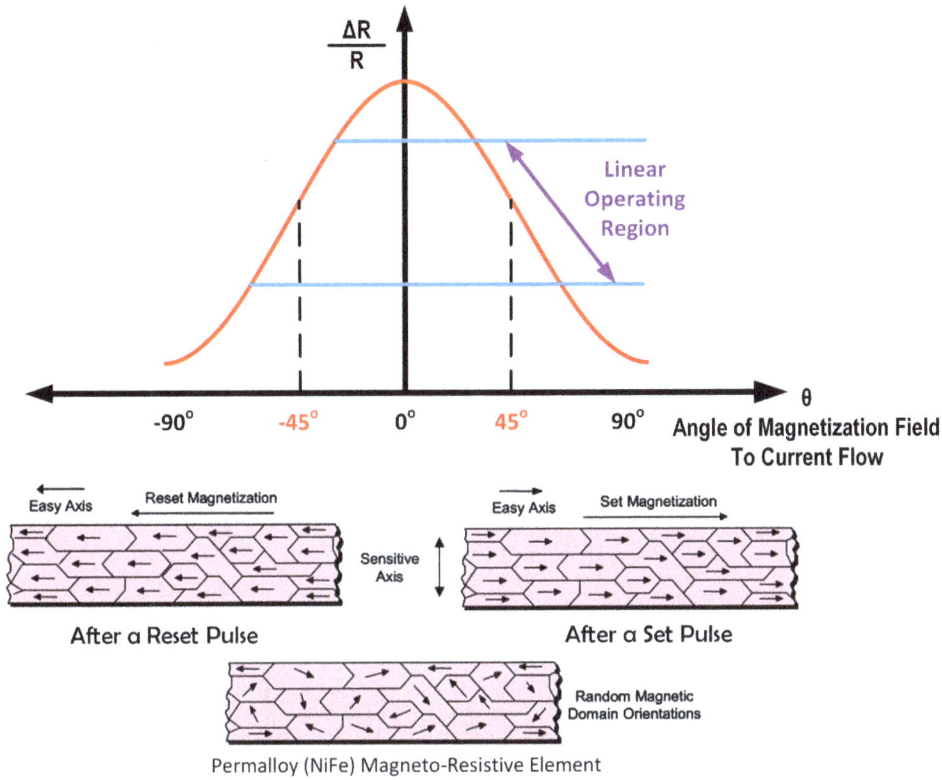

applications. Operational amplifiers (OPA) can be noisier than the sensor element itself (Hadjigeorgiou, 2014). The use of the set/reset coil in a closed-loop architecture can further improve the resolution and suppress the 1/f noise. The AMR sensing element has power spectral density of $15nT / \sqrt{Hz}$ (Tumanski, 2011) as tabulated in Table 1. However, the techniques discussed earlier show that the power spectral density can be as low as $4.5nT / \sqrt{Hz}$ (Hadjigeorgiou et. al, 2017). The AMR sensors have high bandwidth starting from DC and going up to several MHz, but when a closed-loop architecture is used, the frequency is reduced. This bandwidth limitation occurs due to several factors, including the response of the magnetic material in the fast set/reset pulse (Jiles, 2015; Tumanski, 2011) and the set/reset strap that can overheat and permanently damage the sensor. These materials need high magnetic fields in order to rotate and consequently, high current pulses are needed. The method of the closed-loop architecture (Hadjigeorgiou et. al, 2017) is another possible factor. The AMR is a versatile, very widely used magnetic sensor that offers a good compromise between resolution and cost.

Figure 4. The AMR Wheatstone bridge with the Set/Reset and Offset coils (Hadjigeorgiou et al., 2017)

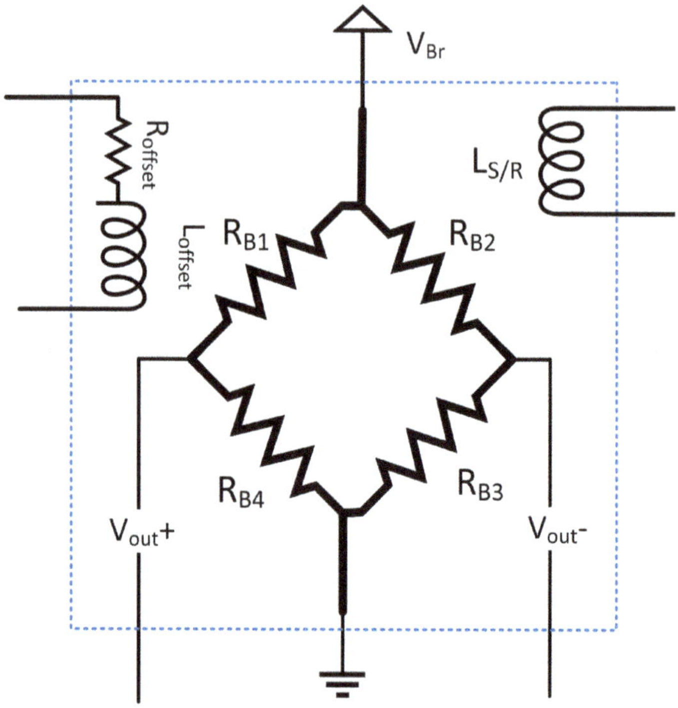

Giant Magnetoresistance (GMR)

Giant Magneto Resistance magnetometers are much more sensitive than AMR magnetometers, but at higher magnetic fields. This type of sensor is easily fabricated on a chip. They are widely used in hard disk drive heads to read the data from the disk plates. The simplest structure of a GMR is constructed by two layers of a magnetic material, which are separated by a conductor. GMR may consist of more magnetic and conductive layers. In each case, a conductor is sandwiched between two magnetic layers. Magnetic layers are constructed by Fe-Co-Ni alloy, or a NiFe pemralloy. Conductive layers are made of Cu or Al. The Giant magneto resistance effect was discovered by the P. Grunberg and A. Fert. Figure 5 provides a simplified cross-section of a GMR.

Figure 5. A Basic GMR structure and its typical response (Grosz et al., 2017)

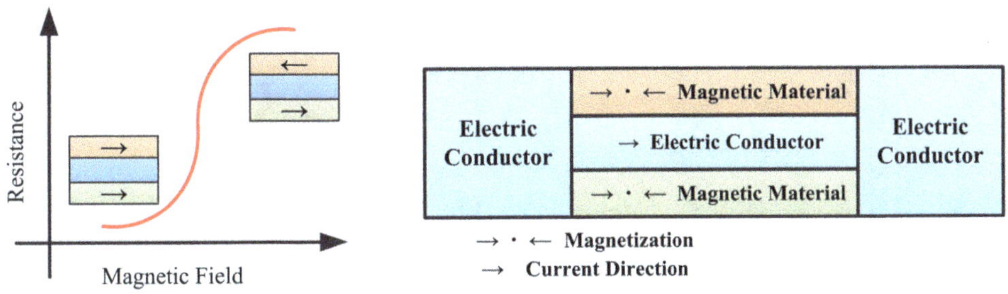

Physical Model

The flow of an electric current in a magnetic multilayer structure, as shown in Figure 5, is influenced by the relative magnetization orientation of the magnetic layers. The resistivity of the magnetic layers is low when the magnetization is parallel to the field and it is higher when the magnetization of the neighboring magnetic layers is antiparallel. This is happening due to the appearance of the effect of the spin-dependent scattering. To observe this phenomenon, the thickness of the separating conductive material must be in the order of nanometers and whilst its thickness is a major factor in sensor's design. When an external magnetic field is applied, a change from antiferromagnetic to ferromagnetic (and vice-versa) coupling can be achieved, which leads to a change in the resistance value. The magnetoresistance (MR) ratio is defined as:

$$\frac{\Delta R}{R} = \frac{R^{\uparrow\downarrow} - R^{\uparrow\uparrow}}{R^{\uparrow\uparrow}} \tag{26}$$

Spin valves are a particular configuration of a sandwich structure. In spin valves, an additional antiferromagnetic (pinning) layer is added to the top or bottom part of the structure, as shown in Figure 5 (right part). In this sort of structures, there is no need of an external excitation to get the antiparallel alignment. In spite of this, the pinned direction (easy axis) is usually fixed by raising the temperature above the knee temperature (at which the antiferromagnetic coupling disappears) and then cooling it within a fixed magnetic field. Obviously, devices based on this phenomenon, have a temperature limitation below the knee temperature. Typical values displayed by spin valves are a MR of 4–20% with saturation fields of 0.8–6 kA/m. For linear applications without excitation, pinned (easy axis) and free layers are preferably arranged in a crossed axis configuration (at 90°), as depicted in Figure 6.

Figure 6. Basic spin value scheme for a multilayer structure in crossed axis configuration (Grosz et al., 2017)

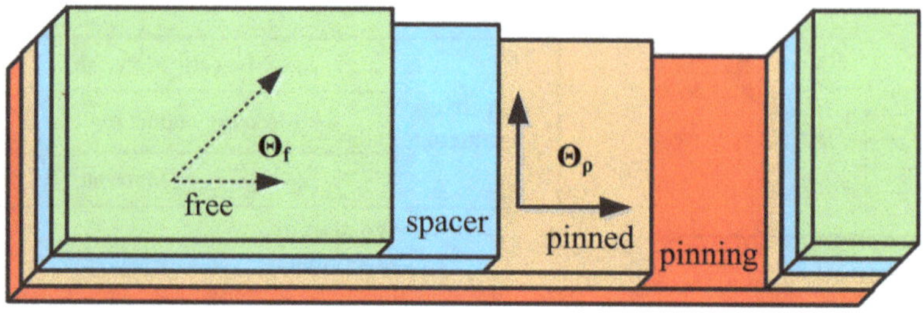

The mathematical expression of the change in resistance for the structure in figure 6 is:

$$\Delta R = \frac{1}{2} i \left(\frac{\Delta R}{R} \right) R_{\blacksquare} \frac{W}{h} \cos\left(\Theta_p - \Theta_f \right) \tag{27}$$

Where $\boldsymbol{\Delta R / R}$ is the maximum MR level (*5-20%*), $\boldsymbol{R_{\blacksquare}}$ is the sensor sheet resistance ($\Omega\blacksquare^{-1}$), $\boldsymbol{W}$ is the width (*m*), $\boldsymbol{h}$ is the thickness (*m*), i is the current passing through the sensor (*A*), and $\boldsymbol{\Theta_p}$ and $\boldsymbol{\Theta_f}$ are the angles of the magnetization of the pinned and free layer (*rads*), respectively (Freitas, 2007).

Both structures that are presented in Figure 5 and Figure 6 can be parts of a Wheatstone bridge. To further improve the sensor's behavior, a proper biasing is needed. Usually, a voltage source is needed to bias the sensor bridge. However, the theoretical analysis and the experiments show that is better to use a constant current source for feeding the bridge instead of a constant voltage. This technique minimizes the temperature drifts of the spin values. Furthermore, the use of AC current for biasing improves the sensor's performance in terms of hysteresis, linearity, offset and noise.

Giant Magneto-Impedance (GMI)

Giant Magnetoimpedance (GMI) effect consists of the huge change of both real and imaginary parts of the impedance upon the application of static magnetic field. The relative change of impedance can reach ratios up to around 600%, with extremely large sensitivities in the very low magnetic field region. The first reports dealing with GMI appeared several decades ago, in 1935 by researches E. P. Harrison et al. (Harrison, 1935). The observation and research about the GMI phenomenon was received with less enthusiasm, probably because of the envisaged lower technological expectations and of the apparent lack of intrinsically new magnetic effects related to its origin. This idea became obsolete and now the interpretation requires a deep understanding of the micromagnetic characteristics of the soft magnetic materials and its dependence on dynamic magnetism. Moreover, the effect has attracted great scientific and technological interest in the last years due to the great impedance changes observed in well suited samples and optimum experimental configurations. Many research teams all over the world investigate GMI and its technological applications. GMI has created a new branch of research merging the micro-magnetics of soft magnets with classical electrodynamics (Grosz, 2017).

Physical Model

The magneto-impedance effect is referring to the complex impedance found in a wire when it is excited with an alternative current. The complex impedance is descripted with the equation:

$$Z = R + jX \tag{28}$$

Where R is the real part or resistance and X is the imaginary resistance (reactance), as shown in Figure 7.

The analytical impedance for a conductor can be calculated by simultaneously solving the deduced Maxwell equation

$$\nabla^2 \boldsymbol{H} - \frac{\mu_0}{\rho}\dot{\overline{}} = \frac{\mu}{\rho}\dot{\mathrm{M}} - grad\, div\, M \tag{29}$$

And the Landau-Lifshitz equation for the motion of the magnetization vector:

Figure 7. Scheme to the definition of impedance

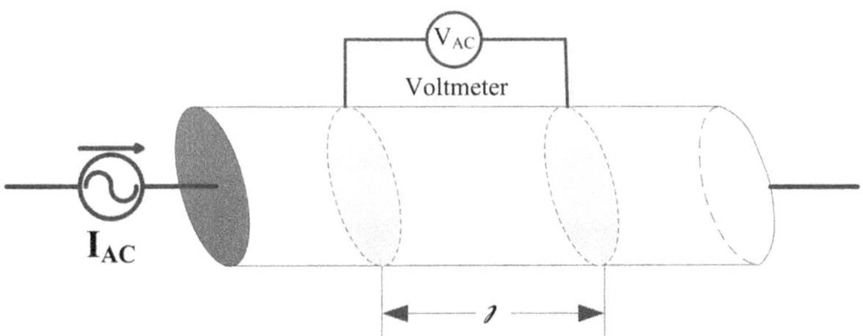

$$\dot{\boldsymbol{M}} = \gamma \boldsymbol{M} \times \boldsymbol{H}_{eff} - \frac{\boldsymbol{a}}{\boldsymbol{M}_S} \boldsymbol{M} \times \dot{\boldsymbol{M}} - \frac{1}{\tau}\left(\boldsymbol{M} - \boldsymbol{M}_0\right) \tag{30}$$

Where γ is the gyromagnetic ratio, $\boldsymbol{M}_S$ the saturation magnetization, $\boldsymbol{M}_0$ the static magnetization, $\boldsymbol{H}_{eff}$ the effective magnetic field, and a is the damping parameter. The expression of the impedance is given by:

$$Z = R_{DC} kt \frac{J_0\left(kr\right)}{2J_1\left(kr\right)} \tag{31}$$

Where $\boldsymbol{J}_0$ and $\boldsymbol{J}_1$ are the Bessel functions of the first kind, $\boldsymbol{r}$ (m) is the radius of the wire, $\boldsymbol{R}_{DC}$ (Ω) is the electrical resistance for a DC and $k = \left(1 + j\right) / \delta_m$ with imaginary unit j. $\boldsymbol{\delta}_m$ (m) is the penetration depth in a magnetic medium, with relative circumferential permeability μ_φ:

$$\delta_m = \frac{c}{\sqrt{4\pi^2 f \sigma \mu_\varphi}} \tag{32}$$

Where $\boldsymbol{c}$ (ms^{-1}) the speed of is light, $\boldsymbol{A}$ (Ωm) the electrical conductivity, and $\boldsymbol{f}$ (Hz) is the frequency of the $\boldsymbol{I}_{AC}$ (A) flowing along the sample (Jordan, 1968). There are two types of magnetic permeability, the longitudinal and the circumferential; in a ribbon, the circumferential is referred as transverse. Figure 8 shows the magnetic domain of a ribbon and Figure 9 shows the longitudinal and the transverse impedance.

Figure 8. Domain wall configuration in a multilayer structure. The magnetization direction is shown in every domain. The AC current flows in the sandwiches metallic layer, producing an AC flux in the surrounding magnetic layers (Grosz, 2017).

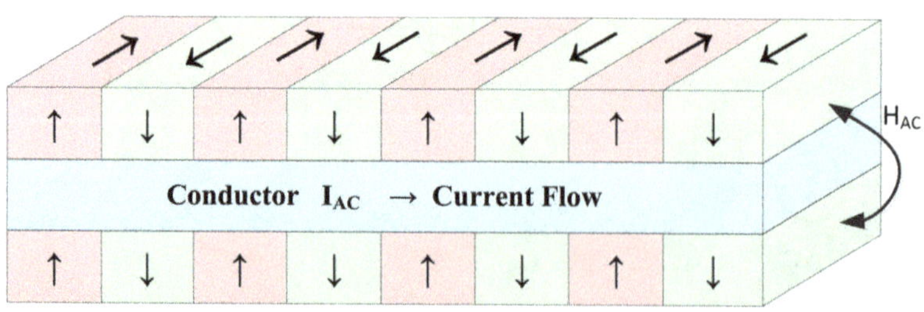

Figure 9. MI field dependences measured in narrow NiFe/Au/NiFe sandwich films with (a) the longitudinal anisotropy and (b) the transverse anisotropy. Dashed curves show the reversed magnetic field behavior which can extracted the hysteresis of the material (Grosz, 2017).

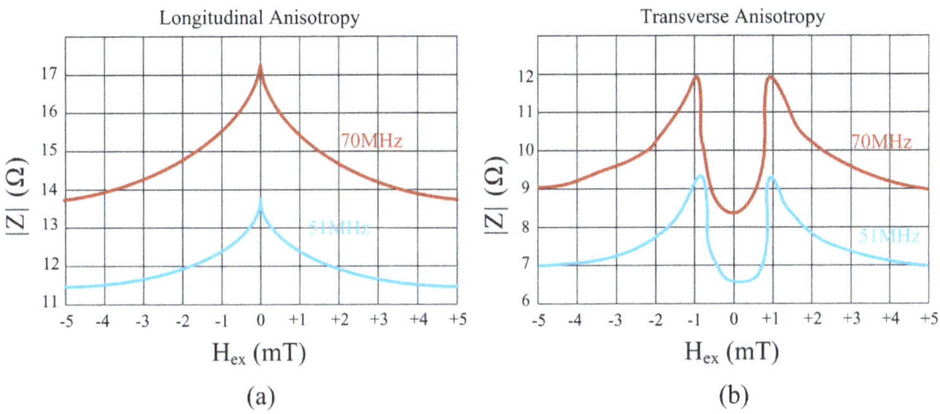

There are many theoretical models, each one for a different frequency rang,e which can be found in bibliography in more detail. These models are the Quasistatic model, the Eddy current model, the Domain wall model, the Electromagnetic model, and the Exchange-conductivity model.

Flux-Gate Sensors

The fluxgate magnetometers received their name from magnetic flux-gating curves. The fluxgate magnetometer is dated back to 1930. It was widely used during the WWII as a trigger in magnetic ship mines. Fluxgate sensors' core is coming in many shapes such as rob, ring and race-track. The design of the core geometry, construction and magnetic material are crucial factors, for low noise performances. There are two types of fluxgate; the parallel and the orthogonal fluxgate. The difference between these types is that the excitation and pick-up coils are parallel to the fluxgate. On the other hand, the orthogonal fluxgate has a direct excitation to the magnetic core and the pick-up coil is orthogonally oriented to the direction of the excitation current. The following figure shows a parallel fluxgate and an orthogonal fluxgate.

Physical Model

The fluxgate sensor is based on two physical phenomena, the induction law and the magnetic core saturation. The output voltage of the pick-up coil on Figure 10a is calculated according to:

$$U_i = -NS\left(\frac{dB_E}{dt} + K\mu_0\mu_r\frac{dH_M}{dt} + K\mu_0 H_M\frac{d\mu_r}{dt}\right) \quad (33)$$

where $\boldsymbol{N}$ is the number of turns of the pick-up coil, $\boldsymbol{S}$ (m^2) is the core cross-sectional area, $\boldsymbol{H}_M$ (Am^{-1}) is the measured external magnetic field intensity, $\boldsymbol{B}_E$ (T) is the alternating excitation flux density in the ferromagnetic core due to the excitation field, μ_0 (Hm^{-1}) is the permeability of vacuum and **K** is a dimensionless coupling coefficient of the core to the field $\boldsymbol{H}_M$ (Am^{-1}). Finally, $\boldsymbol{\mu}_r$ is the relative permeability of the magnetic core varying with the external magnetic field.

The parallel fluxgate sensors operate based on the following principle. A current pass through the excitation coil, creating a magnetic field $\boldsymbol{H}_E$, which encircles the magnetic core. The magnetic field is high enough to force the magnetic core to enter the magnetization saturation region M_s. Figure 11a shows the magnetic flux of one half of the core without an external magnetic field while Figure 11c presents the output voltage. When an external magnetic field $\boldsymbol{H}_{ext}$ is applied, the magnetic flux is different in the two halves of the core as shown in Figure 11b. The output signal is given in Figure 11d.

If the excitation of the core is sinusoidal ($H_e = A\sin(\omega t)$, then the magnetic flux Φ of the core is,

Figure 10. A Parallel Fluxgate on the left (Ripka, 2007) and an Orthogonal Fluxgate on the right (Grosz et al., 2017)

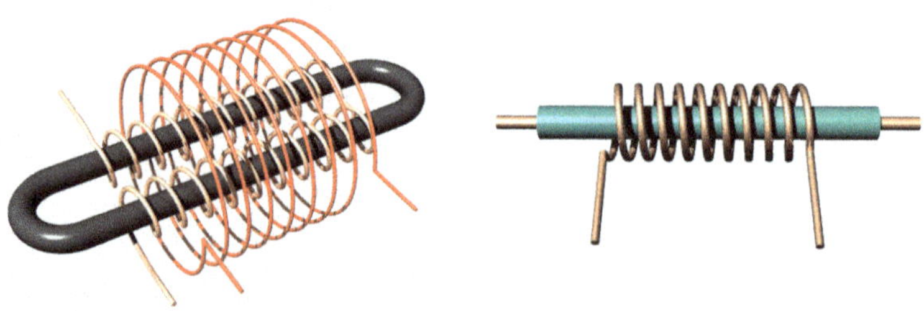

Figure 11. Fluxgate B-H loop without (a), with (b) external magnetic field. At the bottom the excitation magnetic field and the output voltage. (Gordon et al., 1965)

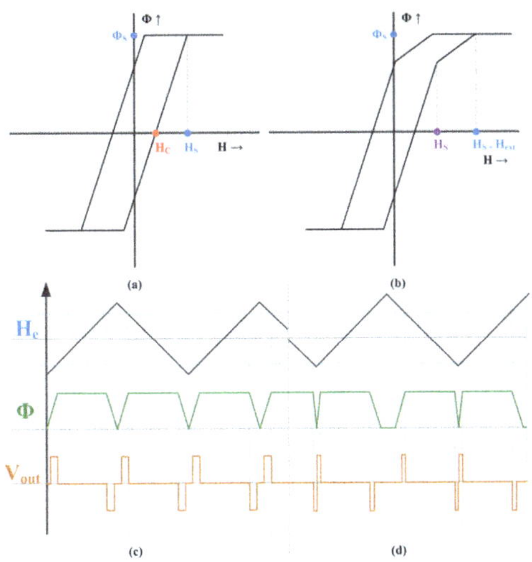

$$= 2S\left(aH_{ext} - bH_{ext}^{3} - \frac{3}{2}bA^{2}H_{ext} + \frac{3}{2}bA^{2}H_{ext}\cos\left(2\omega t\right)\right) \tag{34}$$

where $\boldsymbol{a} > 0\left(Vm / A\right), \boldsymbol{b} > 0 \left(Vm^{3} / A^{3}\right)$, are coefficients, $\boldsymbol{A}$ (Am^{-1}) is the amplitude of the excitation magnetic field, $\boldsymbol{H}_{ext}$ (Am^{-1}) is the external magnetic field and $\boldsymbol{E}$ (rad/s) is the rotating frequency. Considering equation 34, it is clear that the simplest way to extract the external magnetic field (H_{ext}) is the second harmonic. In reality, there are higher-order harmonics present in the output signal, due to the nature of the B-H loop (hysteresis, approach to saturation). Another factor is the non-sinusoidal excitation waveforms which have higher harmonics (Hadjigeorgiou et al., 2015). These effects are sometimes taken into account but most fluxgate magnetometer use only the second harmonic.

Superconducting Quantum Interference Device (SQUID)

Superconducting Quantum Interference Device (SQUID) is a magnetometer that measures the magnetic flux passing throw it. It may be the most sensitive magnetometer that have ever created. There are two types of SQUID; the low temperature which is using liquid Helium for cooling, and the high temperature which is using liquid Nitrogen. Both are based on the Josephson effect (Josephson, 1962). They consist of a superconducting loop interrupted by two resistively shunted Josephson tunnel junctions. The low temperature variation, operating at 4.2 K, is almost noiseless in the order of $\mu\Phi / \sqrt{Hz}$ or $fT / \sqrt{Hz}$. They have large bandwidth and flat frequency response ranging from dc to GHz (Drung et al., 2007). They are excellent for a wide variety of applications, such as e.g. biomagnetism, magnetoencephalography (MEG), calibration of other systems and geophysical exploration. Nevertheless, the cryogenic conditions needed to operate properly are expensive including not only for the installation costs, but also the maintenance costs. (Ripka, 2001; Tumanski, 2011).

Physical Model

SQUID convert magnetic flux into a voltage across the device. The operation of SQUID is based on two physical phenomena: flux quantization in a closed superconducting loop in units of the flux quantum $\Phi_0 = h / 2e = 2.07 \times 10 fTm^2$, and Josephson tunneling. The Josephson Junction (JJ) also called Cooper pairs, have zero spin and follow boson statistics. Figure 12 shows a superconductive ring with a Josephson junction at one end.

Both superconductive branches are in the same quantum state and can be described by a collective superconducting wave function $\Psi = \Psi_0 exp\left(i\phi\right)$, with $\phi\left(x,t\right)$ being the time and space dependent phase and $n_s = \left|\Psi\right|^2$ the Cooper pair density. Whenever two superconductive elements are connected through a thin insulator (weak couple), there is current exchange. In the Josephson tunnel junction (SIS), the thin insulating barrier (I) is placed between two superconductors (S). The current passing from the Josephson junction is described by the first Josephson equation $I_C = I_{C,0} \sin\left(\varphi\right)$,

Figure 12. The SQUID ring is shaped as a square washer with a hole in the center and a slit. The Josephson junctions (purple colure) are located at the outer edge of the slit, as indicated. In the right, an electrical equivalent of the SQUID is presented (Grosz et al., 2017).

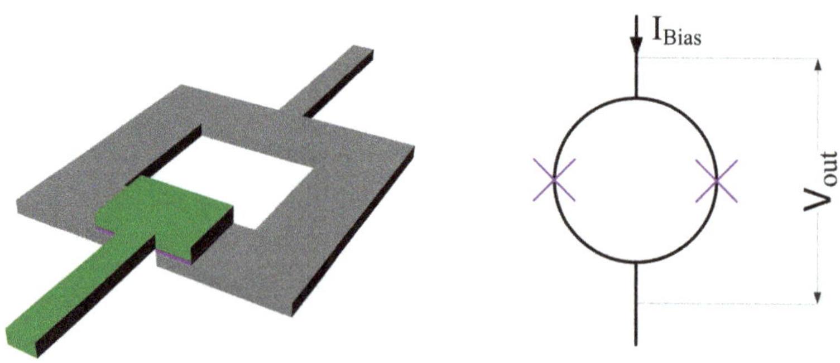

with $I_{C,0}$, the maximum critical current and φ being the difference across the junction. When the maximum critical current is exceeded, the phase difference across the junction increases over time and a DC voltage across the junction appears. It is described by the second Josephson equation:

$$\frac{\partial \varphi}{\partial t} = \frac{2e}{\hbar} V_{DC} = \frac{2\pi}{\Phi_0} V_{DC} \tag{35}$$

Where $\hbar = 1.055 \times 10^{-34}\,\text{Js}$ is the reduced Plank's constant. The current across the Josephson junction is oscillated with the Josephson frequency $2\pi V_{DC} / \Phi_0$. Figure 13 shows this oscillation that can be measured from Figure 13 (Josephson, 1962).

In practice, there is a feedback coil that helps to the linearity of the closed-loop system and the output voltage. However, the electronics needed to be implemented for a linear behavior, introduce new noise sources much higher than the sensor itself.

Magnetic Cleanliness

Magnetic cleanliness also known as magnetic shielding is the ability to remove or block the unwanted magnetic field in a specific region. The magnetic cleanliness can be achieved in what is called Gauss chamber which is a special form of Faraday cage for magnetic field. The Gauss chamber can be passive or active or both.

Figure 13. The Current-Voltage characteristics of a Josephson tunnel junction (Grosz et al., 2017).

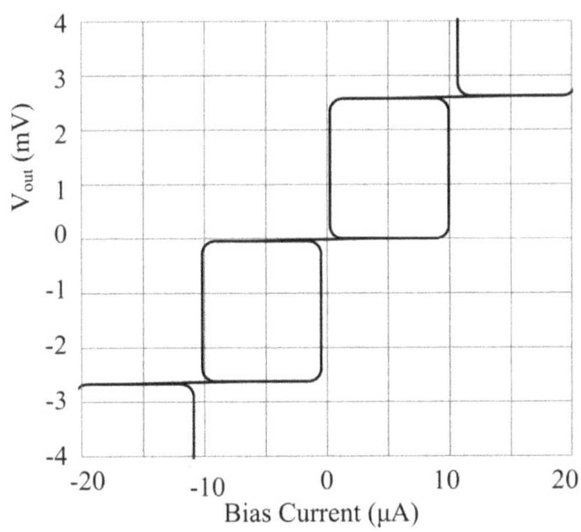

A magnetic field is much harder to prevent from entering a space compared to an electric field. There is no cheap solution that can control the magnetic field in a space, especially for very demanding applications such as very low magnetic field at low frequencies. The magnetic field has the ability to penetrate, even metals, but it depends on the frequency of the magnetic field. A metal cage (Gauss chamber) will drop the amplitude of a magnetic field $\boldsymbol{B}_0$, exponentially. The thickness of each metal needed to reduce the magnetic field to $1/e$ of the original field is called skin depth (δ). The skin depth can be calculated with the following mathematical expression:

$$\delta = \sqrt{\frac{\rho}{\pi f \mu_r \mu_0}} \sqrt{\sqrt{1+\left(2\pi f \rho \varepsilon_r \varepsilon_0\right)^2} + 2\pi f \rho \varepsilon_r \varepsilon_0} \tag{36}$$

Where ρ (Ωm) is the resistivity, μ_r is the relative magnetic permeability, ε_r is the relative permittivity of the sheet material, *f (Hz)* is the frequency of the magnetic field, μ_0 (Hm^{-1}) is the magnetic permeability of vacuum and ε_0 (Fm^{-1}) is the electrical permittivity of vacuum (Jordan, 1968).

The best way to block the magnetic field is to use a superconductor. The reason for that is the property of the superconductor to have zero resistivity. Consequently, even the thinnest superconductors can blog all magnetic field independent of the frequency (Ripka, 2001). However, this type of insulation is extremely expensive not only to build but also to maintain due to the need of cryogenic conditions. A more practical construction of Gauss chamber is made from metal alloys with high magnetic permeability called Mu-Metal. Mu-Metals have relative permeability of the order of 20000 to 200000 compared to pure iron, which exhibits a permeability limited to 5000 (Grosz et al., 2007; Tumanski, 2011). Due to the exponential absorption of the magnetic field, it is preferable that the Gauss chamber is constructed by multiple layers of Mu-Metal instead of a thick sheet. Gauss champers are used in Magnetic Resonance Imaging (MRI's) rooms and Magnetoencephalography (MEG's) rooms (Ishiyama et al., 1991). A magnetic shield's shape is an important parameter in the design phase. For example, a spherical chamber is $4/3$ or 134% more effective than a cylindrical one (Mager, 1970). However, some geometries are not so useful in practice, so every design has its own trade-offs. In order for a cylindrical Gauss chamber to achieve 60 dB reduction at 50 Hz of the magnetic field, a three-layer Mu-Metal is required and each layer needs to be at least 1 mm thick. Figure 14 presents an example of a Mu-metal Gauss chamber.

The passive Gauss chamber sometimes is not good enough and the use of an active magnetic shielding is required. The active shielding is constructed by three

Figure 14. Gauss chamber for instrument calibration (Bartington Inc. c)

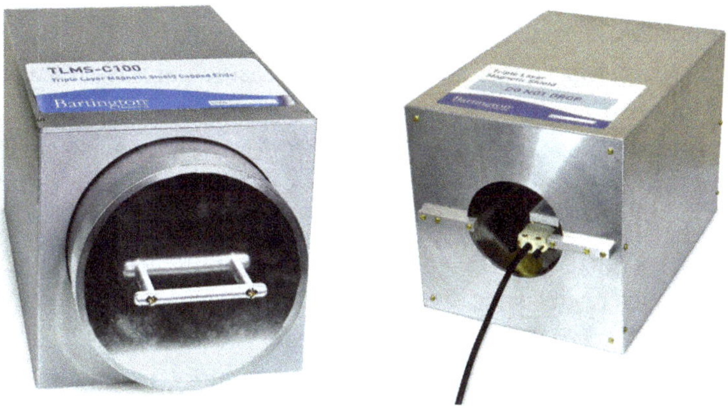

Helmholtz coils, which are orthogonally oriented to each other (Tumanski, 2011). Maxwell coils can also be utilized, but they are rarely used due to the higher construction costs and other functional limitations. Figure 15 shows two types of Helmholtz coils manufactured by Bartington, one with circular coils and one with square coils.

Figure 15. Helmholtz coils for controlling the magnetic field in 3D axes. The (a) has a circular structure (Bartington Inc., (a) and (b) have a square structure (Bartington Inc., b).

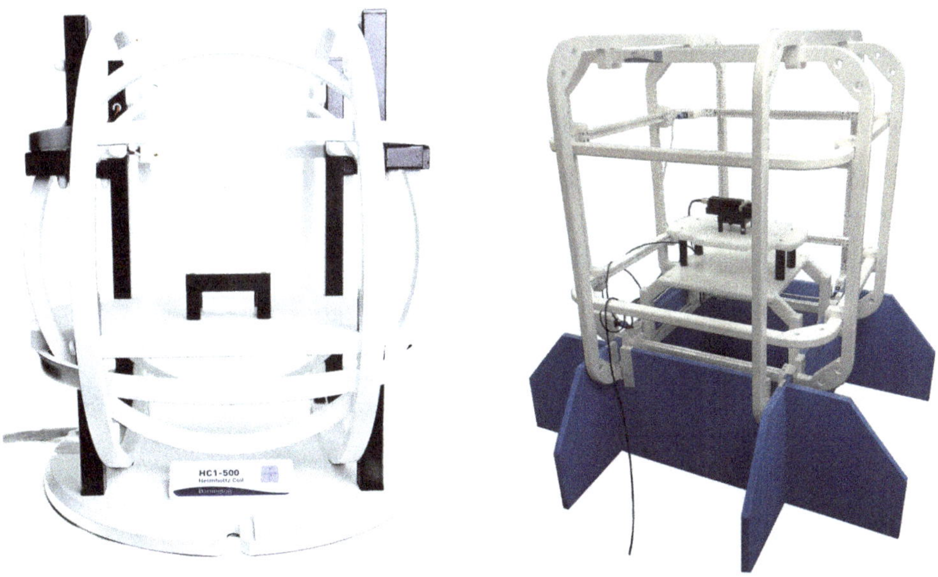

CONCLUSION

This chapter presents a basic overview in the field of magnetic field acquisition. A variety of sensors appropriate for magnetic field measurements is reviewed covering the majority of field applications, with special focus to space missions. These sensors have the advantages that they can easily operate under harsh environmental conditions and elevated temperatures, they are very reliable due to the absence of moving parts and they can be embedded inside building materials. Magnetic sensors are categorized primarily based on their sensing principle, which impacts directly the performance characteristics like measurement range, resolution, frequency response, working temperature and manufacturing cost.

Magnetic Sensors are commonly used for laboratory measurements and for satellite magnetic measurements. Understanding the various types of magnetometers and their limitations is crucial for the interpretation of the satellite magnetometer results. For space use, magnetic sensors must have some unique characteristics. Deep knowledge of sensor's characteristics leads to an optimal selection of the magnetic sensor in accordance to the application requirements. The chapter concludes with a brief discussion about magnetic cleanliness and its implications in space missions.

REFERENCES

Bartington Inc. (n.d.a). *Operation Manual for HC1 Helmholtz Coils*. Retrieved October, 2017, from http://www.bartington.com/Literaturepdf/Operation%20 Manuals/HC1%20Helmholtz%20Coil%20OM3226.pdf

Bartington Inc. (n.d.b). *Operation Manual for HC2 Helmholtz Coils*. Retrieved October, 2017, from http://www.bartington.com/Literaturepdf/Operation%20 Manuals/HC2%20Helmholtz%20Coil%20OM3342.pdf

Bartington Inc. (n.d.c). *Operation Manual for HC1 Helmholtz Coils*. Retrieved October, 2017, from http://www.bartington.com/Literaturepdf/Operation%20 Manuals/HC1%20Helmholtz%20Coil%20OM3226.pdf

Daniel, C. M. (1988). *The Theory of Magnetism I: Statics and Dynamics*. Springer.

Drung, D., Abmann, C., & Beyer, J. (2007). Highly Sensitive and Easy-to-Use SQUID Sensors. *IEEE Transactions on Applied Superconductivity, Volume*, *17*(2), 699–704. doi:10.1109/TASC.2007.897403

Foner, S. (1959). Versatile and Sensitive Vibrating-Sample Magnetometer. *The Review of Scientific Instruments*, *30*(7), 548–557. doi:10.1063/1.1716679

Fraden, J. (2010). *Handbook of Modern Sensors Physics, Designs, and Applications*. New York: Springer-Verlag. doi:10.1007/978-1-4419-6466-3

Fratantuono, L. (2015). *A Reading of Lucretius' De Rerum Natura*. Lexington Books.

Freitas, P. P., Ferreira, R., Cardoso, F., & Cardoso, F. (2007). Magnetoresistive sensors. *Journal of Physics Condensed Matter*, *19*(16), 165221. doi:10.1088/0953-8984/19/16/165221

Gordon, D., Lundsten, R., & Chiarodo, R. (1965). Factors affecting the sensitivity of gamma-level ring-core magnetometers. *IEEE Transactions on Magnetics, Volume*, *1*(4), 330–337. doi:10.1109/TMAG.1965.1062987

Grosz, A., Haji-Sheikh, J. M., & Mukhopadhyay, C. S. (2017). *High Sensitivity Magnetometers*. Cham: Springer. doi:10.1007/978-3-319-34070-8

Hadjigeorgiou, N. (2014). Experimental Study Noise of HMR2003 Sensor with or without the Use of Amplifier AMP04. *Key Engineering Materials*, *605*, 629–632. doi:10.4028/www.scientific.net/KEM.605.629

Hadjigeorgiou, N., Hristoforou, E., & Sotiriadis, P. P. (2017). Closed-loop current-feedback, signal-chopped, low noise AMR sensor with high linearity. *IEEE Modern Circuits and Systems Technologies, 6*.

Hadjigeorgiou, N., Konstantopoulos, C., & Masxas, D. (2015). Fourier Analysis for Orthogonal and Parallel Fluxgate. *Key Engineering Materials*, *644*, 270–273. doi:10.4028/www.scientific.net/KEM.644.270

Hadjigeorgiou, N., Tsalikidou, A.C., Hristoforou, E., & Sotiriadis, P. P. (2017). Highly Linear and Low Noise AMR Sensor Using Closed Loop and Signal-Chopped Architecture. *World Academy of Science, Engineering and Technology, 11*(8).

Harrison, E. P., Turney, G. L., & Rowe, H. (1935). Electrical Properties of Wires of High Permeability. *Nature*, *135*(3423), 961. doi:10.1038/135961a0

Honewell Inc. (2011). *Hall effect sensing and application*. Retrieved September 2017, from www.honeywell.com/sensing

Hristoforou, E., Chiriac, H., & Neagu, M. (1997). A New Magnetic Field Sensor Based on Magnetostrictive Delay Lines. *IEEE Transactions on Instrumentation and Measurement, Vol*, *46*(:2), 632–635. doi:10.1109/19.571942

Hristoforou, E., & Reilly, R. E. (1991). Nonuniformity in amorphous ribbon delay lines after stress and current annealing. *Journal of Physics*, *69*, 5008.

Ishiyama, A., & Hirooka, H. (1991). Magnetic Shielding for MRI Superconducting Magnets. *IEEE Transactions on Magnetics, Volume*, *27*(Issue:2), 1692–1695. doi:10.1109/20.133514

Jiles, D. (2015). *Introduction to Magnetism and Magnetic Materials*. CRC Press.

Jiles, D. C., & Atherton, D. L. (1984). Theory of ferromagnetic hysteresis. *Journal of Applied Physics*, *55*(6), 2115–2120. doi:10.1063/1.333582

Jordan, C. E., & Balmain, G. K. (1968). *Electromagnetic Waves and Radiation Systems*. Englewood Cliffs, NJ: Prentice-Hall, Inc.

Josephson, B. D. (1962). Possible new effects in superconductive tunnelling. *Elsevier Physics Letters, Volume*, *1*(Issue:7), 251–254.

Lowrie, W. (2007). *Fundamentals of Geophysics*. London: Cambridge University Press. doi:10.1017/CBO9780511807107

Mager, J. A. (1970). Magnetic Shields. *IEEE Transactions on Magnetics, Volume*, *6*(Issue:1), 67–75. doi:10.1109/TMAG.1970.1066714

Merrill, R. T., & McElhinny, M. W. (1983). The Earth's magnetic field: Its history, origin and planetary perspective. San Francisco: Academic Press.

Metglas Ind. (2011). *Magnetic Alloy*. Retrieved October, 2017, from www.metglas.com

Prudenziati, M. (1999). *Handbook of Sensors and Actuators. Thick Film Sensors*. Elsevier.

Ripka, P. (1993). Race-track fluxgate sensors. *Sensors and Actuators A. Physical, 37-38*, 417–421.

Ripka, P. (2001). *Magnetic Sensors and Magnetometers*. Boston, MA: Artech House Publishers.

Ripka, P., & Tipek, A. (2007). *Modern Sensors Handbook*. London, UK: ISTE. doi:10.1002/9780470612231

Tumanski, S. (2011). *Handbook of Magnetic Measurements*. CRC Press. doi:10.1201/b10979

William, G. (1991). *De Magnete*. Dover Publications.

KEY TERMS AND DEFINITIONS

Anisotropic Magnetoresistance (AMR): Based magnetometers are used in devices as varied as global positioning systems to provide dead reckoning capability and in automotive ignition systems to provide crankshaft rotational position.

Giant Magneto-Impedance (GMI): Is a physical effect that expresses the large variation in the electrical impedance that occurs in some materials when subject to an external magnetic field. It should not be confused with Giant Magnetoresistance that is a totally different physical phenomenon.

Giant Magnetoresistance (GMR): Is a quantum mechanical magnetoresistance effect observed in multilayers composed of alternating ferromagnetic and non-magnetic conductive layers.

Superconducting Quantum Interference Device (SQUID): Is a very sensitive magnetometer used to measure extremely subtle magnetic fields, based on superconducting loops containing Josephson junctions.

Chapter 8
Progress in Advanced Materials Used in Electromagnetic Interference Shielding for Space Applications

Rafael Vargas-Bernal
Instituto Tecnológico Superior de Irapuato, Mexico

Bárbara Bermúdez-Reyes
Universidad Autónoma de Nuevo León, Mexico

Margarita Tecpoyotl-Torres
Universidad Autónoma del Estado de Morelos, Mexico

ABSTRACT

Aerospace applications experience electromagnetic interference produced by the space environment and by the materials, devices, and systems used in satellites, space shuttles, the international space station, and airplanes. The advanced materials represent a technological possibility to develop coatings that are able to offer a better shielding effectiveness against electromagnetic interference due to the possibility of controlling its electrical and magnetic properties as well as to that the size of the materials is very similar to the electromagnetic waves that it receives. In this chapter, an analysis of progress over advanced materials is presented with the aim of diffusing the role that nanomaterials have had, have and will have to increase the shielding to electromagnetic interference. Nanomaterials will protect aerospace components in the range of Hz to THz, but the huge advantage is that the range of protection can be optimized according to the technical requirements with a considerable weight reduction.

DOI: 10.4018/978-1-5225-5415-8.ch008

INTRODUCTION

Surfaces of aerospace components are subjected to heat, water, ice, electromagnetic waves, corrosion, abrasion, wear, thermal shock and/or fire during their lifetime (Vargas-Bernal, 2016). Materials science and engineering have introduced the use of coatings to protect the surface of the materials under these operating conditions. Materials such as ceramics, polymers, and metals and their combinations either as composites or hybrid materials, as well as functionalized versions of them, have been proposed as different design alternatives. The use of coatings for aerospace industry must offer attractive design solutions, in terms of resistance at climate, durability and reliability during their operation.

The increase in the use of electronic devices in all the applications of high technology vanguard has produced an excessive electromagnetic radiation contamination in our surrounding environment. This situation leads to serious hazards related to the normal operation of electronic devices associated with strategic systems such as nuclear reactors, transformers, aircraft and aerospace systems, control systems and communication systems. Electromagnetic interference can be defined as any electromagnetic emission from a device or system which interferes with the normal operation of another device or system. Thus, electronic systems must have electromagnetic compatibility with their environment or be able to tolerate a fraction of electromagnetic interference and not generate more than a specific amount of interference. Electronic systems experience the effect of indirect lightning strikes which involves damage or failure of their operation due to the application of a near-lightning flash. Among the experiments found are the tripping of circuit breakers due to the physical damage of input or output circuits on the electronic equipment. With the overuse of telecommunication systems, digital systems, and innovative design practices in military, commercial and aerospace systems, the design of electrical products that meet electromagnetic compatibility standards becomes increasingly restrictive. Researchers around the world are continually proposing technological alternatives to minimize the generation of electromagnetic interference from their sources, reducing or eliminating coupling paths by appropriate arrangement, through shielding, filtering and grounding practices, as well as through design hardware with inherent immunity to interference. Three different types of interference can be found on an aircraft: on-board systems (electronic devices used during flight), devices carried by passengers (portable electronic devices that can transmit and receive frequencies), and externally generated interference (lightning strike, high-intensity radiated fields (HIRF), and electromagnetic pulses (EMPs)) (von Klemperer, 2009). The key to achieving successful electromagnetic interference shielding lies in the development of advanced materials and appropriate process technologies.

Electromagnetic shielding is achieved by blocking fields of the same type by means of barriers made of electrical and/or magnetic materials. The amplitude of the reduction of electromagnetic shielding depends on the material used, its thickness, the size of the object to be shielded, the frequencies of the applied fields, as well as the shape, size and orientation of the openings of an object to an electromagnetic field incident. All metals used in the aerospace industry must be lightweight, strong and resistant to corrosion. The most common choices are titanium, aluminum, stainless steel, and nickel. Additionally, these materials must have a high strength-to-weight ratio, be refractory, heat and wear resistant, and flexible with increasing temperature. Materials commonly used for electromagnetic shielding are metals such as copper or nickel, in the form of sheets, screens, or foams. The gaps of these structures must be smaller than the wavelength of the applied electromagnetic fields. More recently, polymers are being combined with electrically conductive or semiconductor materials based on metals or ceramic materials. The substitution of metallic materials by composite materials guarantees to maintain good mechanical and electrical properties as well as a huge weight reduction. For example, metal-clad aramid fibers used in wires and cables for shielding electromagnetic interference in aircraft and satellites, with a weight reduction of up to 80%. The composite materials have been proposed for shielding the electromagnetic interference of electronics and radiation sources mainly by the reflection of electromagnetic radiation at high frequencies. Currently, the design of materials for electromagnetic shielding is associated with the design of functional materials involving distinct kinds of composite materials and hybrid materials where nanomaterials are fully involved.

The purpose of this chapter is to analyze the technological advances that advanced materials are introducing in technical designs, to achieve a successful shielding of electromagnetic interference. In addition, a comparative study of the different materials that have been proposed is carried out in order to guide the researchers and companies in the selection of the most suitable material to achieve the shielding to the electromagnetic interference for diverse applications giving a specific emphasis to space applications such as satellites, aircraft, international space station, etc. The application of macro, micro- and nanomaterials is analyzed and discussed in detail in order to identify the role that advanced materials have had in the optimization of electromagnetic compatibility properties. The chapter has been divided as follows: Firstly, the section called Background will describe the basic concepts associated with the phenomenon of electromagnetic interference. In addition, the importance of designing materials that can reduce the effects of electromagnetic interference in aerospace applications are discussed in the section called Controversies and Problems. Subsequently, the impact of advanced materials on shielding to electromagnetic interference is described in the section called Solutions and Recommendations. In the section called Directions for Future Research, a summary of possible lines of research that will be developed in the near future by researchers around the world is presented. Finally, the Conclusions section discusses the authors' views on the study of advanced materials used in shielding against electromagnetic interference.

BACKGROUND

An electromagnetic radiation can be defined as a kind of radiation such as visible light, radio waves, gamma rays, and/or X-rays, in which electric and magnetic fields vary simultaneously (Tong, 2009). In physics, this type of radiation is an electromagnetic field wave propagating through the free space or through a gaseous, liquid or solid medium carrying radiant energy. This wave can be visualized as a self-propagating transverse oscillating wave from left to right, as shown in Figure 1. The electric and magnetic fields in the electromagnetic waves are always in phase, and both at 90 degrees from each other. The electric field ($\vec{E}$) is in the vertical plane, and the magnetic field ($\vec{B}$) in the horizontal plane. Electromagnetic waves carry energy, momentum or angular momentum away from their source and can impart these quantities to the matter with which they interact. The effects of electromagnetic radiation on materials as well as biological organisms depends on the radiation power and frequency of electromagnetic radiation.

The electric field of electromagnetic radiation in an electrical material induces an electric current that causes a displacement of electric charge within the material to which the radiation is applied (Tong, 2009). The magnetic fields of the electromagnetic radiation generate Eddy currents that cancel the applied magnetic field. Ideally, if the material completely shields electromagnetic radiation, then the internal electromagnetic fields remain inside and the external electromagnetic fields are left out. In real materials, several technical factors limit the ability of electromagnetic shielding among which we can mention: 1) the electrical resistance prevents the complete field cancellation, 2) the ferromagnetic response to low frequency magnetic

Figure 1. The electromagnetic waves composing an electromagnetic radiation

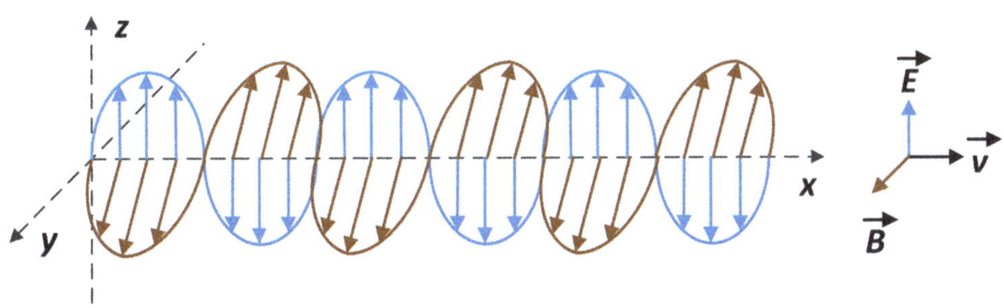

fields is not completely attenuated, 3) high frequency electromagnetic radiation is absorbed into the material unless it is extremely thin.

Electromagnetic shielding allows partial or complete wrapping of electric and magnetic fields generated by an electronic system or to which it may be susceptible from the outside (Tong, 2009). Fortunately, an extensive set of materials have been and are being designed as electromagnetic interference shields to meet the geometries, magnetic permeabilities, and electrical conductivities required. Shielding in addition to reducing internal electromagnetic radiation also allows an isolated grounded reference to attenuate internal crosstalk and global common mode coupling, which in electronic systems decrease the transfer of electromagnetic radiation from the outside to the interior of a system. Also, the shield is used to limit electrostatic dissipation (ESD) and electrostatic potential (ESP) as well as to absorb microwave energy, which also ensure the effectiveness of electromagnetic shielding.

Electromagnetic compatibility (EMC) is the property of a material, device or system to be compatible with its electromagnetic environment and does not emit levels of electromagnetic energy that cause electromagnetic interference (EMI) in other neighboring materials, devices or systems. Any material, device or system can be vulnerable to EMI if the electromagnetic energy levels in its environment exceed the electromagnetic immunity or resistance to the electromagnetic fields to which it was designed or tested. Forms of electromagnetic energy that can cause electromagnetic interference are conducted, radiated, and by electrostatic discharge (ESD). The natural space environment such as radiation, plasma, and cosmic rays, or the occurrence of geomagnetic storms can cause potential disruptive effects that may lead to electromagnetic incompatibility. The unwanted electromagnetic interaction in aerospace applications has increased due to the excessive increase in the number of electronic systems using electrical signals with higher bandwidths, as these are accommodated in a very small physical space along with other equipment. Among the effects that the electromagnetic incompatibility has on the space systems are the temporary interruptions of telemetry, noisy data from onboard science, permanent damage of the power supplies, accidental mistake of the circuit protection devices, false commands, instability of the power distribution systems, etc. In order to ensure the electromagnetic compatibility of space systems it is necessary to develop coatings with advanced materials that produce a shielding to electromagnetic interference in the devices or systems. Materials that have been used to produce electromagnetic interference shielding are of a different nature: metals, ceramics, polymers, and composites. Their operation has been optimized through the use of alloys or combinations of these materials (composites, advanced materials or functional materials), using more than one material of the same group or at least two materials of different nature. Additionally, with the introduction of nanomaterials in the development of high added value applications, and their multiple properties such as high surface to volume ratio, good electrical properties, excellent mechanical properties, interesting magnetic properties, and outstanding thermal properties, different options for achieving electromagnetic compatibility for various technological applications have been proposed by different researchers around the world.

Shielding effectiveness (*SE*) can be quantified by means of the attenuation of the electric and/or magnetic fields as well as the resistance to the waves in the plane produced by the material used in the shielding (Tong, 2009). The attenuation of the electromagnetic field depends on the frequency, the distance from the shield to the generating source, the thickness of the shielding material, and the material used as shielding. Shielding effectiveness is usually expressed in decibels (dB) as a function of the logarithm of the ratio of the intensity of the incident (or input) electromagnetic field to the intensity of the output electromagnetic field. Shielding effectiveness using mathematical expressions can be defined as:

$$SE\left(dB\right) = 20log\left(\frac{E_{out}}{E_{in}}\right) \quad (1)$$

$$SE\left(dB\right) = 20log\left(\frac{B_{out}}{B_{in}}\right) \quad (2)$$

$$SE\left(dB\right) = 20log\left(\frac{F_{out}}{F_{in}}\right) \quad (3)$$

where E represents the magnitude of the electric field, B the magnitude of the magnetic field, F the strength of the plane wave field, and the *out* and *in* subscripts are used to denote the output and input, respectively. Electromagnetic interferences present three mechanisms that determine the effectiveness of a shield: 1) reflection from the front surface of the material, 2) absorption inside the material, and 3) reflection from the back surface of the shield to the front, as illustrated in Figure 2.

The total shielding effectiveness of a material (*SE*) can be mathematically defined as the sum of the absorption factor (*A*), the reflection factor (*R*), and the correction factor (or multiple reflection factor, *B*), all expressed in dB, as follows:

$$SE = A + B + R \quad (4)$$

Figure 2. Multiple reflections of electromagnetic waves presented by a coating for shielding to electromagnetic interference

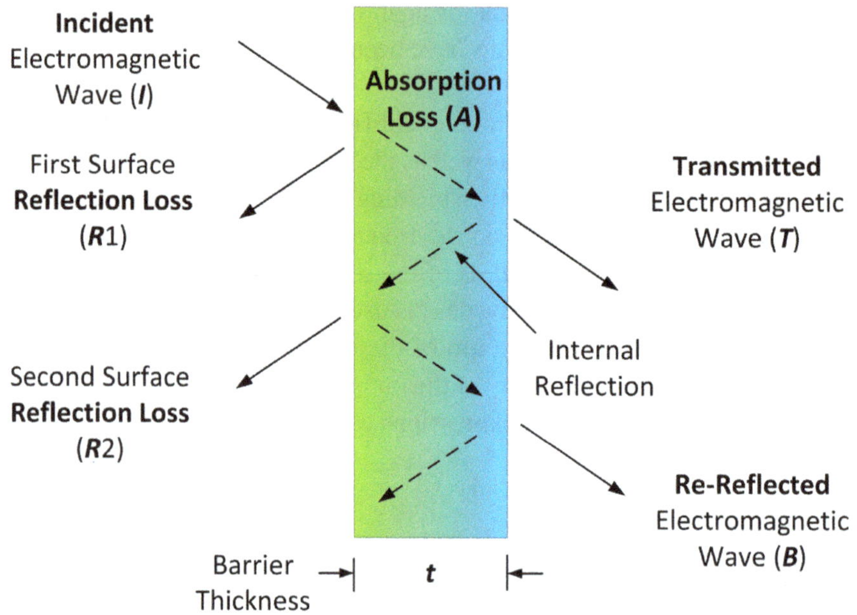

The multiple reflection factor *B* is considered insignificant when the absorption loss *A* is greater than 10 dB. In practical calculations, the *B* factor can also be ignored for electric fields and plane waves. The absorption loss *A* is a function of the physical characteristics of the shield and it is independent of the type of electromagnetic source applied (Tong, 2009). In mathematical terms, this can be expressed in dB as:

$$A = 20\left(\frac{t}{\delta}\right) log\left(e\right) = 8.686\left(\frac{t}{\delta}\right) = 15.395t\sqrt{f\mu\sigma} \quad (5)$$

where *t* is the thickness of the shield in cm, δ is defined as skin depth or distance required for the wave to be attenuated to 1/*e* and mathematically expressed as $\delta = \frac{1}{\sqrt{\pi f \mu \sigma}}$, *f* is the frequency in MHz, μ is the magnetic permeability, σ is the electrical conductivity. According to the mathematical expression of the skin depth, this decreases with the increase in the frequency, electrical conductivity and/or magnetic permeability of the electromagnetic radiation. The multi-reflection correction factor can be positive or negative and becomes insignificant when the absorption loss A is greater than 6 dB (Tong, 2009). This can be expressed mathematically in simplified form as:

$$B = 20log\left(1 - e^{-\frac{2t}{\delta}}\right). \tag{6}$$

The loss by reflection is produced due to the relative mismatch between the incident radiation and the surface impedance of the shield. This mathematically can be expressed as the sum of three terms: 1) reflection of electric fields (R_E), 2) reflection of magnetic fields (R_B), and 3) reflection of plane wave fields (R_p), as follows:

$$R_E = 321.8 + 10log\left(\frac{\sigma}{f^3 r^2 \mu}\right) \tag{7}$$

$$R_B = 14.6 + 10log\left(\frac{fr^2\sigma}{\mu}\right) \tag{8}$$

$$R_p = 168 - 10log\left(\frac{f\mu}{\sigma}\right) \tag{9}$$

where σ is the relative conductivity related to copper, f is the frequency in Hz, μ is the relative permeability related to free space, and r is the distance from the source to the shield in m.

The absorption loss is proportional to the product $\sigma_r\mu_r$, whereas the loss by reflection is proportional to the ratio σ_r/μ_r, where σ_r is the volumetric electrical conductivity relative to copper and μ_r is the relative magnetic permeability (Chung, 2000). Electrical metals such as silver, copper, gold and aluminum mainly exhibit the reflection mechanism due to their high electrical conductivity. On the other hand, magnetic materials such as superpermalloy (nickel-iron-molybdenum alloy) or mu-metal (nickel-iron alloy) mainly exhibit the absorption mechanism due to their high magnetic permeability. The mechanism of reflection loss decreases with increasing frequency, while the mechanism of absorption loss increases with frequency. The mechanism of multiple reflections involves the presence of a large surface or interface area, which can be achieved through foamed or porous materials. The mechanism of multiple reflections involves the presence of a large surface or interface area, which can be achieved through foamed or porous materials commonly presented in composite materials. This loss mechanism can be considered insignificant when the distance between the reflecting surfaces or interfaces is comparatively large with the skin depth. In order to illustrate the behavior of the metallic materials with respect to the mechanisms of electromagnetic shielding, the behavior of coatings with a thickness of one thousandth of an inch separated by a length of 1 meter of the source of electromagnetic radiation, as shown in Figure 3.

Figure 3. Prediction of the shielding effectiveness for metallic materials such as copper, nickel and stainless steel 4340 for different frequencies: (a) absorption loss (A), (b) multiple reflection loss (B), (c) electric field reflection loss (R_E) (d) magnetic field reflection loss (R_B), (e) plane wave reflection loss (R_P), and (f) total shielding effectiveness (SE)

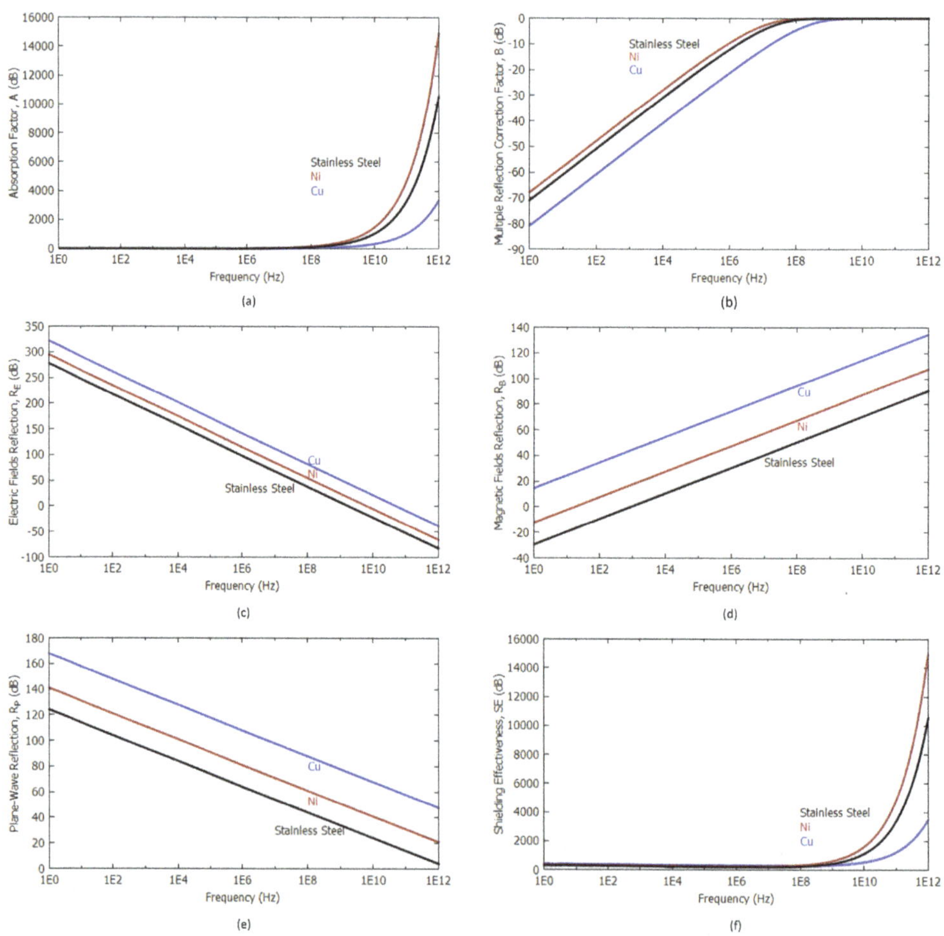

SHIELDING EFFECTIVENESS

Issues, Controversies and Problems

Metals are by far the most common materials for shielding from electromagnetic interference (Chung, 2000). They mainly operate by the reflection mechanism because their electrical carriers are the free electrons found in them. The sheets of metal used as shielding coatings are bulky, and are made by electroplating, electroless plating, or vacuum deposition. The coatings are made of bulky materials, fibers or particles. These tend to experience light wear and scratch resistance. The mechanism of absorption to the electromagnetic radiation can be presented in the metals, when fillers with high dielectric constants and high magnetic permeabilities are aggregated. In the manufacturing process, electromagnetic shield designs based on metals always contain joints, seams, openings or holes for access and ventilation, or they allow the attachment of wires or pipes for signaling or services, which reduce the effectiveness of electromagnetic shielding. The metal-based packaging of a system due to its design tends to cause loss of radiation and decreases the effectiveness of electromagnetic shielding.

In order to meet the technical requirements of polymer materials used for aviation components, mechanical properties (specific strength, creep resistance), thermal properties (glass transition temperature, melting temperature, long-term service temperature, short-term service temperature, negative service temperatures, linear thermal expansion coefficient), electrical properties (surface resistance, specific volumetric resistance, dielectric strength, dissipation factor, comparative tracking index), radiation resistance (ionizing radiation, electromagnetic radiation, ultraviolet radiation), combustibility (fire load, smoke, toxicity, degassing) and chemical resistance must be controlled. In order to achieve these properties it is necessary to add to the polymeric materials some suitable filler in the necessary proportions.

The efficiency of a material used as shielding to electromagnetic interference is directly related to the ratio surface to volume of the material, and which is greater with the use of nanomaterials as electromagnetic fillers. The unit size of the filler should be comparable to or less than the skin depth of the material used in the shielding (Chung, 2000). A filler of unit size of 1 micron or less is typically preferred, although small unit size is not commonly available for most fillers and the dispersion of these fillers in a composite or hybrid material is very difficult when the unit size of the filler is smaller. Metal-coated polymer particles or fibers are used as materials for electromagnetic shielding; however, they do not have a high shielding efficiency because most polymers regularly exhibit diamagnetic behavior.

At the volumetric level, metals have better shielding effectiveness than carbon based materials due to their high electrical conductivity (Chung, 2000). However, carbon based materials have better oxidation resistance and thermal stability than metals. In addition, fibers with small diameter are more attractive for shielding than particles because of their high aspect ratio. Particularly, electronic enclosures and aircraft use polymer matrix composite materials reinforced with carbon-based continuous fibers coated with nickel or interleaved with others based on metals to increase electrical conductivity.

Solutions and Recommendations

Polymer matrix composites containing electrically conductive fillers are attractive to electromagnetic shielding because of their processability, which helps to reduce or eliminate the seams, joints, openings or voids used in the design of the encapsulation of the system being shielded. Polymer matrices are regularly electrically insulating and do not contribute to electromagnetic shielding, since they can affect the connectivity of the electric filler to be incorporated. Greater connectivity between the fillers improves the effectiveness of the material to be used in the shielding. The use of electrically conductive polymeric materials has the advantage of being able to connect the added fillers without touching one another, and therefore, with improved connectivity it is possible to increase the efficiency of the shielding. Low filler content is desirable for several reasons: cost savings, weight reduction, high strength and ductility, and improved processability. Electrically conductive filler is most effective when it has small unit size, high electrical conductivity, and a high aspect ratio.

The effectiveness of the electromagnetic shielding of a composite material can be mathematically determined by means of the Simon formula expressed as:

$$SE = 50 + 10log\left(\frac{\sigma}{f}\right) + 1.7t\sqrt{\sigma f} \tag{10}$$

where *SE* is given in dB, σ is the bulk conductivity in S/cm, *t* is the thickness of the sample in cm, and *f* is the frequency of the applied electromagnetic radiation in MHz. Shielding effectiveness of a composite material based on polymer and carbon nanotubes has been determined using the Simon formula for materials with different thicknesses (from 10 to 200 microns) with an electrical conductivity of 300 S/cm (see Figure 4 (a)), and for materials with different electrical conductivities (from 1 S/cm to 100 S/cm) with a thickness of 50 microns (see Figure 4 (b)). The shielding effectiveness against electromagnetic interference of a composite material depends on the aspect ratio, electrical conductivity, dielectric constant, and magnetic permeability of the filler used.

Figure 4. Shielding effectiveness for a composite material based on polymer and carbon nanotubes: (a) for different coating thicknesses with the same electrical volumetric conductivity, and (b) for different electrical conductivities with the same thickness

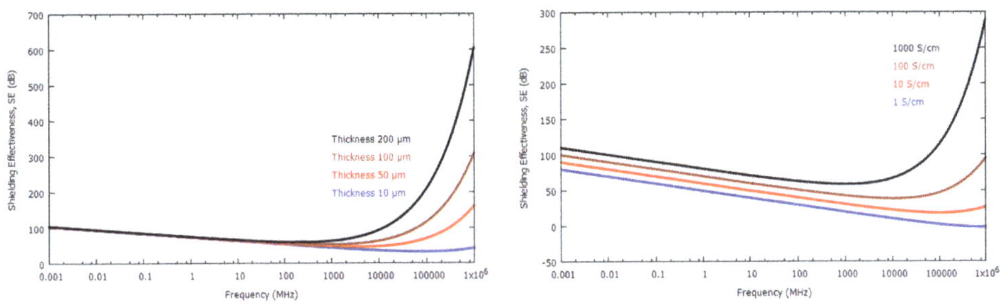

(a) (b)

The aerospace industry requires high performance plastics with ease of processing, reduced weight, good chemical resistance, inherent flame resistance, good radiation resistance, and low outgassing in vacuum. A high-performance plastic can be defined as a plastic that meets much higher requirements than standard or engineering plastics. These plastics exhibit good stability to temperature, chemical resistance, superior mechanical properties, production quantity, and price. A summary of the classification of plastics that can be used to design advanced materials for electromagnetic shielding is illustrated in Figure 5. Advanced materials for electromagnetic shielding are used in high speed and/or high altitude applications where electronic equipment is vulnerable to interference for applications in manned and unmanned aircraft (both civilian and military) with fixed or rotating wings, for aerospace applications, missiles, rockets, dirigibles, and racing vehicles. The physical properties of high performance plastics can be improved by the addition of stabilizers or reinforcing materials. Particularly, different fillers may be added to favor shielding to electromagnetic interference by any of the aforementioned mechanisms.

Electromagnetic Compatibility (EMC) Tests

To approve the use of military and aerospace components in a specific application, they must be tested for electromagnetic compatibility (EMC) (Pettit, 2017). Electromagnetic emissions and immunity are called susceptibility in the military and aerospace fields. Emission tests are run to determine how much electromagnetic energy the product emits. Immunity tests are run to determine that the product will operate as intended while subjected to various sources of electromagnetic energy in its intended operating environment. A higher level of protection from

Figure 5. Classification of engineering plastics

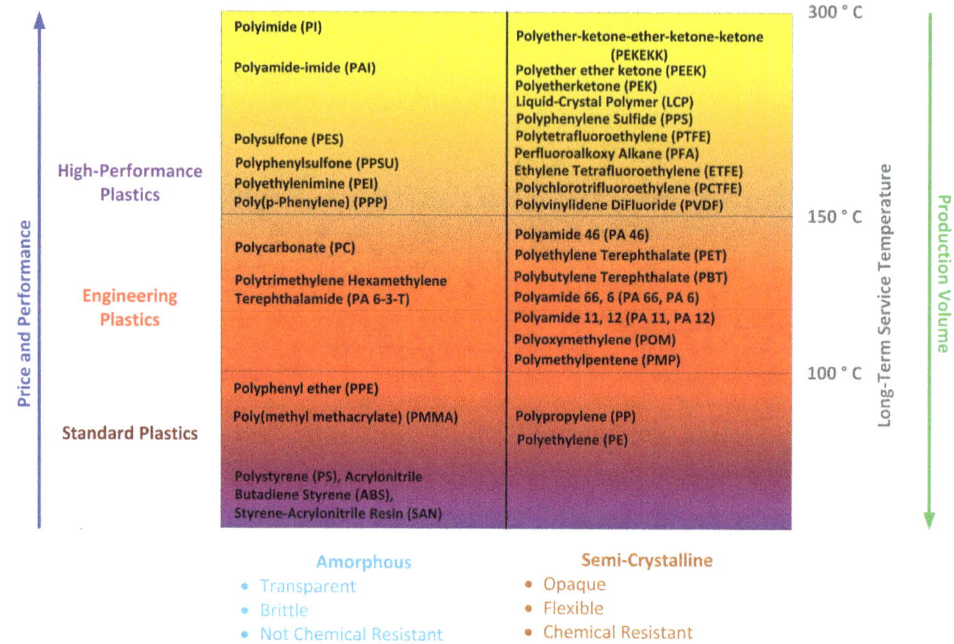

interference in military and aerospace fields is necessary with respect to commercial field. Whether these specified test levels are adequate or not is a subject open to debate. Although the standards, including limits and test methods may differ, all electromagnetic compatibility test standards have a few issues in common. The MIL-STD-461G standard (MIL-STD-461G, 2015) provides requirements at the equipment and subsystem level. It divides test requirements into 4 basic types: conducted emissions (CE), conducted susceptibility (CS), radiated emissions (RE) and radiated susceptibility (RS). In particular, susceptibility tests are necessary for these types of coatings. Some examples of section of the standard that can be applied are described below. CS101 conducted susceptibility is applied from 30 Hz to 150 kHz for equipment and subsystem AC and DC power input leads. CS103, CS104 and CS105 conducted susceptibility, antenna port, intermodulation, rejection of undesired signals and cross-modulation are applied according to the range of operating frequencies of the systems. CS109 conducted susceptibility in structure current is applicable from 60 Hz to 100 kHz. CS114 conducted susceptibility in bulk cable injection is applicable from 10 KHz to 200 MHz for all electrical cables interfacing with the EUT (Equipment Under Test) enclosures. CS115 conducted susceptibility for bulk cable injection and impulse excitation. CS117 conducted susceptibility for Lightning induced transients, cables and power leads. For space systems including launch vehicles are applied CS101, CS114, CS115, and CS117. The ISO 14302 standard (ISO 14302, 2002) provides performance requirements for the purpose of ensuring space systems electromagnetic compatibility (EMC). The ISO 24637 standard (ISO 24637, 2009) provides specific requirements for the equipment-level electromagnetic interference (EMI) test report for integration analysis of complex space systems.

Practical Guidelines for Selecting Materials Used as Coatings

Materials having constituents with atomic/molecular and nuclear cross sections, and therefore with lower atomic mass, have a better effectiveness to electromagnetic shielding (Wilson, 1995). A spacecraft designed with high performance structural material, such as a high hydrogen content carbon composite material, has effective radiation shielding properties (Musgrave, 2009). Traditionally, metals have been used as materials for passive-type electromagnetic shielding for space applications; however, these materials are very heavy. Recently, highly hydrogenated light polymeric materials have been introduced (Shavers, 2004). An attractive technological alternative to passive coatings are active coatings whose surfaces can deflect electromagnetic waves, which can be reached thanks to nanomaterials (Spillantini, 2007). Coatings for electromagnetic shielding must meet other properties such as withstanding thermal shocks and hazardous environments, compatibility with the substrate, as well as being chemically inert. Currently, a lot of coating materials options are available for simple metals, metal alloys or metal based compounds; polymer matrix composite materials reinforced with ceramic materials or metals; metal-reinforced ceramic matrix composites; hybrid materials or heterostructures based on combinations of the foregoing materials. The coating will be exposed to radiated electromagnetic interference so that its performance will depend on either the purpose of the space component or system or the nature of the resulting effect. The greater the value of the shielding effectiveness (SE) of the coating, the more effective it is to withstand electromagnetic interference (EMI). For example, at an altitude of 100 miles with orbit inclinations of 57° and 90°, electromagnetic fields as much as 200 V/m for the frequency range from 2 to 3 GHz imply a shielding effectiveness of not less than 20 dB. According to ASTM ES7-83 (ASTM ES7-83, 1988), a material used as a coating for electromagnetic shielding should have high shielding (SE) effectiveness for electromagnetic signals in the frequency range 30 MHz to 18 GHz.

Shielding Materials Based on Carbon Nanomaterials

Common materials used as shielding to electromagnetic interference are metals. The main carbon fillers that have been used for the development of materials for electromagnetic shielding are carbon black, carbon fibers, carbon nanotubes, and graphene (Thomassin, 2013). High performance polymers are specially designed by different companies to withstand extreme temperatures for the manufacture of aeronautical parts such as bearings, cable covers, seals and foams, plastic structures for passenger seats, etc. Currently, composite materials based on polymers, in addition to providing shielding, they possess interesting properties such as low density, corrosion resistance, flexibility and processability excellent. Particularly, graphene represents very promising conductive electrical filler for the fabrication of materials for shielding to electromagnetic interference, due to its high electrical conductivity, high specific surface area as well as a high aspect ratio.

One of the most prominent materials for the development of electromagnetic interference shielding is the polymer-nanomaterial carbon composite foam. This material has a low density, and a mechanical compressibility which allows an adjustable shielding to electromagnetic interference. Through the foaming it is possible to reduce the density of the shielding material, and it is possible to increase the attenuation to the microwaves through the promotion of multiple reflections within the porous material. Carbon nanomaterials offer conductive electrical dissipation as well as dispersion and multiple reflections towards applied electromagnetic waves. To form a conductive electrical network, high contents of carbon nanomaterials in the order of 5 to 30 weight percent should be added to polymers to form composite materials with shielding capabilities. The polymer foams commonly used for the shielding materials are polypropylene (PP), polymethylmethacrylate (PMMA), polystyrene (PS), polycaprolactone (PCL), polyurethane (PU), polyvinylidene fluoride (PVDF), polyimide (PI), and poly(dimethyl-siloxane) (PDMS), as it is summarized in Table 1.

Most of these materials cannot be used in applications in the electronics industry where in addition to the electromagnetic field; properties such as high heat resistance, excellent flame retardancy, and low smoke generation are required. The fillers to be added to the composites must have electrical and/or magnetic properties to be used in materials for electromagnetic shielding as noted above. For example, carbon nanomaterials have high electrical conductivities because they exhibit mechanisms of electric charge transport in both ballistic and diffusion forms, as shown in Figure 6 (Vargas-Bernal, 2018).

The integrated circuits are sources of electromagnetic interference which must be packaged with coatings to shield this interference (Vargas-Bernal, 2015). Composite materials based on polymers and carbon nanotubes or polymers and

Table 1. Materials for electromagnetic shielding based on polymers and carbon nanomaterials.

Material	EMI SE (dB)	Density (g/cm^3)	Carbon nanomaterial content wt %	Reference
PMMA/Graphene	13-19	0.79	5	Zhang, 2011
PS/Graphene	17	0.27	30	Yan, 2012
PI/Graphene	17-21	0.28	1-16	Li, 2015
PI/Graphene	24	0.40	1-4	Li, 2016
PU/Graphene	15-22.5	0.027-0.030	10	Shen, 2016
PP/Carbon Fiber	17-24	1.8	7-8.5 vol%	Ameli, 2013
PP/Carbon Nanotubes	1-35	1.75	1-7.5 vol%	Al-Saleh, 2009
PVA/EVA/Paraffin /Graphene	27	-	1-40%	Song, 2014a
PDMS/Graphene/Carbon Nanotube	75	0.97	0.28-2.7	Sun, 2016
PDMS/Graphene/Carbon Nanotube	55	1.2	5	Kong, 2014
Epoxy Resin/Carbon Fiber/Carbonyl Iron Powder	8.6-53.9	-	-	Hu, 2015
PVDF/Carbonyl Iron Powder	20	0.9	50 vol%	Joseph, 2013
Wax/Highly Ordered Porous Carbon (HOPC)	50	-	20	Song, 2014b
PMMA/Fe_3O_4/Carbon Nanotube	13-50	0.22-0.38	7	Zhang, 2017a
Poly(3,4-ethylenedioxythiophene) -poly(4-styrenesulfonate)/Graphene/Carbon Nanotube/Fe_2O_3	90-130	-	-	Lee, 2017

Figure 6. Mechanisms of transport in carbon nanomaterials: (a) ballistic and (b) diffusive (Vargas-Bernal, 2018)

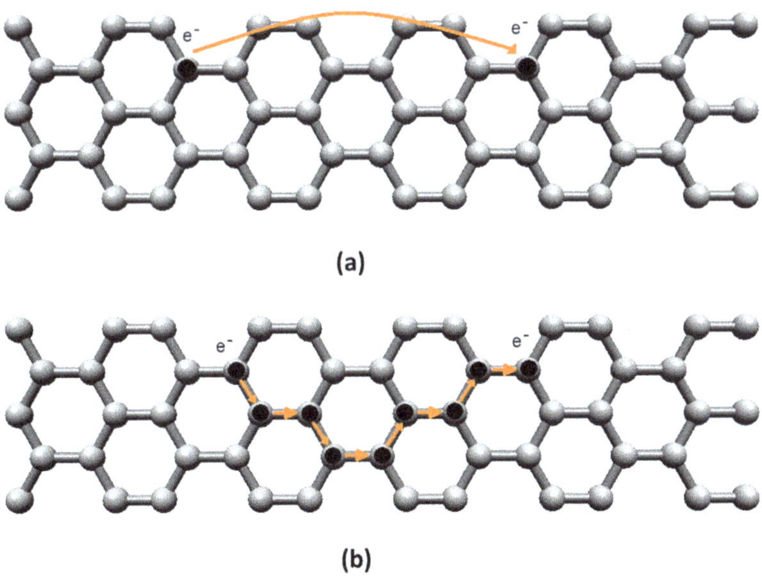

graphene have been proposed to perform this function. A high effective shielding involves the reflection of free mobile charge carriers from outside, the absorption of mobile charge carriers produced inside as well as the alignment of the electric and magnetic dipoles within the shielding material. An excessively large proportion of the electromagnetic radiation found in aerospace applications is produced by electronic circuits or electrical systems included in satellites, space shuttles, the international space station, and in airplanes. The shielding in these applications reduces the coupling between the signals, crosstalk between electrical components, noise in wires or cables, interference produced by analog and/or digital communication systems, etc. The frequency of operation of the devices in the integrated circuits as well as the frequency of the communication systems is in the order of the THz, and therefore materials in the size of nanometers or picometers must be developed to achieve the shielding to the electromagnetic interference required (Liu, 2014).

Carbon nanomaterials increase the effectiveness of electromagnetic shielding in a composite material through the absorption mechanism as the main shielding mechanism and reflection is the secondary shielding mechanism due to its electrical properties (Al-Saleh, 2009). In addition, the multiple reflection mechanism presented on the inner surfaces of the carbon nanomaterials and between the outer surfaces of nanomaterials decreases the shielding effectiveness to electromagnetic interference of the resulting composite material. In particular, composite materials based on polypropylene and multi-wall carbon nanotubes increase their shielding effectiveness by increasing the content of carbon nanotubes and the thickness of the shielding material (Al-Saleh, 2009). Injection molding of composite foams based on polypropylene and carbon fibers can be used to design electromagnetic shielding materials for applications in the electronics, aerospace and automotive industries (Ameli, 2013). These materials have improved shielding effectiveness thanks to the increase in electrical conductivity and dielectric permittivity by the addition of carbon based materials. Composite materials including carbon nanomaterials are chemically inert, water repellent and have the self-cleaning ability that can be used in the manufacture of cost effective and flexible electromagnetic shielding materials (Liu, 2014). In addition to high frequency electronic applications it is possible to design shields for quasi-optical components used in sensing and imaging systems, such as polarizers and filters.

Another of the properties that must be satisfied by the composite materials applied to electromagnetic shielding is the thermal stability that can be achieved by using high performance polymers in the design of composite materials such as polysulfone (PSU) (Nayak, 2015). In this work, the performance in electromagnetic shielding of composite materials based on PSU and carbon nanomaterials was verified for three different combinations: PSU and carbon nanotubes, PSU and carbon fibers, and PSU reinforced with carbon nanotubes and carbon nanotubes.

The composite material with PSU and carbon fiber presented greater effectiveness to the shielding. The composite material with PSU/carbon nanotubes/carbon fiber had the best thermal stability.

Composite materials based on epoxy resin and multiple wall carbon nanotubes at 10 wt% as a filler had similar shielding properties than those of a metallic aluminum enclosure at frequencies in GHz (Micheli, 2011). The dominant mechanism of the composite material is electromagnetic absorption due to high electrical conductivity, which makes it attractive for use in electromagnetic shielding. In addition, the multifunctional properties of carbon nanotubes allow high thermal conductivity and flame retardancy, which produce an interesting combination for space applications. These materials can attenuate the electric field within a satellite and thereby improve its immunity against electromagnetic interference (Micheli, 2013). Other applications that can benefit from electromagnetic absorption are the anechoic chambers, the protection against lightning strikes, and the insulation of thermal blankets. Metal nanoparticles can be incorporated into these composite materials to tune the electromagnetic reflectivity of satellite surfaces in specific frequency ranges especially for those surfaces that must act as antenna radiation patterns for communication systems (Micheli, 2017). The metals studied were silver (Ag), cobalt (Co), nickel (Ni), titanium (Ti), zinc (Zn), copper (Cu), iron (Fe), boron (B), bismuth (Bi), and hafnium (Hf). The results showed that the materials with higher electrical permittivity (those that formed ceramic materials) present a higher electrical conductivity which favors the improvement in the properties of electromagnetic shielding.

The use of sandwich structures based on two-dimensional structures, films or thin sheets with embedded nanoscale materials has allowed the development of materials with higher electromagnetic shielding effectiveness (Song, 2014a). Graphene thanks to its two-dimensional structure favors the absorption mechanism with the increase in the thickness of the shielding material as well as the electrical conductivity of the hybrid material formed between the paraffin and the polymer-graphene composite material (whose dominant mechanism is reflection).

Graphene-carbon nanotubes/poly(dimethyl siloxane) composites have exceptional microwave shielding because their constituent materials with multiple scales favor percolation and high porosity increases the effectiveness of electromagnetic shielding (Sun, 2016). The shielding effectiveness values achieved by these materials are high, and these materials have low density and are very attractive for aerospace components. Thanks to the high porosity achieved, the microwaves can be dissipated and attenuated in the nanoscale materials through the dissipation of surface currents and polarization in the electric field. These materials are electromagnetic wave absorbers in civil and military applications in the range of gigahertz due to their intensified interfacial polarization and impedance matching that can be optimized by the dispersion of the filler into 3D hybrids (Kong, 2014).

In addition to carbon nanomaterials, metal layers are used as coatings of carbon nanomaterials, metal nanoparticles or metal oxides are co-added to composites, and even intrinsic conductive polymers are used to modify carbon nanomaterials or co-added directly to the composite (Thomassin, 2013). Materials such as carbonyl iron powder (CIP) have been used as metallic magnetic fillers with stable shielding properties by absorption (at 50 vol%, shielding effectiveness of 20 dB) in PVDF-carbon nanotubes composites particularly in the X-band (8.0-12.0 GHz) for medical, spatial and military applications (Joseph, 2013). Carbonyl iron powder-carbon fiber/epoxy resin composites made of layers have a better shielding effectiveness when the layer angle is very small in the X-band because the combined effect of both fillers cancels the effects of absorption and reflection with the increase in the angle (Hu, 2015).

PMMA/Fe_3O_4@MWCNTs composite foams benefit from the microcellular structure and porosity to increase shielding effectiveness to electromagnetic interference significantly, which makes them attractive for use in the electronics and aerospace industries (Zhang, 2017a). These foams absorb electromagnetic radiation with less reflection. Ideal fillers for electromagnetic shielding must provide both dielectric loss and magnetic loss. The property of electromagnetic absorption of the carbon nanomaterials is very poor, therefore, when adding magnetic ceramic materials such as Fe_3O_4, in addition to increase the electrical conductivity (reflection mechanism) also favors the absorption of the electromagnetic waves.

Thin and flexible films made from polymer-based composites such as poly(3,4-ethylenedioxythiophene)-poly(4-styrenesulfonate) and reinforced with nanomaterials based on graphene, carbon nanotubes and iron oxide (Fe_2O_3) have shown excellent shielding effectiveness (Lee, 2017). These materials have a seamlessly networked hetero-nanostructure with a low electrical contact resistance and high electrical conductivity that optimizing the shielding effectiveness. In addition, they possess high cyclic stability under repetitive bending conditions, which is required in some structural applications.

Shielding Materials Using Ceramic Matrices

In aerospace applications, high temperature components require advanced materials with high mechanical resistance, oxidation resistance, low density, good thermo-chemical stability, as well as electromagnetic shielding capabilities derived from their electrical properties. Carbides and ternary nitrides arranged in layers with chemical formula M_3BX_2 (312 phases) and M_2BX (211 phases), where M is a transition metal,

B is an element of groups IIIB or IVB, and *X* is either C and/or N are good electrical conductors. These can be used as electrical fillers in composite materials for shielding to electromagnetic interference. Ceramic composites such as SiC_f/SiC require the use of fillers such as Ti_3SiC_2 to optimize dielectric properties in electromagnetic shielding applications in the temperature range of 25-600 °C at 8.2-12.4 GHz (Mu, 2015). The dominant mechanism of effectiveness to electromagnetic shielding is absorption, which depends directly on temperature (high temperatures are preferred) because the electrical conductivity of the composite is increased when dielectric filler is used.

The electromagnetic interference shielding of composites based on carbon fiber reinforced silicon carbide (C/SiC) is enhanced with the introduction of carbon nanotubes on the carbon fibers by electrophoretic deposition (EPD) thanks to the achieved electrical conductivity (Mei, 2016). These materials meet the specifications of electromagnetic shielding due to the absorption and reflection mechanisms for commercial aeronautical and aerospace military applications. In addition, ceramic-based composites have low density, high strength, high fracture toughness and high oxidation resistance, which are additional supplementary requirements that are required for some aerospace components.

Carbon fiber reinforced multilayered pyrocarbon-silicon carbide matrix composites represent an interesting alternative to develop materials for electromagnetic shielding through of the control of the number PyC-SiC sequences by means of layer-by-layer deposition (Jia, 2017). Shielding effectiveness of the composites is increased with the increment of PyC-SiC sequences number due to the increasing electrical conductivity and polarization of the multilayered matrix. The high shielding effectiveness combined with low density and good mechanical properties of $C/(PyC\text{-}SiC)_n$ composites allow the manufacture of lightweight and high-performance structural components.

Shielding Materials Using Biomimetic Materials

Artificial hybrid films similar to nacre composed of two-dimensional reduced graphene oxide (RGO) and magnetic graphene (MG) as inorganic building blocks as well as polyvinyl alcohol (PVA) as a glue have been developed as materials for shielding against electromagnetic pollution (Yuan, 2014). Magnetic graphene is obtained by decorating graphene with magnetic nanoparticles such as gamma iron oxide Fe_2O_3, which absorb microwaves. Graphene oxide produces dielectric loss while magnetic graphene produces magnetic loss, therefore, the resulting composite material shows higher shielding effectiveness. These materials can be considered promising candidates (thin, flexible and lightweight) for electromagnetic shielding in aviation and aerospace industry applications.

The influence of nanometric electrical and/or magnetic ceramic fillers has been studied through the manufacture of paraffin-based composites and nanospheres made of α-Fe_2O_3, γ-Fe_2O_3 and Fe_3O_4 (Wu, 2015). Composite materials formed with the first type of nanosphere presented better shielding to electromagnetic interference than composite materials based on the last two types of nanospheres. The mechanism favored was electromagnetic absorption, and the resulting materials are lightweight for use in practical applications.

Highly ordered porous carbon (HOPC) behaves as an effective electromagnetic absorber thanks to the large surface area and the three-dimensional conductive interconnections even when it is included in a composite as filler in low percentages (Song, 2014b). These materials have a competitive performance for electromagnetic attenuation and shielding when they are compared with carbon nanotube- and graphene-based composites with a much higher filler content. The three-dimensional conductive framework allows the dissipation of the electrical current induced by the microwave which attenuates the electromagnetic waves and generates an ideal impedance match condition required for producing the electromagnetic shielding.

Composite materials based on bagasse fiber and polyaniline (PANI) uniformly deposited on the fiber have structural, electrical and electromagnetic properties, which make them attractive for electromagnetic shielding. The dominant mechanism is the electromagnetic absorption derived from the use of a core-shell architecture, which improves the electrical conductivity of the resulting material (Zhang, 2017b).

Coatings used as electromagnetic interference shields are more easily processed in films and if these are flexible and show electrical conductivity then these are highly desirable. Two-dimensional materials such as carbides and nitrides ($Ti_3C_2T_x$) known as MXenes have high electrical conductivity and a hydrophilic surface that can be exploited in the design of polymer matrix composites for electromagnetic shielding (Shahzad, 2016). If the films were standing suspended the dominant mechanism is the multiple reflection, however, being incorporated in a composite material, the dominant mechanisms are the simple absorption and reflection. The authors used as a polymer matrix sodium alginate, a material that allowed to reach levels of shielding effectiveness of 90 dB, at using 10 wt% of the MXene.

A comprehensive study of two-dimensional materials that can be used either in composite materials or in hybrid materials to develop materials for electromagnetic shielding has recently been published (Vargas, 2017). The author of this paper describes the electrical properties of these materials specifically for the design of integrated circuits, but these materials can be exploited in space applications. Unfortunately, the study of these materials is in its infancy and much effort must be made in the coming decades to apply them at a practical level.

DIRECTIONS FOR FUTURE RESEARCH

The advancement in electromagnetic shielding materials has been quite good so far, however, as described above, material engineering researchers around the world will continue to develop new materials, which are based on different combinations of available materials where it will be sought to control either the magnetic and/or electric effect on them. Many of the advanced materials that have been developed and are being developed will definitely drive the manufacture of different parts of systems for space applications in the coming decades.

In the space area, the main applications of the advanced materials are in the satellites (antenna covers, sliding elements, bearings) and in the construction and insulation components (tube holders, fixing elements, torque cylinder, radar covers, sealing rings, wire coils). In the aviation area, the main applications of functional materials are in aerostructures (boxes, panels, door fairings, cable ducts, slats and flaps, wings, fuselage components), components (fasteners, bearings, fuel and refueling systems, seals, bushings), equipment (air management, power and thermal management, electrical landing system, sensors, de-icing, flight control, door opening/closing control, retraction system, engine control), cabin interior (galley, cabin lighting, seat, cooling systems, oxygen systems, drinking water systems, vacuum waste systems, cargo equipment), and propulsion systems (bearings for engine guide vanes, turbines, propeller system).

Among the materials for electromagnetic shielding that will continue as a matrix material are standard polymers (for large-scale commercial applications, materials not suitable for space applications), engineering polymers (multifunctional applications for internal aerospace components) and high-performance polymers (thermal and high-temperature applications, for internal and external aerospace components). Ceramic matrices will continue to be used in the design of structural materials for aerospace components. In the case of filler materials for electromagnetic shielding, there is a wide range of possibilities including carbon-based materials (carbon nanofibers, carbon nanotubes, graphene, pyrocarbon, highly ordered porous carbon (HOPC)) and their derivatives such as functionalized, coated or sandwich-like materials, ceramic (Fe_2O_3, Fe_3O_4) and metallic (silver (Ag), cobalt (Co), nickel (Ni), titanium (Ti), zinc (Zn), copper (Cu), iron (Fe), boron (B), bismuth (Bi), and hafnium (Hf)) materials in morphologies such as bars, spheres, and hollow structures, and two-dimensional materials such as Mxenes (Mo_2TiC_2, Cr_2TiC_2, Ti_3AlC_2, Ti_4AlN_3), transition metal chalcogenides (MoS_2, WS_2, $MoSe_2$, WSe_2, $MoTe_2$), or metals (boron, tin, silicon, arsenic, antimony, phosphorus, germanium). The use of hybrid structures, i.e., where inorganic and organic materials work together, will continue to produce materials for shielding to electromagnetic interference, even seeking to exploit mimetics to imitate nature. Heterostructures as well as three-dimensional structures, where two-dimensional and unidirectional fills are incorporated together, represent some of the options for electromagnetic shielding materials that must be fully exploited.

CONCLUSION

The contribution that advanced materials have made to electromagnetic shielding is considerable, with the introduction of nanomaterials especially at high frequencies, because nanometer size is directly correlated with immunity at operating frequencies in the GHz range and THz. Although metallic materials continue to be widely regarded as the materials for electromagnetic shielding par excellence, advanced materials have achieved levels of shielding effectiveness that make them ideal candidates to reduce weight and provide the required functionality of metals, especially at high frequencies. In the coming decades, researchers around the world will continue to design materials for electromagnetic shielding by exploiting both electrical and magnetic properties to find the best way to exploit the mechanisms presented by a coating used in the shield. Despite all the advances made to date, the area of electromagnetic compatibility through shielding to electromagnetic interference cannot yet establish a cooking recipe that can completely solve all the problems that are in practice, but the contribution of advanced materials is undoubtedly an important study topic for current and future scientific research.

ACKNOWLEDGMENT

This work is carried out as part of the activities of the first two authors within the National Thematic Network of Aeronautics of Mexico supported economically by CONACYT. The first author is grateful for the support of the Instituto Tecnológico Superior de Irapuato (ITESI) to carry out this research in advanced materials applied in the electronics industry.

REFERENCES

Al-Saleh, M. H., & Sundararaj, U. (2009). Electromagnetic Interference Shielding Mechanisms of CNT/Polymer Composites. *Carbon*, *47*(7), 1738–1746. doi:10.1016/j.carbon.2009.02.030

Ameli, A., Jung, P. U., & Park, C. B. (2013). Electrical Properties and Electromagnetic Interference Shielding Effectiveness of Polypropylene/Carbon Fiber Composite Foams. *Carbon*, *60*, 379–391. doi:10.1016/j.carbon.2013.04.050

ASTM ES7-83. (1988). *Test Method for Electromagnetic Shielding Effectiveness of Planar Materials, American Society for Testing and Materials.*

Chung, D. D. L. (2000). Materials for Electromagnetic Interference Shielding. *Journal of Materials Engineering and Performance*, *9*(3), 350–354. doi:10.1361/105994900770346042

Hu, T., Wang, J., Wang, J., & Chen, R. (2015). Electromagnetic Interference Shielding Properties of Carbonyl Iron Powder-Carbon Fiber felt/Epoxy Resin Composites with Different Layer Angle. *Materials Letters*, *142*, 242–245. doi:10.1016/j.matlet.2014.12.026

ISO 14302. (2002). *Space Systems – Electromagnetic Compatibility Requirements.* The International Organization for Standarization (ISO).

ISO 24637. (2009). *Space Systems – Electromagnetic Interference (EMI) Test reporting requirements.* The International Organization for Standarization (ISO).

Jia, Y., Li, K., Xue, L., Ren, J., Zhang, S., & Li, H. (2017). Mechanical and Electromagnetic Shielding Performance of Carbon Fiber Reinforced Multilayered $(PyC\text{-}SiC)_n$ Matrix Composites. *Carbon*, *111*, 299–308. doi:10.1016/j.carbon.2016.10.004

Joseph, N., & Sebastian, M. T. (2013). Electromagnetic Interference Shielding Nature of PVDF-Carbonyl Iron Composites. *Materials Letters*, *90*, 64–67. doi:10.1016/j.matlet.2012.09.014

Kong, L., Yin, X., Yuan, X., Zhang, Y., Liu, X., Cheng, L., & Zhang, L. (2014). Electromagnetic Wave Absorption Properties of Graphene modified with Carbon Nanotube/Poly(Dimethyl Siloxane) Composites. *Carbon*, *73*, 185–193. doi:10.1016/j.carbon.2014.02.054

Lee, S.-H., Kang, D., & Oh, I.-K. (2017). Multilayered Graphene-Carbon Nanotube-Iron Oxide Three-Dimensional Heterostructure for Flexible Electromagnetic Interference Shielding Film. *Carbon*, *111*, 248–257. doi:10.1016/j.carbon.2016.10.003

Li, Y., Pei, X., Shen, B., Zhai, W., Zhang, L., & Zheng, W. (2015). Polyimide/Graphene Composite Foam Sheets with Ultrahigh Thermostability for Electromagnetic Interference Shielding. *RSC Advances*, *5*(31), 24342–24351. doi:10.1039/C4RA16421K

Li, Y., Shen, B., Pei, X., Zhang, Y., Yi, D., Zhai, W., & Zheng, W. et al. (2016). Ultrathin Carbon Foams for Effective Electromagnetic Interference Shielding. *Carbon*, *100*, 375–385. doi:10.1016/j.carbon.2016.01.030

Liu, L., Das, A., & Megaridis, C. M. (2014). Terahertz Shielding of Carbon Nanomaterials and Their Composites – A Review and Applications. *Carbon*, *69*, 1–16. doi:10.1016/j.carbon.2013.12.021

Mei, H., Han, D., Xiao, S., Ji, T., Tang, J., & Cheng, L. (2016). Improvement of the Electromagnetic Shielding Properties of C/SiC Composites by Electrophoretic Deposition of Carbon Nanotube on Carbon Fibers. *Carbon*, *109*, 149–153. doi:10.1016/j.carbon.2016.07.070

Micheli, D., Apollo, C., Pastore, R., Morles, R. B., Laurenzi, S., & Marchetti, M. (2011). Nanostructured Composite Materials for Electromagnetic Interference Shielding Applications. *Acta Astronautica*, *69*(9-10), 747–757. doi:10.1016/j.actaastro.2011.06.004

Micheli, D., Pastore, R., Delfini, A., Giusti, A., Vricella, A., Santoni, F., & Vasilyeva, E. et al. (2017). Electromagnetic Characterization of Advanced Nanostructured Materials and Multilayer Design Optimization for Metrological and Low Radar Observability Applications. *Acta Astronautica*, *134*, 33–40. doi:10.1016/j.actaastro.2017.01.044

Micheli, D., Pastore, R., Gradoni, G., Primiani, V. M., Moglie, F., & Marchetti, M. (2013). Reduction of Satellite Electromagnetic Scattering by Carbon Nanostructured Multilayers. *Acta Astronautica*, *88*, 61–73. doi:10.1016/j.actaastro.2013.03.003

MIL-STD-451G. (2015). *Requirements for the Control of Electromagnetic Interference Characteristics of Subsystems and Equipment.* Department of Defense Interface Standard, Department of Defense, United States of America.

Mu, Y., Zhou, W., Wan, F., Ding, D., Hu, Y., & Luo, F. (2015). High-Temperature Dielectric and Electromagnetic Interference Shielding Properties of SiC_f/SiC Composites using Ti_3SiC_2 as Inert Filler. *Composites. Part A, Applied Science and Manufacturing*, *77*, 195–203. doi:10.1016/j.compositesa.2015.07.004

Musgrave, G., Larsen, A., & Sgobba, T. (2009). *Safety Design for Space Systems.* Oxford, UK: Butterworth-Heinemann.

Nayak, L., Pradhan, R. R., Khastgir, D., & Chaki, T. K. (2015). Thermally Stable Electromagnetic Interference Shielding Material from Polysulfone Nanocomposites: Comparison on Carbon Nanotube and Nanofiber Reinforcement. *Polymer Composites*, *36*(3), 566–575. doi:10.1002/pc.22973

Pettit, G. (2017). Summary of Military and Aerospace EMC Tests. In *2017 EMC Testing Guide*. Retrieved from www.interferencetechnology.com

Shahzad, F., Alhabeb, M., Hatter, C. B., Anasori, B., Hong, S. M., Koo, C. M., & Gogotsi, Y. (2016). Electromagnetic Interference Shielding with 2D Transition Metal Carbides (MXenes). *Science*, *353*(6304), 1137–1140. doi:10.1126/science.aag2421 PMID:27609888

Shavers, M.R., & Cucinotta. (2004). Implementation of ALARA Radiation Protection on the ISS through Polyethylene Shielding Augmentation of the Service Module Crew Quarters. *Adv. Space Res.*, *34*(6), 1333-1337. doi: /10.1016/j.asr.2003.10.051

Shen, B., Li, Y., Zhai, W., & Zheng, W. (2016). Compressible Graphene-Coated Polymer Foams with Ultralow Density for Adjustable Electromagnetic Interference (EMI) Shielding. *ACS Applied Materials & Interfaces*, *8*(12), 8050–8057. doi:10.1021/acsami.5b11715 PMID:26974443

Song, W. L., Cao, M.-S., Fan, L.-Z., Lu, M.-M., Li, Y., Wang, C.-Y., & Ju, H.-F. (2014b). Highly Ordered Porous Carbon/Wax Composites for Effective Electromagnetic Attenuation and Shielding. *Carbon*, *77*, 130–142. doi:10.1016/j.carbon.2014.05.014

Song, W.-L., Cao, M.-S., Lu, M.-M., Bi, S., Wang, C.-Y., Liu, J., & Fan, L.-Z. et al. (2014a). Flexible Graphene/Polymer Composite Films in Sandwich Structures for Effective Electromagnetic Interference Shielding. *Carbon*, *66*, 67–76. doi:10.1016/j.carbon.2013.08.043

Spillantini, P., Casolino, M., Durante, M., Mueller-Mellin, R., Reitz, G., Rossi, L., & Sorbi, M. et al. (2007). Shelding from Cosmic Radition for Interplanetary Missions: Active and Passive Methods. *Radiation Measurements*, *42*(1), 14–23. doi:10.1016/j.radmeas.2006.04.028

Sun, X., Liu, X., Shen, X., Wu, Y., Wang, Z., & Kim, J.-K. (2016). Graphene foam/Carbon Nanotube/Poly(dimethyl siloxane) Composites for Exceptional Microwave Shielding. *Composites. Part A, Applied Science and Manufacturing*, *85*, 199–206. doi:10.1016/j.compositesa.2016.03.009

Thomassin, J.-M., Jérôme, C., Pardoen, T., Bailly, C., Huynen, I., & Detrembleur, C. (2013). Polymer/Carbon based Composites as Electromagnetic Interference (EMI) Shielding Materials. *Materials Science and Engineering R Reports*, *74*(7), 211–232. doi:10.1016/j.mser.2013.06.001

Tong, X. C. (2009). *Advanced Materials and Design for Electromagnetic Interference Shielding*. Boca Raton, FL: CRC Press.

Vargas-Bernal, R. (2015). Performance Analysis of Electromagnetic Interference Shielding based on Carbon Nanomaterials used in AMS/RF IC Design. In M. Fakhfakh, E. Tlelo-Cuautle, & M. H. Fino (Eds.), *Performance Optimization Techniques in Analog, Mixed-Signal, and Radio-Frequency Circuit Design* (pp. 268–294). Hershey, PA: IGI Global; doi:10.4018/978-1-4666-6627-6.ch011

Vargas-Bernal, R. (2016). Advances in Functional Nanocoatings Applied in the Aerospace Industry. In A. Zuzuarregui & M. C. Morant-Miñana (Eds.), *Research Perspectives on Functional Micro- and Nanoscale Coatings* (pp. 318–340). Hershey, PA: IGI Global; doi:10.4018/978-1-5225-0066-7.ch012

Vargas-Bernal, R. (2017). The Next Generation of Nanomaterials for Designing Analog Integrated Circuits. In E. Tlelo-Cuautle, M. Fakhfakh, & L. G. de la Fraga (Eds.), *Analog Circuits: Fundamentals, Synthesis and Performance* (pp. 103–121). Nova Science Publishers.

Vargas-Bernal, R. (2018). Mechanisms of Electrical Conductivity in Carbon Nanotubes and Graphene. In M. Khosrow-Pour (Ed.), *Encyclopedia of Information Science and Technology* (4th ed.; pp. 2673–2684). Hershey, PA: IGI Global; doi:10.4018/978-1-5225-2255-3.ch233

Von Klemperer, C. J., & Maharaj, D. (2009). Composite Electromagnetic Interference Shielding Materials for Aerospace Applications. *Composite Structures*, *91*(4), 467–472. doi:10.1016/j.compstruct.2009.04.013

Wilson, J. W., Kim, M., Schimmeling, W., Badavi, F. F., Thibeault, S. A., Cucinotta, F. A., & Kiefer, R. et al. (1995). Issues in Space Radiation Protection: Galactic Cosmic Rays. *Health Physics*, *68*(1), 50–58. doi:10.1097/00004032-199501000-00006 PMID:7989194

Wu, H., Wu, G., & Wang, L. (2015). Peculiar Porous α-Fe_2O_3, γ-Fe_2O_3 and Fe_3O_4 Nanospheres: Facile Synthesis and Electromagnetic Properties. *Powder Technology*, *269*, 443–451. doi:10.1016/j.powtec.2014.09.045

Yan, D.-X., Ren, P. G., Pang, H., Fu, Q., Yang, M.-B., & Li, Z. M. (2012). Efficient Electromagnetic Interference Shielding of Lightweight Graphene/Polystyrene Composite. *Journal of Materials Chemistry*, *22*(36), 18772–18774. doi:10.1039/c2jm32692b

Yuan, B., Bao, C., Qian, X., Song, L., Tai, Q., Liew, K. M., & Hu, Y. (2014). Design of Artificial Nacre-like Hybrid Films as Shielding to Mitigate Electromagnetic Pollution. *Carbon*, *75*, 178–189. doi:10.1016/j.carbon.2014.03.051

Zhang, H., Zhang, G., Li, J., Fan, X., Jing, Z., Li, J., & Shi, X. (2017a). Lightweight, Multifunctional Microcellular PMMA/Fe_3O_4@MWCNTs Nanocomposite Foams with Efficient Electromagnetic Interference Shielding. *Composites. Part A, Applied Science and Manufacturing*, *100*, 128–138. doi:10.1016/j.compositesa.2017.05.009

Zhang, H.-B., Yan, Q., Zheng, W.-G., He, Z., & Yu, Z. Z. (2011). Tough Graphene-Polymer Microcellular Foams for Electromagnetic Interference Shielding. *ACS Applied Materials & Interfaces*, *3*(3), 918–924. doi:10.1021/am200021v PMID:21366239

Zhang, Y., Qiu, M., Yu, Y., Wen, B., & Cheng, L. (2017b). A Novel Polyaniline-Coated Bagasse Fiber Composite with Core-Shell Heterostructure provides Effective Electromagnetic Shielding Performance. *ACS Applied Materials & Interfaces*, *9*(1), 809–818. doi:10.1021/acsami.6b11989 PMID:27982585

ADDITIONAL READING

Chung, D. D. L. (2001). Electromagnetic Interference Shielding Effectiveness of Carbon Materials. *Carbon*, *39*(2), 279–285. doi:10.1016/S0008-6223(00)00184-6

D'Aloia, A. G., Marra, F., Tamburrano, A., De Bellis, G., & Sarto, M. S. (2014). Electromagnetic Absorbing Properties of Graphene-Polymer Composite Shields. *Carbon*, *73*, 175–184. doi:10.1016/j.carbon.2014.02.053

Eddib, A. A., & Chung, D. D. L. (2017). The Importance of the Electrical Contact between Specimen and Testing Fixture in Evaluating the Electromagnetic Interference Shielding Effectiveness of Carbon Materials. *Carbon*, *117*, 427–436. doi:10.1016/j.carbon.2017.02.091

Hsiao, S.-T., Ma, C.-C. M., Tien, H.-W., Liao, W.-H., Wang, Y.-S., Li, S.-M., & Huang, Y. C. (2013). Using a Non-Covalent Modification to Prepare a High Electromagnetic Interference Shielding Performance Graphene Nanosheet/Water-Borne Polyurethane Composite. *Carbon*, *60*, 57–66. doi:10.1016/j.carbon.2013.03.056

Huang, Y., Li, N., Ma, Y., Du, F., Li, F., He, X., & Chen, Y. et al. (2007). The Influence of Single-Walled Carbon Nanotube Structure on the Electromagnetic Interference Shielding Efficiency of Its Epoxy Composites. *Carbon*, *45*(8), 1614–1621. doi:10.1016/j.carbon.2007.04.016

Kuang, T., Chang, L., Chen, F., Sheng, Y., Fu, D., & Peng, X. (2016). Facile Preparation of Lightweight High-Strength Biodegradable Polymer/Multi-Walled Carbon Nanotubes Nanocomposite Foams for Electromagnetic Interference Shielding. *Carbon*, *105*, 305–313. doi:10.1016/j.carbon.2016.04.052

Liu, X., Yin, X., Kong, L., Li, Q., Liu, Y., Duan, W., & Cheng, L. et al. (2014). Fabrication and Electromagnetic Interference Shielding Effectiveness of Carbon Nanotube reinforced Carbon Fiber/Pyrolytic Carbon Composites. *Carbon, 68*, 501–510. doi:10.1016/j.carbon.2013.11.027

Liu, Z., Bai, G., Huang, Y., Ma, Y., Du, F., Li, F., & Chen, Y. et al. (2007). Reflection and Absorption Contributions to the Electromagnetic Interference Shielding of Single-Walled Carbon Nanotube/Polyurethane Composites. *Carbon, 45*(4), 821–827. doi:10.1016/j.carbon.2006.11.020

Micheli, D., Vricella, A., Pastore, R., Delfini, A., Giusti, A., Albano, M., & Primiani, V. M. et al. (2016). Ballistic and Electromagnetic Shielding Behaviour of Multifunctional Kevlar Fiber Reinforced Epoxy Composites modified by Carbon Nanotubes. *Carbon, 104*, 141–156. doi:10.1016/j.carbon.2016.03.059

Singh, A. P., Garg, P., Alam, F., Singh, K., Mathur, R. B., Tandon, R. P., & Dhawan, S. K. et al. (2012). Phenolic Resin-based Composite Sheets filled with Mixtures of Reduced Graphene Oxide, γ-Fe_2O_3 and Carbon Fibers for Excellent Electromagnetic Interference Shielding in the X-Band. *Carbon, 50*(10), 3868–3875. doi:10.1016/j.carbon.2012.04.030

Yang, S., Lozano, K., Lomeli, A., Foltz, H. D., & Jones, R. (2005). Electromagnetic Interference Shielding Effectiveness of Carbon Nanofiber/LCP Composites. *Composites. Part A, Applied Science and Manufacturing, 36*(5), 691–697. doi:10.1016/j.compositesa.2004.07.009

KEY TERMS AND DEFINITIONS

Advanced Materials: Novel materials or those obtained by modifying existing materials with superior performance in one or more properties that are critical to an application under consideration.

Carbon Nanomaterials: Nanostructures of carbon such as fullerenes, carbon nanotubes, nanofibers, and graphene with unique physicochemical properties with multiple technological applications.

Composites: Materials made from two or more constituents materials with significantly different physical and/or chemical properties, that when are combined, produce a material with different characteristics from the individual components.

Electromagnetic Effectiveness: The sum of the reflection, absorption, and re-reflection losses that presents an electromagnetic barrier when it blocks and/or isolate electromagnetic waves expressed in dB.

Electromagnetic Shielding: The action of reducing and/or isolating the electromagnetic field is a space by blocking the electromagnetic waves with barriers made of conductive and/or magnetic materials.

Graphene: A two-dimensional, crystalline allotrope of carbon whose atoms are densely packed in a regular sp^2-bonded atomic-scale chicken wire (hexagonal) pattern composed by a one-atom thick layer of graphite.

Hybrid Materials: Composite materials consisting of at least 2 materials at the nanometric or molecular level where at least one is inorganic in nature and another is organic, and its mixing leads to a more homogeneous material showing characteristics of the original two phases or even new properties.

Mxenes: Materials consist of few atoms thick layers of transition metal carbides, nitrides, or carbonitrides.

Permeability: The measure of the ability of a material to support the formation of a magnetic field within itself.

Two-Dimensional Materials: Materials of a single layer of atoms or with a thickness of a few nanometers or less.

Compilation of References

AC 20-158A. (2014). *The certification of aircraft electrical and electronic systems for operation in the high-intensity radiated fields (HIRF) environment.* Advisory U.S. Department of Transportation Circular Federal Aviation Administration.

Acuna, M. H. (2002). Space-based magnetometers. *The Review of Scientific Instruments*, *73*(11), 3717–3736. doi:10.1063/1.1510570

Adams, J. W., Cruz, J., & Melquist, D. (1992). Comparison measurements of currents induced by radiation and injection. *IEEE Transactions on Electromagnetic Compatibility*, *34*(3), 360–362. doi:10.1109/15.155856

Agrawal, A. K., Price, H. J., & Gurbaxani, S. H. (1980). Transient response of multiconductor transmission lines excited by a nonuniform electromagnetic field. *IEEE Transactions on Electromagnetic Compatibility*, *EMC-22*(2), 119–129. doi:10.1109/TEMC.1980.303824

Airbus Defence & Space SAS. (2014). *Medium Power PCDU, POW2 datasheet.* Author.

Al-Saleh, M. H., & Sundararaj, U. (2009). Electromagnetic Interference Shielding Mechanisms of CNT/Polymer Composites. *Carbon*, *47*(7), 1738–1746. doi:10.1016/j.carbon.2009.02.030

Ameli, A., Jung, P. U., & Park, C. B. (2013). Electrical Properties and Electromagnetic Interference Shielding Effectiveness of Polypropylene/Carbon Fiber Composite Foams. *Carbon*, *60*, 379–391. doi:10.1016/j.carbon.2013.04.050

Antonucci, F., Armano, M., Audley, H., Auger, G., Benedetti, M., Binetruy, P., & Zweifel, P. et al. (2011). LISA Pathfinder: Mission and status. *Classical and Quantum Gravity*, *28*(9), 9. doi:10.1088/0264-9381/28/9/094001

ASTM ES7-83. (1988). *Test Method for Electromagnetic Shielding Effectiveness of Planar Materials, American Society for Testing and Materials.*

Baklezos, A. T., Nikolopoulos, C. D., Capsalis, C. N., & Tsatalas, S. (2017, March). Effect of LVDS link speed and pattern length on spectrum measurements of a Spacewire harness. In *Antenna Technology: Small Antennas, Innovative Structures, and Applications (iWAT), 2017 International Workshop on* (pp. 38-41). IEEE. doi:10.1109/IWAT.2017.7915291

Baklezos, A. T., Nikolopoulos, C. N., & Capsalis, C. N. (2017). An equivalent dipole method with novel measurement positioning for modeling electric emissions in space missions. *Electromagnetics*, *37*(7). doi:10.1080/02726343.2017.1376902

Bartington Inc. (n.d.a). *Operation Manual for HC1 Helmholtz Coils*. Retrieved October, 2017, from http://www.bartington.com/Literaturepdf/Operation%20Manuals/HC1%20Helmholtz%20 Coil%20OM3226.pdf

Bartington Inc. (n.d.b). *Operation Manual for HC2 Helmholtz Coils*. Retrieved October, 2017, from http://www.bartington.com/Literaturepdf/Operation%20Manuals/HC2%20Helmholtz%20 Coil%20OM3342.pdf

Bartington Inc. (n.d.c). *Operation Manual for HC1 Helmholtz Coils*. Retrieved October, 2017, from http://www.bartington.com/Literaturepdf/Operation%20Manuals/HC1%20Helmholtz%20 Coil%20OM3226.pdf

Baum, C. E., Liu, T. K., & Tesche, F. M. (1978). On the analysis of general multiconductor transmission-line networks. *Interaction Note*, *350*(6), 467–547.

Bayle, O., Lorenzoni, L., Blancquaert, T., Langlois, S., Walloschek, T., Portigliotti, S., & Capuano, M. (2011). *Exomars Entry Descent and Landing Demonstrator Mission and Design Overview*. NASA Solar System.

Benkhoff, J., Casteren, J., Hayakawa, H., Fujimoto, M., Laakso, H., Novara, M., & Ziethe, R. et al. (2010). BepiColombo—Comprehensive exploration of Mercury: Mission overview and science goals. *Planetary and Space Science*, *58*(1), 2–20. doi:10.1016/j.pss.2009.09.020

Berenger, J. P. (1994). A perfectly matched layer for the absorption of electromagnetic waves. *Journal of Computational Physics*, *114*(2), 185–200. doi:10.1006/jcph.1994.1159

Berenger, J. P. (2000). A multiwire formalism for the FDTD method. *IEEE Transactions on Electromagnetic Compatibility*, *42*(3), 257–264. doi:10.1109/15.865332

Besnier, P., & Demoulin, B. (2011). *Electromagnetic reverberation chambers*. John Wiley & Sons.

BIPM, I., IFCC, I., IUPAC, I., & ISO, O. (2008). Evaluation of measurement data—guide for the expression of uncertainty in measurement. JCGM 100: 2008. *Citado en las*, 167.

Bojarski, N. (1982). Low frequency inverse scattering. *IEEE Transactions on Antennas and Propagation*, *30*(4), 775–778. doi:10.1109/TAP.1982.1142846

Boschetti, D., Gervasio, G., & Marziali, I. (2012, May). Montecarlo approach for magnetic cleanliness evaluation on spacecraft. In *Aerospace EMC, 2012 Proceedings ESA Workshop on* (pp. 1-3). IEEE.

Brownlee, J. (2011). *Clever algorithms: Nature-inspired programming recipes*. Jason Brownlee.

Bureau International des Poids et Mesures. (2008). Evaluation of measurement data - Guide to the expression of uncertainty in measurement. GUM 1995 with minor corrections. *JCGM*, *100*, 2008.

Cabello, M. R., Angulo, L. D., Bretones, A. R., Martin, R. G., Garcia, S. G., & Alvarez, J. (2017a, May). A novel subgriding scheme for arbitrarily dispersive thin-layer modeling. In *Numerical Electromagnetic and Multiphysics Modeling and Optimization for RF, Microwave, and Terahertz Applications (NEMO), 2017 IEEE MTT-S International Conference on* (pp. 266-268). IEEE. doi:10.1109/NEMO.2017.7964255

Cabello, M. (2017d). *Subcell FDTD techniques for electromagnetic compatibility assessment in aeronautics (PhD dissertation)*. University of Granada. Retrieved from http://www.sembahome.org/wp-content/uploads/2017/PhD_dissertation_MRCabello_definprint.pdf

Cabello, M. R., Angulo, L. D., Alvarez, J., Bretones, A. R., Gutierrez, G. G., & Garcia, S. G. (2015, August). A New Efficient and Stable 3D Conformal FDTD. *IEEE Microwave and Wireless Components Letters*, *26*(8), 553–555. doi:10.1109/LMWC.2016.2588579

Cabello, M. R., Angulo, L., Alvarez, J., Flintoft, I. D., Bourke, S., Dawson, J. F., & Gómez Martín, R. (2017b, May). A Hybrid Crank Nicolson FDTD Subgridding Boundary Condition for Lossy Thin-Layer Modeling. *IEEE Transactions on Microwave Theory and Techniques*, *65*(5), 1397–1406. doi:10.1109/TMTT.2016.2637348

Cabello, M. R., Fernández, S., Pous, M., Pascual-Gil, E., Angulo, L. D., López, P., & Garcia, S. G. et al. (2017c). SIVA UAV: A Case Study for the EMC Analysis of Composite Air Vehicles. *IEEE Transactions on Electromagnetic Compatibility*, *59*(4), 1103–1113. doi:10.1109/TEMC.2017.2648507

Carrubba, E., Junge, A., Marliani, F., & Monorchio, A. (2012, May). Particle swarm optimization to solve multiple dipole modelling problems in space applications. In *Aerospace EMC, 2012 Proceedings ESA Workshop on* (pp. 1-6). IEEE.

Chen, D. X., & Gu, C. (2005). AC susceptibilities of conducting cylinders and their applications in electromagnetic measurements. *IEEE Transactions on Magnetics*, *41*(9), 2436–2446. doi:10.1109/TMAG.2005.855450

Chung, D. D. L. (2000). Materials for Electromagnetic Interference Shielding. *Journal of Materials Engineering and Performance*, *9*(3), 350–354. doi:10.1361/105994900770346042

Colby, J., (2011). *Application Note: ESD and Surge Circuit Protection*. Chicago: Littelfuse, Inc.

Cook, B., & Walker, P. (2007). Reducing Electromagnetic Emissions From SpaceWire. In *DAta Systems in Aerospace*. Naples: DASIA.

Daniel, C. M. (1988). *The Theory of Magnetism I: Statics and Dynamics*. Springer.

Danzmann, K. (2008). LISA mission overview. *Advances in Space Research*, *25*(6), 1129–1136. doi:10.1016/S0273-1177(99)00973-4

Dawar, D., & Ludwig, S. A. (2014, December). Differential evolution with dither and annealed scale factor. In *Differential Evolution (SDE), 2014 IEEE Symposium on* (pp. 1-8). IEEE. doi:10.1109/SDE.2014.7031528

Department of Defense, United States of America. (1996). *MIL-STD-88G3, Test Method Standard: Microcircuits*. Author.

Department of Defense, United States of America. (2015). *MIL-STD-461G: Requirements for the Control of Electromagnetic Interference Characteristics of Subsystems and Equipment*. Author.

Department of the Air Force, United States of America. (1987). *MIL-STD-1541A (USAF), Electromagnetic compatibility, requirements for space systems*. Author.

Dey, S., & Mittra, R. (1997). A locally conformal finite-difference time-domain (FDTD) algorithm for modeling three-dimensional perfectly conducting objects. *IEEE Microwave and Guided Wave Letters*, *7*(9), 273–275. doi:10.1109/75.622536

Drăgoi, E. N., Curteanu, S., & Vlad, D. (2012, October). Differential evolution applications in electromagnetics. In *Electrical and Power Engineering (EPE), 2012 International Conference and Exposition on* (pp. 636-640). IEEE. doi:10.1109/ICEPE.2012.6463801

Drinkwater, M. R., Floberghagen, R., Haagmans, R., Muzi, D., & Popescu, A. (2003). GOCE: ESA's first Earth Explorer Core mission. In Earth gravity field from space—From sensors to earth sciences, (pp. 419-432). Springer Netherlands.

Drung, D., Abmann, C., & Beyer, J. (2007). Highly Sensitive and Easy-to-Use SQUID Sensors. *IEEE Transactions on Applied Superconductivity, Volume*, *17*(2), 699–704. doi:10.1109/TASC.2007.897403

Duffy, A. P., Martin, A. J. M., Orlandi, A., Antonini, G., Benson, T. M., & Woolfson, M. S. (2006). Feature Selective Validation (FSV) for validation of computational electromagnetics (CEM). Part I – The FSV method. *IEEE Trans. on Electromagnetic. Compatibility*, *48*, 449–459.

Duffy, A. P., Martin, A. J., Orlandi, A., Antonini, G., Benson, T. M., & Woolfson, M. S. (2006, August). Feature selective validation (FSV) for validation of computational electromagnetics (CEM). Part I-The FSV method. *IEEE Transactions on Electromagnetic Compatibility*, *48*(3), 449–459. doi:10.1109/TEMC.2006.879358

Dumond, O., & Berge, R. (2012, May). Determination of the magnetic moment with spherical measurements and spherical harmonics modelling. In *Aerospace EMC, 2012 Proceedings ESA Workshop on* (pp. 1-5). IEEE.

Dunn, B. D. (2016). *Materials and processes - for Spacecraft and High Reliability Applications*. New York: Springer.

Duvvury, C., & Gossner, H. (2015). *System level ESD CO-Design*. Wiley, IEEE Press. doi:10.1002/9781118861899

ECSS, C. (2012). *European Cooperation for Space Standardization. ECSS-E-HB-20-07A. Space engineering – Electromagnetic compatibility Handbook*. ESA Publication.

ECSS-E50-12A vs ECSS-E-ST-50-12C, what's the difference? | STAR-Dundee. (2017). Retrieved 15 June 2017, from https://www.star-dundee.com/knowledge-base/ecss-e50-12a-vs-ecss-e-st-50-12c-whats-difference

ECSS-E-ST-50-12C Rev. 1 Working Group. (2015). *ECSS-E-ST-50-12C Rev. 1 Draft (DIR3), Space Engineering – SpaceWire- Links, nodes, routers and networks, ESA-ESTEC*. Noordwijk, The Netherlands: Requirements & Standards Division.

Edelvik, F. (2002). *Hybrid solvers for the Maxwell equations in time-domain* (Doctoral dissertation). Acta Universitatis Upsaliensis.

Enouf, K., & Hermant, S. (2016, October). Compact, impedance-matched SpaceWire connector development—"MicroMach SpaceWire": SpaceWire components, long paper. In SpaceWire Conference (SpaceWire), 2016 International (pp. 1-7). IEEE.

ESA. (2014). *PLATO EMC Specification, ESA-PLATO-ESTEC-MIS-SP-002*. ESA TEC-EEE.

ESA. (2015). *THOR Mission Requirements Document ESA-THOR-EST-MIS-RS-001, ESA-ESTEC*. ESA THOR Study Team.

Escoubet, C. P., Schmidt, R., & Goldstein, M. L. (1997). Cluster-Science and mission overview. In The Cluster and Phoenix Missions, (pp. 11-32). Springer Netherlands. doi:10.1007/978-94-011-5666-0_1

Eurocae ED-105A. (2013). *Aircraft Lightning Test Methods. Revision A*. Author.

Eurocae ED-107A. (2010). *Guide to certification of aircraft in a high-intensity radiated field (HIRF) environment*. Author.

Eurocae ED-14G. (2011). *Environmental conditions and test procedures for airborne equipment*. Author.

European Cooperation for Space Standardization (ECSS) Secretariat. (2003). *ESCC Detail Specification No. 3902/003, Cable, "Spacewire", Round, Quad Using Symmetric Cables, FlexiblE, -200 TO +180°C, ESA-ESTEC*. Noordwijk, The Netherlands: Requirements & Standards Division.

European Cooperation for Space Standardization (ECSS) Secretariat. (2008). *ECSS- E-ST-50-12C, Space Engineering – SpaceWire- Links, nodes, routers and networks, ESA-ESTEC*. Noordwijk, The Netherlands: Requirements & Standards Division.

European Cooperation for Space Standardization (ECSS) Secretariat. (2008). *ECSS- E-ST-50-12C, Space Engineering – SpaceWire- Links, nodes, routers and networks, ESA-ESTEC*. Requirements & Standards Division.

European Cooperation for Space Standardization (ECSS) Secretariat. (2012). *ECSS-E-HB-20-07A. Space engineering – Electromagnetic compatibility Handbook*. ESA-ESTEC. Noordwijk, The Netherlands: Requirements & Standards Division.

European Cooperation for Space Standardization (ECSS) Secretariat. (2012). *ECSS-E-ST-20-07, C Rev. 1. Space engineering, Electromagnetic compatibility*. Author.

European Cooperation for Space Standardization (ECSS) Secretariat. (2012). *ECSS-E-ST-20-07C Rev. 1, Space Engineering –Electromagnetic compatibility, ESA-ESTEC*. Requirements & Standards Division.

European Cooperation for Space Standardization (ECSS) Secretariat. (2014). *ESCC Detail Specification No. 3902/003, Cable, "Spacewire", Round, Quad Using Symmetric Cables, Flexible, -200 To +180 °c, Esa-Estec*. Noordwijk, The Netherlands: Requirements & Standards Division.

European Cooperation for Space Standardization (ECSS). (2008). *ECSS-E-ST-20-06C: Space Engineering Spacecraft charging*. Author.

European Cooperation for Space Standardization (ECSS). (2012). *ECSS-E-ST-20-07C Rev. 1: Space engineering, Electromagnetic compatibility*. Author.

Feliziani, M., & Maradei, F. (2002). Full-wave analysis of shielded cable configurations by the FDTD method. *IEEE Transactions on Magnetics*, *38*(2), 761–764. doi:10.1109/20.996197

Foner, S. (1959). Versatile and Sensitive Vibrating-Sample Magnetometer. *The Review of Scientific Instruments*, *30*(7), 548–557. doi:10.1063/1.1716679

Fraden, J. (2010). *Handbook of Modern Sensors Physics, Designs, and Applications*. New York: Springer-Verlag. doi:10.1007/978-1-4419-6466-3

Fratantuono, L. (2015). *A Reading of Lucretius' De Rerum Natura*. Lexington Books.

Freitas, P. P., Ferreira, R., Cardoso, F., & Cardoso, F. (2007). Magnetoresistive sensors. *Journal of Physics Condensed Matter*, *19*(16), 165221. doi:10.1088/0953-8984/19/16/165221

Friis-Christensen, E., Lühr, H., Knudsen, D., & Haagmans, R. (2008). Swarm–an Earth observation mission investigating geospace. *Advances in Space Research*, *41*(1), 210–216. doi:10.1016/j.asr.2006.10.008

GarciaS. G. (2013). *SEMBA*. Retrieved from http://www.sembahome.org

Garcia, S. G., Perez, I. V., Martin, R. G., & Olmedo, B. G. (1999). BiPML: A PML to match waves in bianisotropic media. *Microwave and Optical Technology Letters*, *20*(1), 44–48. doi:10.1002/(SICI)1098-2760(19990105)20:1<44::AID-MOP12>3.0.CO;2-W

Garcia, S. G., Rubio, R. G., Bretones, A. R., & Lopez, R. G. (2007, November). Revisiting the Stability of Crank-Nicolson and ADI-FDTD. *IEEE Transactions on Antennas and Propagation*, *55*(11), 3199–3203. doi:10.1109/TAP.2007.908794

Garrett, H. B., & Whittlesey, A. C. (2011). Guide to mitigating spacecraft charging effects. Jet Propulsion Laboratory, California Institute of Technology, JPL Space Science and Technology Series.

Gil, E. P., & Gutierrez, G. G. (2011). Simplification and cleaning of complex CAD models for EMC simulations. *Int. Symp. Electromagnetic Compatibility (EMC Europe).*

Gkatzianas, M., & Tsiboukis, T. (2003). Thin-slot/thin-layer subcell FDTD algorithms for EM penetration through apertures. *Electromagnetics*, *23*(2), 119–133. doi:10.1080/02726340390159469

Gordon, D., Lundsten, R., & Chiarodo, R. (1965). Factors affecting the sensitivity of gamma-level ring-core magnetometers. *IEEE Transactions on Magnetics, Volume*, *1*(4), 330–337. doi:10.1109/TMAG.1965.1062987

Grassi, F., Spadacini, G., Marliani, F., & Pignari, S. A. (2008). Use of double bulk current injection for susceptibility testing of avionics. *IEEE Transactions on Electromagnetic Compatibility*, *50*(3), 524–535. doi:10.1109/TEMC.2008.926810

Grosz, A., Haji-Sheikh, J. M., & Mukhopadhyay, C. S. (2017). *High Sensitivity Magnetometers*. Cham: Springer. doi:10.1007/978-3-319-34070-8

Guiffaut, C., Reineix, A., & Pecqueux, B. (2012). New oblique thin wire formalism in the FDTD method with multiwire junctions. *IEEE Transactions on Antennas and Propagation*, *60*(3), 1458–1466. doi:10.1109/TAP.2011.2180304

Gustavsen, B. (2006, July). Improving the pole relocating properties of vector fitting. *IEEE Transactions on Power Delivery*, *21*(3), 1587–1592. doi:10.1109/TPWRD.2005.860281

Gutierrez, G. G., Romero, D. M., Cabello, M. R., Pascual-Gil, E., Angulo, L. D., Gomez, D. G., & Garcia, S. G. (2016). On the design of aircraft electrical structure networks. *IEEE Transactions on Electromagnetic Compatibility*, *58*(2), 401–408. doi:10.1109/TEMC.2016.2514379

Güttler, C., Hillenmaier, O., & Auster, U. (2009). Measurement of magnetic field emissions at low frequencies. *ESA Workshop on Aerospace EMC*, 30.

Hadjigeorgiou, N., Hristoforou, E., & Sotiriadis, P. P. (2017). Closed-loop current-feedback, signal-chopped, low noise AMR sensor with high linearity. *IEEE Modern Circuits and Systems Technologies, 6*.

Hadjigeorgiou, N., Tsalikidou, A.C., Hristoforou, E., & Sotiriadis, P. P. (2017). Highly Linear and Low Noise AMR Sensor Using Closed Loop and Signal-Chopped Architecture. *World Academy of Science, Engineering and Technology, 11*(8).

Hadjigeorgiou, N. (2014). Experimental Study Noise of HMR2003 Sensor with or without the Use of Amplifier AMP04. *Key Engineering Materials*, *605*, 629–632. doi:10.4028/www.scientific.net/KEM.605.629

Hadjigeorgiou, N., Konstantopoulos, C., & Masxas, D. (2015). Fourier Analysis for Orthogonal and Parallel Fluxgate. *Key Engineering Materials*, *644*, 270–273. doi:10.4028/www.scientific.net/KEM.644.270

Han, Dutton & Fan. (2006, March). Model dispersive media in finite-difference time-domain method with complex-conjugate pole-residue pairs. *IEEE Microwave and Wireless Components Letters*.

Hao, Y., & Mittra, R. (2009). *FDTD Modeling of Metamaterials Theory and Application*. Artech House.

Harrison, E. P., Turney, G. L., & Rowe, H. (1935). Electrical Properties of Wires of High Permeability. *Nature*, *135*(3423), 961. doi:10.1038/135961a0

Hill, D. A. (1992). Currents induced on multiconductor transmission lines by radiation and injection. *IEEE Transactions on Electromagnetic Compatibility*, *34*(4), 445–450. doi:10.1109/15.179277

Hill, D. A. (1998). Plane wave integral representation for fields in reverberation chambers. *IEEE Transactions on Electromagnetic Compatibility*, *40*(3), 209–217. doi:10.1109/15.709418

Hill, D. A. (2009). *Electromagnetic fields in cavities: deterministic and statistical theories* (Vol. 35). John Wiley & Sons. doi:10.1002/9780470495056

HIRFSE. (2008). *HIRF -SE project EU FP7, 2008-2013*. Retrieved from http://HIRFse.axessim.eu/

Hoang, V. D., Hernández, D. C., & Jo, K. H. (2014, April). Simple and efficient method for calibration of a camera and 2D laser rangefinder. In *Asian Conference on Intelligent Information and Database Systems* (pp. 561-570). Springer. doi:10.1007/978-3-319-05476-6_57

Holland, R., & Simpson, L. (1981). Finite-difference analysis of EMP coupling to thin struts and wires. *IEEE Transactions on Electromagnetic Compatibility*, *EMC-23*(2), 88–97. doi:10.1109/TEMC.1981.303899

Honewell Inc. (2011). *Hall effect sensing and application*. Retrieved September 2017, from www.honeywell.com/sensing

Hristoforou, E., Chiriac, H., & Neagu, M. (1997). A New Magnetic Field Sensor Based on Magnetostrictive Delay Lines. *IEEE Transactions on Instrumentation and Measurement, Vol*, *46*(:2), 632–635. doi:10.1109/19.571942

Hristoforou, E., & Reilly, R. E. (1991). Nonuniformity in amorphous ribbon delay lines after stress and current annealing. *Journal of Physics*, *69*, 5008.

Hu, T., Wang, J., Wang, J., & Chen, R. (2015). Electromagnetic Interference Shielding Properties of Carbonyl Iron Powder-Carbon Fiber felt/Epoxy Resin Composites with Different Layer Angle. *Materials Letters*, *142*, 242–245. doi:10.1016/j.matlet.2014.12.026

IEC 61000-4-21. (2011). *Electromagnetic compatibility (EMC) - Part 4-21: Testing and measurement techniques - Reverberation chamber test method*. Author.

IEEE Standard P1597. (2008). *Standard for Validation of Computational Electromagnetics Computer Modeling and Simulation – Parts 1&2*. IEEE.

IEEE Std 1597.1. (2008). *IEEE Standard for Validation of Computational Electromagnetics Computer Modeling and Simulations, IEEE 1597.1-2008*. IEEE.

IEEE Std 1597.2. (2010). *IEEE Recommended Practice for Validation of Computational Electromagnetics Computer Modeling and Simulations, IEEE Standard 1597.2-2010*. IEEE.

IEEE Std 299.1. (2014). *IEEE Standard Method for Measuring the Shielding Effectiveness of Enclosures and Boxes Having all Dimensions between 0.1 m and 2 m. IEEE Std 299.1-2013*. IEEE.

International Electrotechnical Commission (IEC). (2009). *IEC 61000-4-2:2009 - Electromagnetic compatibility (EMC). Testing and Measurement Techniques. Electrostatic Discharge Immunity Test*. Author.

International Electro-technical Commission, IEC 61000-4-2. (2008). *Electromagnetic compatibility (EMC) - Part 4-2: Testing and measurement techniques - Electrostatic discharge immunity test*. Author.

International Standardization Organization (ISO). (2002). *ISO 14302:2002 Space systems -- Electromagnetic compatibility requirements*. ISO.

International Standardization Organization (ISO). (2002). *Space Systems – Electromagnetic Compatibility Requirements*. Author.

Ishiyama, A., & Hirooka, H. (1991). Magnetic Shielding for MRI Superconducting Magnets. *IEEE Transactions on Magnetics, Volume*, *27*(Issue:2), 1692–1695. doi:10.1109/20.133514

ISO 14302. (2002). *Space Systems – Electromagnetic Compatibility Requirements*. The International Organization for Standarization (ISO).

ISO 24637. (2009). *Space Systems – Electromagnetic Interference (EMI) Test reporting requirements*. The International Organization for Standarization (ISO).

Jackson, J. D. (1999). Classical Electrodynamics. John Wiley & Sons.

Jackson, J. D. (1999). *Classical Electrodynamics*. New York: John Wiley and Sons.

Jauregui, R., Pous, M., & Silva, F. (2014b, September). Use of reference limits in the Feature Selective Validation (FSV) method. *IEEE 2014 International Symposium on Electromagnetic Compatibility*, 1031-1036.

Jauregui, R., Aragon, M., & Silva, F. (2013, February). The role of uncertainty in the feature selective validation (FSV) method. *IEEE Transactions on Electromagnetic Compatibility*, *55*(1), 217–220. doi:10.1109/TEMC.2012.2227262

Jauregui, R., Zhang, G., Rojas-Mora, J., Ventosa, O., Silva, F., Duffy, A. P., & Sasse, H. (2014a, August). Analyzing transient phenomena in the time domain using the feature selective validation (FSV) method. *IEEE Transactions on Electromagnetic Compatibility*, *56*(4), 825–834. doi:10.1109/TEMC.2013.2246167

Jauregui, T. R., Silva, F., Orlandi, A., Sasse, H., & Duffy, A. (2010, April). Factors influencing the successful validation of transient phenomenon modelling. *2010 IEEE Asia-Pacific Symposium on Electromagnetic Compatibility (APEMC)*, 338-341.

Javor, K. (1997, August). On field-to-wire coupling versus conducted injection techniques. In *Electromagnetic Compatibility, 1997. IEEE 1997 International Symposium on* (pp. 479-487). IEEE.

Jiang, R. H-C. (2012). *System Level ESD/EMI Issues and Protection Design.* Academic Press.

Jia, Y., Li, K., Xue, L., Ren, J., Zhang, S., & Li, H. (2017). Mechanical and Electromagnetic Shielding Performance of Carbon Fiber Reinforced Multilayered $(PyC\text{-}SiC)_n$ Matrix Composites. *Carbon, 111*, 299–308. doi:10.1016/j.carbon.2016.10.004

Jiles, D. (2015). *Introduction to Magnetism and Magnetic Materials.* CRC Press.

Jiles, D. C., & Atherton, D. L. (1984). Theory of ferromagnetic hysteresis. *Journal of Applied Physics*, *55*(6), 2115–2120. doi:10.1063/1.333582

Jordan, C. E., & Balmain, G. K. (1968). *Electromagnetic Waves and Radiation Systems.* Englewood Cliffs, NJ: Prentice-Hall, Inc.

Joseph, N., & Sebastian, M. T. (2013). Electromagnetic Interference Shielding Nature of PVDF-Carbonyl Iron Composites. *Materials Letters*, *90*, 64–67. doi:10.1016/j.matlet.2012.09.014

Josephson, B. D. (1962). Possible new effects in superconductive tunnelling. *Elsevier Physics Letters, Volume*, *1*(Issue:7), 251–254.

Junge, A., & Marliani, F. (2011, August). Prediction of DC magnetic fields for magnetic cleanliness on spacecraft. In *Electromagnetic Compatibility (EMC), 2011 IEEE International Symposium on* (pp. 834-839). IEEE. doi:10.1109/ISEMC.2011.6038424

Kakarakis, S. D. J., Kapsalis, N. C., & Capsalis, C. N. (2015). A semianalytical heuristic approach for prediction of eut's multiple dipole model by reducing the number of heuristics. *IEEE Transactions on Electromagnetic Compatibility*, *57*(1), 87–92. doi:10.1109/TEMC.2014.2358712

Kakarakis, S. D. J., Spantideas, S. T., Kapsalis, N. C., Capsalis, C. N., & Junge, A. (2017). A Software-Based Calibration Technique for Characterizing the Magnetic Signature of EUTs in Measuring Facilities. *IEEE Transactions on Electromagnetic Compatibility*, *59*(2), 334–341. doi:10.1109/TEMC.2016.2615132

Kapsalis, N. C., Kakarakis, S. D. J., & Capsalis, C. N. (2012). Prediction of multiple magnetic dipole model parameters from near field measurements employing stochastic algorithms. *Progress In Electromagnetics Research Letters*, *34*, 111–122. doi:10.2528/PIERL12030905

Kennedy, J. (2011). Particle swarm optimization. In Encyclopedia of machine learning (pp. 760-766). Springer US.

Klingler, M., Szelag, M., & Heddebaut, M. (1994). Double bulk current injection: A possible substitute to field-to-wire coupling. *Proc. EUROEM*, 94, 1249-1256.

Knoepfel, H. E. (2008). *Magnetic fields: a comprehensive theoretical treatise for practical use.* John Wiley & Sons.

Kobidze, G. (2010). Implementation of collocated surface impedance boundary conditions in FDTD. *IEEE Transactions on Antennas and Propagation*, *58*(7), 2394–2403. doi:10.1109/TAP.2010.2048859

Kong, L., Yin, X., Yuan, X., Zhang, Y., Liu, X., Cheng, L., & Zhang, L. (2014). Electromagnetic Wave Absorption Properties of Graphene modified with Carbon Nanotube/Poly(Dimethyl Siloxane) Composites. *Carbon*, *73*, 185–193. doi:10.1016/j.carbon.2014.02.054

Koutantos, G. I., Nikolopoulos, C. D., Baklezos, A. T., & Capsalis, C. N. (2016). On the Modeling of ELF Electric Fields for Space Mission Equipment. *IEEE Transactions on Electromagnetic Compatibility*, *58*(8), 1782–1786.

Kunz, K. S., Steich, D. J., & Luebbers, R. J. (1997, May). *Determining an Aircraft's Attenuation Characteristics of HIRF Testing*. SAE Technical Paper.

Kunz, S., Steich, D. J., & Luebbers, R. J. (1993, August). Low Frequency Shielding Effects of a Conducting Shell with an Aperture: Response of an Internal Wire. *IEEE Transactions on Electromagnetic Compatibility*, *34*(3), 370–373. doi:10.1109/15.155859

Lai, S. T. (2011). *Spacecraft and human electrostatic discharge: a comparison of the two phenomena, Fundamentals of Spacecraft Charging: Spacecraft Interactions with Space Plasmas*. Princeton, NJ: Princeton University Press.

Ledfelt, G. (2001). *Hybrid time-domain methods and wire models for computational electromagnetics* (Doctoral dissertation). Numerisk analys och datalogi.

Lee, S.-H., Kang, D., & Oh, I.-K. (2017). Multilayered Graphene-Carbon Nanotube-Iron Oxide Three-Dimensional Heterostructure for Flexible Electromagnetic Interference Shielding Film. *Carbon*, *111*, 248–257. doi:10.1016/j.carbon.2016.10.003

Leininger, M. M., Thurecht, F., Pfeiffer, E., & Ruddle, A. (2012, May). Advanced grounding methods in the presence of carbon fibre reinforced plastic structures. In *Aerospace EMC, 2012 Proceedings ESA Workshop on* (pp. 1-6). IEEE.

Li, P., Li, Y., Jiang, L. J., & Hu, J. (2013). A wide-band equivalent source reconstruction method exploiting the Stoer-Bulirsch algorithm with the adaptive frequency sampling. *IEEE Transactions on Antennas and Propagation*, *61*(10), 5338–5343. doi:10.1109/TAP.2013.2274032

Liu, L., Das, A., & Megaridis, C. M. (2014). Terahertz Shielding of Carbon Nanomaterials and Their Composites – A Review and Applications. *Carbon*, *69*, 1–16. doi:10.1016/j.carbon.2013.12.021

Li, Y., Pei, X., Shen, B., Zhai, W., Zhang, L., & Zheng, W. (2015). Polyimide/Graphene Composite Foam Sheets with Ultrahigh Thermostability for Electromagnetic Interference Shielding. *RSC Advances*, *5*(31), 24342–24351. doi:10.1039/C4RA16421K

Li, Y., Shen, B., Pei, X., Zhang, Y., Yi, D., Zhai, W., & Zheng, W. et al. (2016). Ultrathin Carbon Foams for Effective Electromagnetic Interference Shielding. *Carbon*, *100*, 375–385. doi:10.1016/j.carbon.2016.01.030

Lowrie, W. (2007). *Fundamentals of Geophysics*. London: Cambridge University Press. doi:10.1017/CBO9780511807107

Mager, J. A. (1970). Magnetic Shields. *IEEE Transactions on Magnetics, Volume*, *6*(Issue:1), 67–75. doi:10.1109/TMAG.1970.1066714

Mehlem, K., Wiegand, A., & Weickert, S. (2012, May). New developments in magnetostatic cleanliness modeling. In *Aerospace EMC, 2012 Proceedings ESA Workshop on* (pp. 1-6). IEEE.

Mehlem, K. (1978). Multiple magnetic dipole modeling and field prediction of satellites. *IEEE Transactions on Magnetics*, *14*(5), 1064–1071. doi:10.1109/TMAG.1978.1059983

Mei, H., Han, D., Xiao, S., Ji, T., Tang, J., & Cheng, L. (2016). Improvement of the Electromagnetic Shielding Properties of C/SiC Composites by Electrophoretic Deposition of Carbon Nanotube on Carbon Fibers. *Carbon*, *109*, 149–153. doi:10.1016/j.carbon.2016.07.070

Merrill, R. T., & McElhinny, M. W. (1983). The Earth's magnetic field: Its history, origin and planetary perspective. San Francisco: Academic Press.

Metglas Ind. (2011). *Magnetic Alloy*. Retrieved October, 2017, from www.metglas.com

Meyer, M., Flourens, F., Rouquette, J. A., & Delnevo, A. (2008, September). Modeling of lightning indirect effects in CFRP aircraft. *2008 International Symposium on Electromagnetic Compatibility (EMC Europe)*, 1-5. doi:10.1109/EMCEUROPE.2008.4786842

Micheli, D., Apollo, C., Pastore, R., Morles, R. B., Laurenzi, S., & Marchetti, M. (2011). Nanostructured Composite Materials for Electromagnetic Interference Shielding Applications. *Acta Astronautica*, *69*(9-10), 747–757. doi:10.1016/j.actaastro.2011.06.004

Micheli, D., Pastore, R., Delfini, A., Giusti, A., Vricella, A., Santoni, F., & Vasilyeva, E. et al. (2017). Electromagnetic Characterization of Advanced Nanostructured Materials and Multilayer Design Optimization for Metrological and Low Radar Observability Applications. *Acta Astronautica*, *134*, 33–40. doi:10.1016/j.actaastro.2017.01.044

Micheli, D., Pastore, R., Gradoni, G., Primiani, V. M., Moglie, F., & Marchetti, M. (2013). Reduction of Satellite Electromagnetic Scattering by Carbon Nanostructured Multilayers. *Acta Astronautica*, *88*, 61–73. doi:10.1016/j.actaastro.2013.03.003

Mikki, S. M., & Antar, Y. M. (2012). Near-field analysis of electromagnetic interactions in antenna arrays through equivalent dipole models. *IEEE Transactions on Antennas and Propagation*, *60*(3), 1381–1389. doi:10.1109/TAP.2011.2180318

MIL-HDBK-5H. (1998). Department of defense test method standard: Metallic Materials and Elements for Aerospace Vehicle Structures. *Section, 5*(1).

MIL-STD-202H. (2015, April). *Department of defense test method standard: Test method standard electronic and electrical component parts.*

MIL-STD-451G. (2015). *Requirements for the Control of Electromagnetic Interference Characteristics of Subsystems and Equipment.* Department of Defense Interface Standard, Department of Defense, United States of America.

MIL-STD-461G. (2015). *Department of defense test method standard: Requirements for the control of electromagnetic interference characteristics of subsystems and equipment.*

MIL-STD-464C. (2010). *Department of defense test method standard: electromagnetic environmental effects requirements for systems.*

Mittra, R. (2016). *Computational Electromagnetics*. Springer-Verlag.

Moglie, F., & Pastore, A. P. (2006). FDTD analysis of plane wave superposition to simulate susceptibility tests in reverberation chambers. *IEEE Transactions on Electromagnetic Compatibility*, *48*(1), 195–202. doi:10.1109/TEMC.2006.870793

Müller, D., Marsden, R. G., Cyr, O. S., & Gilbert, H. R. (2013). Solar orbiter. *Solar Physics*, *285*, 25–70. doi:10.1007/s11207-012-0085-7

Musgrave, G., Larsen, A., & Sgobba, T. (2009). *Safety Design for Space Systems*. Oxford, UK: Butterworth-Heinemann.

Musmann, G. (Ed.). (2010). *Fluxgate magnetometers for space research*. BoD–Books on Demand.

Mu, Y., Zhou, W., Wan, F., Ding, D., Hu, Y., & Luo, F. (2015). High-Temperature Dielectric and Electromagnetic Interference Shielding Properties of SiC_f/SiC Composites using Ti_3SiC_2 as Inert Filler. *Composites. Part A, Applied Science and Manufacturing*, *77*, 195–203. doi:10.1016/j.compositesa.2015.07.004

National Aeronautics and Space Administration (NASA). (1984). *NASA Technical Paper 2361: Design Guidelines for Assessing and Controlling Spacecraft Charging Effects*. NASA.

National Aeronautics and Space Administration (NASA). (1993a). *NASA Preferred Reliability Practices, Practice No. PD-AP-1301: Surface Charging/ESD Analysis*. NASA.

National Aeronautics and Space Administration (NASA). (1993b). *NASA Preferred Reliability Practices, Practice No. PT-TE-1414: Electrostatic Discharge (ESD)*. NASA.

National Aeronautics and Space Administration (NASA). (1999). *NASA Handbook on electrical transients Specification, Measurement, and Control of Electrical Switching Transients, NASA/CR-1999-209574*. NASA.

National Aeronautics and Space Administration (NASA). (2007). *Low Earth Orbit Spacecraft Charging Design Handbook, NASA-HDBK-4006*. NASA.

National Aeronautics and Space Administration. (1995). *Marshall Space Flight Center Electromagnetic Compatibility Design and Interference Control (MEDIC) Handbook*. Marshall Space Flight Center.

Nayak, L., Pradhan, R. R., Khastgir, D., & Chaki, T. K. (2015). Thermally Stable Electromagnetic Interference Shielding Material from Polysulfone Nanocomposites: Comparison on Carbon Nanotube and Nanofiber Reinforcement. *Polymer Composites*, *36*(3), 566–575. doi:10.1002/pc.22973

Nayyeri, V., Soleimani, M., & Ramahi, O. M. (2013). Modeling graphene in the finite-difference time-domain method using a surface boundary condition. *IEEE Transactions on Antennas and Propagation*, *61*(8), 4176–4182. doi:10.1109/TAP.2013.2260517

Nicoletto, M., Boschetti, D., & Savi, P. (2014, September). High Speed Digital Lines routed on non-metallic Spacecraft structures. In *Electromagnetic Compatibility (EMC Europe), 2014 International Symposium on* (pp. 785-789). IEEE. doi:10.1109/EMCEurope.2014.6931011

Obiekezie, C. S., Thomas, D. W., Nothofer, A., Greedy, S., Arnaut, L. R., & Sewell, P. (2014). Complex locations of equivalent dipoles for improved characterization of radiated emissions. *IEEE Transactions on Electromagnetic Compatibility*, *56*(5), 1087–1094. doi:10.1109/TEMC.2014.2313406

Oliveri, G., Rocca, P., & Massa, A. (2012, March). Differential evolution as applied to electromagnetics: Advances, comparisons, and applications. In *Antennas and Propagation (EUCAP), 2012 6th European Conference on* (pp. 3058-3059). IEEE.

Orlandi, A., Duffy, A. P., Archambeault, B., Antonini, G., Coleby, D. E., & Connor, S. (2006). Feature selective validation (FSV) for validation of computational electromagnetics (CEM). part II-assessment of FSV performance. *IEEE Transactions on Electromagnetic Compatibility*, *48*(3), 460–467. doi:10.1109/TEMC.2006.879360

Parkes, S. (2012). SpaceWire User's Guide. STAR-Dundee Limited.

Parmantier J., Issac F., Gobin V. (2012, December). Indirect Effects of Lightning on Aircraft and Rotorcraft. *Journal of Aerospace Lab., ONERA*, 1-27.

Paul, C. R. (1992). *Introduction to electromagnetic compatibility (EMC)* (2nd ed.). Introduction to Electromagnetic Compatibility.

Paul, C. R. (2006). *Introduction to electromagnetic compatibility* (Vol. 184). John Wiley & Sons.

Pelissou, P., Daout, B., & Wolf, K.-F. J. (2015). Improved ESD test method for testing spacecraft equipment. *Asia Pacific Symposium on Electromagnetic Compatibility (APEMC 2015)*. Doi:10.1109/APEMC.2015.7175316

Pettit, G. (2017). Summary of Military and Aerospace EMC Tests. In *2017 EMC Testing Guide*. Retrieved from www.interferencetechnology.com

Pisacane, V. L. (2016). *The Space Environment and its Effects on Space Systems*. AIAA Education Series. doi:10.2514/4.103537

Primdahl, F. (1989). *A Pedestrian's Approach to Magnetic Cleanliness*. Danish Space Research Institute.

Prudenziati, M. (1999). *Handbook of Sensors and Actuators. Thick Film Sensors*. Elsevier.

Rachidi, F. (2012, August). A Review of Field-to-Transmission Line Coupling Models With Special Emphasis to Lightning-Induced Voltages on Overhead Lines. *IEEE Transactions on Electromagnetic Compatibility*, *54*(4), 898–911. doi:10.1109/TEMC.2011.2181519

Radasky, W. (2014) Electromagnetic Warfare is Here. *IEEE Spectrum*. Retrieved from http://spectrum.ieee.org/aerospace/military/electromagnetic-warfare-is-here

Radasky, W. A., Baum, C. E., & Wik, M. W. (2004). Introduction to the special issue on high-power electromagnetics (HPEM) and intentional electromagnetic interference (IEMI). *IEEE Transactions on Electromagnetic Compatibility*, *46*(3), 314–321. doi:10.1109/TEMC.2004.831899

Rasek & Gabrisak. (2011, April). Wire bundle currents for high intensity radiated fields (HIRF) and indirect effects of lightning (IEL) with focus on bulk current injection (BCI) test. *21st International Conference Radioelektronika.*

Rasek, G. A., & Loos, S. E. (2008, August). Correlation of direct current injection (DCI) and free-field illumination for HIRF certification. *IEEE Transactions on Electromagnetic Compatibility*, *50*(3), 499–503. doi:10.1109/TEMC.2008.926872

Rasek, S., Tobola, R., & Loos, T. (2015, February 3). HIRF transfer function observations: Notes on results versus requirements and certification approach. *IEEE Transactions on Electromagnetic Compatibility*, *57*(2), 195–202. doi:10.1109/TEMC.2014.2381095

Rierson, L. K. (1997). *Determining an Aircraft's Attenuation Characteristics of HIRF Testing* (No. 971488). SAE Technical Paper.

Ripka, P. (1993). Race-track fluxgate sensors. *Sensors and Actuators A. Physical*, *37-38*, 417–421.

Ripka, P. (2001). *Magnetic Sensors and Magnetometers*. Boston, MA: Artech House Publishers.

Ripka, P., & Tipek, A. (2007). *Modern Sensors Handbook*. London, UK: ISTE. doi:10.1002/9780470612231

Romero, S. F., Morales, A. L., Cancela, M. A., & Gutierrez, G. G. (2012, September). Validation procedure of low level coupling tests on real aircraft structure. In *Electromagnetic Compatibility (EMC EUROPE), 2012 International Symposium on* (pp. 1-6). IEEE. doi:10.1109/EMCEurope.2012.6396706

RTCA/DO-160 G. (2014). *Environmental Conditions and Test Procedures for Airborne Equipment*.

Sarto, M. S. (1999). A new model for the FDTD analysis of the shielding performances of thin composite structures. *IEEE Transactions on Electromagnetic Compatibility*, *41*(4), 298–306. doi:10.1109/15.809798

Sarto, M. S., Greco, S., & Tamburrano, A. (2014). Shielding effectiveness of protective metallic wire meshes: EM modeling and validation. *IEEE Transactions on Electromagnetic Compatibility*, *56*(3), 615–621. doi:10.1109/TEMC.2013.2292715

Scheeres, D. J., Marzari, F., Tomasella, L., & Vanzani, V. (1998). ROSETTA mission: Satellite orbits around a cometary nucleus. *Planetary and Space Science*, *46*(6), 649–671. doi:10.1016/S0032-0633(97)00200-6

Schmidt, S., & Lazzi, G. (2004). Use of the FDTD thin-strut formalism for biomedical telemetry coil designs. *IEEE Transactions on Microwave Theory and Techniques*, *52*(8), 1952–1956. doi:10.1109/TMTT.2004.832019

Schneider. (2010). *Understanding the Finite-Difference Time-Domain Method.* Retrieved from http://www.eecs.wsu.edu/~schneidj/ufdtd/ufdtd.pdf

Schroder, Bruns, Kucera, & Schuster. (2014, February). Analysis of High Intensity Radiated Field Coupling into Aircraft Using the Method of Moments. *IEEE Trans. Electromagnetic Compatibility, 56*(1).

Schuster, C., Christ, A., & Fichtner, W. (2000). Review of FDTD time-stepping schemes for efficient simulation of electric conductive media. *Microwave and Optical Technology Letters*, *25*(1), 16–21. doi:10.1002/(SICI)1098-2760(20000405)25:1<16::AID-MOP6>3.0.CO;2-O

Secretariat, E. C. S. S. (2012). *ECSS-E-HB-20-07A, Space Engineering – Electromagnetic Compatibility Handbook, ESA-ESTEC.* Noordwijk, The Netherlands: Requirements & Standards Division.

Secretariat, E. C. S. S. (2012). *ECSS-E-ST-20-07C Rev. 1, Space Engineering – Electromagnetic Compatibility, ESA-ESTEC.* Requirements & Standards Division.

Shahzad, F., Alhabeb, M., Hatter, C. B., Anasori, B., Hong, S. M., Koo, C. M., & Gogotsi, Y. (2016). Electromagnetic Interference Shielding with 2D Transition Metal Carbides (MXenes). *Science*, *353*(6304), 1137–1140. doi:10.1126/science.aag2421 PMID:27609888

Shavers, M.R., & Cucinotta. (2004). Implementation of ALARA Radiation Protection on the ISS through Polyethylene Shielding Augmentation of the Service Module Crew Quarters. *Adv. Space Res.*, *34*(6), 1333-1337. doi: /10.1016/j.asr.2003.10.051

Shen, B., Li, Y., Zhai, W., & Zheng, W. (2016). Compressible Graphene-Coated Polymer Foams with Ultralow Density for Adjustable Electromagnetic Interference (EMI) Shielding. *ACS Applied Materials & Interfaces*, *8*(12), 8050–8057. doi:10.1021/acsami.5b11715 PMID:26974443

Song, W. L., Cao, M.-S., Fan, L.-Z., Lu, M.-M., Li, Y., Wang, C.-Y., & Ju, H.-F. (2014b). Highly Ordered Porous Carbon/Wax Composites for Effective Electromagnetic Attenuation and Shielding. *Carbon*, *77*, 130–142. doi:10.1016/j.carbon.2014.05.014

Song, W.-L., Cao, M.-S., Lu, M.-M., Bi, S., Wang, C.-Y., Liu, J., & Fan, L.-Z. et al. (2014a). Flexible Graphene/Polymer Composite Films in Sandwich Structures for Effective Electromagnetic Interference Shielding. *Carbon*, *66*, 67–76. doi:10.1016/j.carbon.2013.08.043

SpaceWire - The Standard. (2017). Retrieved 15 June 2017, from http://spacewire.esa.int/content/Standard/Standard.php

Spadacini, G., & Pignari, S. A. (2004). A bulk current injection test conforming to statistical properties of radiation-induced effects. *IEEE Transactions on Electromagnetic Compatibility*, *46*(3), 446–458. doi:10.1109/TEMC.2004.831896

Spantideas, S. T., Kapsalis, N. C., Kakarakis, S. D., & Capsalis, C. N. (2016, May). A novel technique for accurate extrapolation of complex magnetic sources. In *Aerospace EMC (Aerospace EMC), 2016 ESA Workshop on* (pp. 1-5). IEEE. doi:10.1109/AeroEMC.2016.7504599

Spantideas, S. T., Kakarakis, S. D. J., Kapsalis, N. C., & Capsalis, C. N. (2016). Theoretical methods for studying distance and frequency scaling for AC magnetic fields in satellite missions. *IEEE Transactions on Magnetics*, *52*(4), 1–5. doi:10.1109/TMAG.2015.2496901

Spantideas, S., Kapsalis, N. C., Kakarakis, S. D. J., & Capsalis, C. N. (2014). A method of predicting composite magnetic sources employing particle swarm optimization. *Progress In Electromagnetics Research M*, *39*, 161–170. doi:10.2528/PIERM14092902

Spillantini, P., Casolino, M., Durante, M., Mueller-Mellin, R., Reitz, G., Rossi, L., & Sorbi, M. et al. (2007). Shelding from Cosmic Radition for Interplanetary Missions: Active and Passive Methods. *Radiation Measurements*, *42*(1), 14–23. doi:10.1016/j.radmeas.2006.04.028

STANAG 4370 Ed 5. (2014). NATO Standardization Agreements.

Storn, R., & Price, K. (1995). *Differential evolution–a simple and efficient adaptive scheme for global optimization over continuous spaces: technical report TR-95-012*. Berkeley, CA: International Computer Science.

Sun, X., Liu, X., Shen, X., Wu, Y., Wang, Z., & Kim, J.-K. (2016). Graphene foam/Carbon Nanotube/Poly(dimethyl siloxane) Composites for Exceptional Microwave Shielding. *Composites. Part A, Applied Science and Manufacturing*, *85*, 199–206. doi:10.1016/j.compositesa.2016.03.009

Taflove & Hagness. (2005). *Computational Electrodynamics The Finite- Differences Time Domain Method*. Norwood, MA: Artech House.

Taflove, A., Oskooi, A., & Johnson, S. G. (Eds.). (2013). *Advances in FDTD computational electrodynamics: photonics and nanotechnology*. Artech House.

Telecommunications Industry Association. (2001). *ANSI/TIA/EIA-644, Electrical Characteristics of Low Voltage Differential Signaling (LVDS) Interface Circuits*. Arlington, VA: Standards and Technology Department.

Thomassin, J.-M., Jérôme, C., Pardoen, T., Bailly, C., Huynen, I., & Detrembleur, C. (2013). Polymer/Carbon based Composites as Electromagnetic Interference (EMI) Shielding Materials. *Materials Science and Engineering R Reports*, *74*(7), 211–232. doi:10.1016/j.mser.2013.06.001

Tong, X. C. (2009). *Advanced Materials and Design for Electromagnetic Interference Shielding*. Boca Raton, FL: CRC Press.

Tsitsas, N. L. (2013). A Low-Frequency Electromagnetic Near-Field Inverse Problem for A Spherical Scatterer. *Journal of Computational Mathematics*, *31*(5), 439–448. doi:10.4208/jcm.1304-m4388

Tumanski, S. (2011). *Handbook of Magnetic Measurements*. CRC Press. doi:10.1201/b10979

US Department of Defense. MIL-STD-461-F. (2007). Requirement for the control of electromagnetic interference characteristics of subsystems and equipment. Author.

Vance, E. F. (1975). Shielding effectiveness of braided-wire shields. *IEEE Transactions on Electromagnetic Compatibility*, *EMC-17*(2), 71–77. doi:10.1109/TEMC.1975.303389

Vargas-Bernal, R. (2015). Performance Analysis of Electromagnetic Interference Shielding based on Carbon Nanomaterials used in AMS/RF IC Design. In M. Fakhfakh, E. Tlelo-Cuautle, & M. H. Fino (Eds.), *Performance Optimization Techniques in Analog, Mixed-Signal, and Radio-Frequency Circuit Design* (pp. 268–294). Hershey, PA: IGI Global; doi:10.4018/978-1-4666-6627-6.ch011

Vargas-Bernal, R. (2016). Advances in Functional Nanocoatings Applied in the Aerospace Industry. In A. Zuzuarregui & M. C. Morant-Miñana (Eds.), *Research Perspectives on Functional Micro- and Nanoscale Coatings* (pp. 318–340). Hershey, PA: IGI Global; doi:10.4018/978-1-5225-0066-7.ch012

Vargas-Bernal, R. (2017). The Next Generation of Nanomaterials for Designing Analog Integrated Circuits. In E. Tlelo-Cuautle, M. Fakhfakh, & L. G. de la Fraga (Eds.), *Analog Circuits: Fundamentals, Synthesis and Performance* (pp. 103–121). Nova Science Publishers.

Vargas-Bernal, R. (2018). Mechanisms of Electrical Conductivity in Carbon Nanotubes and Graphene. In M. Khosrow-Pour (Ed.), *Encyclopedia of Information Science and Technology* (4th ed.; pp. 2673–2684). Hershey, PA: IGI Global; doi:10.4018/978-1-5225-2255-3.ch233

Von Klemperer, C. J., & Maharaj, D. (2009). Composite Electromagnetic Interference Shielding Materials for Aerospace Applications. *Composite Structures*, *91*(4), 467–472. doi:10.1016/j.compstruct.2009.04.013

Wang, Y. M., & Chew, W. C. (1989). An iterative solution of the two-dimensional electromagnetic inverse scattering problem. *International Journal of Imaging Systems and Technology*, *1*(1), 100–108. doi:10.1002/ima.1850010111

Weikert, S., Mehlem, K., & Wiegand, A. (2012, May). Spacecraft magnetic cleanliness prediction and control. In *Aerospace EMC, 2012 Proceedings ESA Workshop on* (pp. 1-5). IEEE.

William, G. (1991). *De Magnete*. Dover Publications.

Wilson, J. W., Kim, M., Schimmeling, W., Badavi, F. F., Thibeault, S. A., Cucinotta, F. A., & Kiefer, R. et al. (1995). Issues in Space Radiation Protection: Galactic Cosmic Rays. *Health Physics*, *68*(1), 50–58. doi:10.1097/00004032-199501000-00006 PMID:7989194

Wu, H., Wu, G., & Wang, L. (2015). Peculiar Porous α-Fe_2O_3, γ-Fe_2O_3 and Fe_3O_4 Nanospheres: Facile Synthesis and Electromagnetic Properties. *Powder Technology*, *269*, 443–451. doi:10.1016/j.powtec.2014.09.045

Yan, D.-X., Ren, P. G., Pang, H., Fu, Q., Yang, M.-B., & Li, Z. M. (2012). Efficient Electromagnetic Interference Shielding of Lightweight Graphene/Polystyrene Composite. *Journal of Materials Chemistry*, *22*(36), 18772–18774. doi:10.1039/c2jm32692b

Yang, Y., Chen, R. S., & Yung, E. K. (2006). The unconditionally stable Crank Nicolson FDTD method for three-dimensional Maxwell's equations. *Microwave and Optical Technology Letters*, *48*(8), 1619–1622. doi:10.1002/mop.21684

Yatsenko, V. V., Tretyakov, S. A., Maslovski, S. I., & Sochava, A. A. (2000). Higher order impedance boundary conditions for sparse wire grids. *IEEE Transactions on Antennas and Propagation*, *48*(5), 720–727. doi:10.1109/8.855490

Yee, K. (1966, May). Numerical solution of initial boundary value problems involving Maxwell's equations in isotropic media. *IEEE Transactions on Antennas and Propagation*, *14*, 202–307.

Yu, Mittra, Su, Liu, &Yang (2006). *Parallel Finite-Difference Time-Domain Method.* Artech House, Inc.

Yuan, B., Bao, C., Qian, X., Song, L., Tai, Q., Liew, K. M., & Hu, Y. (2014). Design of Artificial Nacre-like Hybrid Films as Shielding to Mitigate Electromagnetic Pollution. *Carbon*, *75*, 178–189. doi:10.1016/j.carbon.2014.03.051

Zaiping, N., & Yerong, Z. (1998). Hybrid Born iterative method in low-frequency inverse scattering problem. *IEEE Transactions on Geoscience and Remote Sensing*, *36*(3), 749–753. doi:10.1109/36.673668

Zhang, H.-B., Yan, Q., Zheng, W.-G., He, Z., & Yu, Z. Z. (2011). Tough Graphene-Polymer Microcellular Foams for Electromagnetic Interference Shielding. *ACS Applied Materials & Interfaces*, *3*(3), 918–924. doi:10.1021/am200021v PMID:21366239

Zhang, H., Zhang, G., Li, J., Fan, X., Jing, Z., Li, J., & Shi, X. (2017a). Lightweight, Multifunctional Microcellular PMMA/Fe_3O_4@MWCNTs Nanocomposite Foams with Efficient Electromagnetic Interference Shielding. *Composites. Part A, Applied Science and Manufacturing*, *100*, 128–138. doi:10.1016/j.compositesa.2017.05.009

Zhang, Y., Qiu, M., Yu, Y., Wen, B., & Cheng, L. (2017b). A Novel Polyaniline-Coated Bagasse Fiber Composite with Core-Shell Heterostructure provides Effective Electromagnetic Shielding Performance. *ACS Applied Materials & Interfaces*, *9*(1), 809–818. doi:10.1021/acsami.6b11989 PMID:27982585

Zivanovic, S. S., Yee, K. S., & Mei, K. K. (1991). A subgridding method for the time-domain finite-difference method to solve Maxwell's equations. *IEEE Transactions on Microwave Theory and Techniques*, *39*(3), 471–479. doi:10.1109/22.75289

About the Contributors

Christos Nikolopoulos was born in Athens, Greece, in 1981. He received the B.S. degree in Physics, M.Sc in Telecom Engineering from the National & Kapodistrian University of Athens (UOA) in 2006 and 2011 respectively, also he holds a Ph.D from National Technical University of Athens since 2013. He is currently working at the Wireless and Long Distance Communication Lab., School of Electrical and Computer Engineering, National Technical University of Athens as an Associate Researcher. His main research interests are in the field of Electromagnetic Compatibility, Inverse EM Scattering, Smart Antennas Design and Mobile Communications. He participates as a member in various workgroups in Hellenic Ministries. He is a licensed Physicist-Radioelectrologist Grade A' since 2014. Past positions include Hellenic Telecommunications Organization and various research projects in collaboration with National and Kapodistrian University of Athens, National Technical University of Athens, European Research Agencies, European Space Agency, and the Hellenic Ministry of Finance. He also acts as a consultant to EMC issues in EMTech SPACE P.C.

* * *

Jesus Alvarez was born in Leon, Spain. He received the B.Sc. degree in 2001 from the University of Cantabria, Spain, the M.Sc. degree from the University Carlos III of Madrid, Spain, in 2008, and Ph.D. from University of Granada, Spain, in 2013. Since 2006 he has been working as a RCS, Antenna and EMC Engineer with Airbus Defence and Space. His current research interests include computational electrodynamics in time domain, method of moments, and fast algorithms for integral equations in frequency domain and computational electromagnetics applied to electromagnetic compatibility, antenna and radar cross section.

Manuel Añón-Cancela was born in Carballo-A Coruña. He received the M.Sc. degree in physics from the Universidade de Santiago de Compostela, Santiago de Compostela, Spain (www.usc.es), in 1989. He joined www.INTA.es in 1991 as

E3 Test Engineer in the Electromagnetic Compatibility Area, being involved in several national and international projects for the aerospace and defence sectors. He is currently the Head of the Electromagnetic Compatibility Area at INTA.Mr. Añón-Cancela is a member of national and international standardization groups for military and civilian E3 specs.

Anargyros T. Baklezos was born in Lamia, Greece, in 1981. He holds a B.S. degree in Physics and an M.Sc in Telecom Engineering from the National & Kapodistrian University of Athens (UOA). He is currently working towards a PhD degree in Intelligent Antenna Systems from the School of Electrical and Computer Engineering of National Technical University of Athens (NTUA). His main research interests are in the field of electromagnetic compatibility, smart antennas and biomedical antenna design. He has taken part in number of research projects in collaboration with European Research Agencies and the Hellenic Ministry of Finance. He is a published author in international peer reviewed journals and scientific books.

Bárbara Bermúdez-Reyes studied Materials Engineering at Instituto Tecnologico de Morelia (Morelia's Technologic Institute.) In 2004, she received the M. S. degree in Materials Science in Centro de Investigacion y Estudios Avanzados (Research and Advance Studies Center from National Polytechnic Institute), and in 2009 she received her Ph. D. degree in Metallurgic and Materials Science in Universidad Michoacana de San Nicolas de Hidalgo. In 2011, she held a postdoctoral in materials design at the Universidad Autonoma de Sinaloa (Autonomous University of Sinaloa). From 2009 to 2010, she was lecturer in Aeronautics Engineering Program from National Institute Polytechnic. Currently, she is a Research Professor in Aeronautics Engineering Research and Innovation Center of Autonomous University of Nuevo León. Her current research areas are: Protective aerospace Materials and satellite structures. She is Point of Contact between University Space Engineering (UNISEC) and UNISEC- Mexico.

Miguel Ruiz Cabello Nuñez was born in Granada, Spain. He received the B.Sc. in 2008, M.Sc. in physics in 2010, and Ph.D in 2017 all from the University of Granada, Granada, Spain. He is currently working in the Department of Electromagnetism and Matter Physics, working on HPC time domain solvers for electromagnetic analysis and design.He has participated in the 7PM EU project HIRF-SE as a Developer of the SEMBAUGRFDTD solver.

Christos N. Capsalis was born in Nafplion, Greece, in 1956. He received the Diploma in electrical and mechanical engineering from the National Technical University of Athens (NTUA), in 1979, the B.S. degree in economics from the University of Athens, in 1983, and the Ph.D. degree in electrical engineering from NTUA, in 1985. He is currently a Professor in the Department of Electrical and Computer Engineering, NTUA. His current scientific activity concerns satellite and mobile communications, antenna theory and design, and electromagnetic compatibility.

Luis Diaz Angulo received the M.Sc. degree in physics in 2008, the Ph.D. degree in 2014, and the M.Sc. degree in electronics engineering in 2015, all from the University of Granada, Granada, Spain.He is currently an Interim Teacher in the Department of Electromagnetism, University of Granada. He works in time-domain numerical methods applied to electromagnetism, especially discontinuous Galerkin and finite-difference methods. His research interests include applications of numerical methods in EMC, terahertz technologies, GPR imaging, and bioelectromagnetics.

David Escot-Bocanegra was born in Madrid, Spain. He received the M.Sc. and Ph.D. degrees in telecommunication engineering from the University of Alcala, Madrid, Spain, in 2002 and 2012, respectively. From 2002 to 2003, he worked as a junior Researcher at Telecom Bretagne, Brest, France. In 2004, he joined INTA, Spain, first as a Researcher and currently as a Scientist in the Radiofrequency Area. He has published numerous refereed journal articles and has been the principal investigator in several national and international research projects. His research interests include radar cross section, computational electromagnetics, materials characterization, and EMC simulations.

Mireya Fernández Chimeno received the M.Sc. and Ph.D. degrees in electrical engineering (Ingenierí de Telecomunicación, specialization in electronics) from the Technical University of Catalunya (UPC BarcelonaTech), Barcelona, Spain, in 1990 and 1996, respectively.Since 1996, she has been an Associate Professor in the Department of Electronic Engineering, UPC BarcelonaTech. She belongs to the Biomedical and Electronics Instrumentation Group (http://ieb.eel.upc.edu/), where she develops her research and is responsible of the electrical safety area. She is the Quality Manager of the electromagnetic compatibility group at UPC BarcelonaTech. She was a Visiting Researcher at the Health Sciences Department de la Universidad de York (2003). She was the Associate Dean (Academic Laboratories) at the School of Electrical and Telecom Engineering (ETSETB), UPC, (1996-2000). She is co-author of more than 100 scientific and academic publications (papers, congress communications and books) and 6 patents. Her research interests include the design and development of biomedical instrumentation and the electromagnetic compatibility.

Mario Fernández Pantoja received the B.S., M.S. and Ph.D. degree in 1996, 1998 and 2001, respectively, in electrical engineering from the University of Granada, Granada, Spain.Between 1997 and 2001, he was an Assistant Professor at the University of Jaen, Jaén, Spain, and then he joined the University of Granada, where, in 2004 and 2016, he was appointed as an Associate Professor and Full Professor, respectively. He has been a Guest Researcher at Dipartimento Ingenieria dell. Informazione, University of Pisa, Pisa, Italy, in the Antenna and Electromagnetics Group at Denmark Technical Univ., and in the CEARL, Pennsylvania State University, State College, PA, USA. His research is focused mainly on the areas of time-domain analysis of electromagnetic radiation and scattering problems, optimization methods applied to antenna design, Terahertz technology, and nanoelectromagnetics.

Sergio Fernandez Romero was born in Valencia, Spain, in 1978. He received the M.S. degree in telecommunication engineering from the University of Valencia, Spain, in January 2002, and the Ph.D. degree in telecommunication engineering from University of Alcala, Madrid, Spain, in 2016. Since 2003, he has been with the electromagnetic compatibility (EMC) on Aircraft Test Group, National Institute for Aerospace Technology, Torrejon de Ardoz, Madrid, Spain. His current research interest include reverberation chambers, EMC and shielding effectiveness measurements and some related topics such as time domain measurements and reverberant microwave propagation models. Dr. Romero is a member of the International Electrotechnical Commission joint task force on reverberation chambers.

Salvador G. Garcia received the M.S. and Ph.D. degrees (with extraordinary award) in physics from the University of Granada, Granada, Spain, in 1989 and 1994, respectively.He is currently a Full Professor in the Department of Electromagnetism and Matter Physics, University of Granada. He has published more 80 refereed journal articles and book chapters and leaded several national and international projects. He has received grants to stay as a Visiting Scholar at the University of Duisburg (1997), the Institute of Mobile and Satellite Communication Techniques (1998), the University of Wisconsin-Madison (2001), and the University of Kentucky (2005). His current research interests include computational electromagnetics, electromagnetic compatibility, RCS, and antennas.

Rafael Gómez Martín received the M.Sc. degree in physics from the University of Seville, Seville, Spain, in 1971, and the Ph.D. (cum laude) degree in physics from the University of Granada, Granada, Spain, in 1974.From 1971 to 1974, he was with the University of Santiago de Compostela, Santiago, Spain, as an Assistant Professor. From 1975 to 1985, he was an Associate Professor at the University of Granada, where he became a Full Professor in 1986. He is currently Professor Emeritus in the Department of Electromagnetism and Matter Physics.

Guadalupe G. Gutierrez was born in Leon, Spain, in 1979. She received the M.Sc. degree in physics from the University of Salamanca, Spain, in 2002. Since 2003, she has worked in different areas as an EME and Antennas Engineer at Airbus. Her main research activities concern electromagnetic simulations on aircrafts.

Neoclis, G. Hadjigeorgiou is a PhD candidate at the School of Electrical and Computer Engineering, National Technical University of Athens (NTUA), Greece. He has had his Master Degree on the subject Automation Systems at the School of Mechanical Engineering, National Technical University of Athens. His Master thesis was "Detection of the position of sea vessel with magnetic techniques." in 2013. He has also had his Bachelor degree at the School of Electrical and Computer Engineering, of the National Technical University of Athens. He has his Bachelor thesis on "Acceleration of the search of the cyclic representation of posets with GPU" in 2011.

Ricardo Jauregui received the B.Sc. in Electrical Engineering from Los Andes University, Venezuela, in 2001, M.Sc. degrees in Mobile Communication from Desuto University, Bilbao-Spain, in 2003 and the Ph.D. degree from Polytechnic University of Catalonia (UPC), Barcelona-Spain, in 2010. He worked for 7 years in Electromagnetic Compatibility Group at the Universitat Politecnica de Catalunya (GCEM-UPC), being responsible for the electromagnetic simulation and validation methods. In 2012 he is certifying as Project Manager Professional by the PMI and in 2013, Roche Diagnostic hires him to coordinate development of diabetes Software Project in Spain. In 2015, Ricardo founds his own medical device Company (DyCare), where he is in charge of aligning technology-related decisions and the objectives of the organization.

Nikolaos C. Kapsalis was born in Athens in 1985. He finished his studies in Computer Science in the Economical University of Athens in 2008. He received his Ph.D. from the National Technical University of Athens in 2015, and he is a member of the Wireless and Long Distance Communications Laboratory. His scientific interests include smart antennas, electromagnetic interference and resource allocation.

Eleni P. Nicolopoulou (PhD) was born in 1987 in Athens, Greece, where she completed her school education. She received her diploma and the PhD degree in Electrical and Computer Engineering from the School of Electrical and Computer Engineering of the National Technical University of Athens in 2010 and 2016, respectively. She speaks English (Certificate of Proficiency in English, 2001) and German (Kleines Deutsches Sprachdiplom, 2010). Career: Since 2010 she is a laboratory assistant at the High Voltage Laboratory of the National Technical University of Athens. Author Activities: She is author of 1 diploma thesis, 1 doctoral thesis, 5 papers in international scientific journals and 18 papers in conference proceedings. Other Activities: Member of the Technical Chamber of Greece (TEE). She participates in research activities of the NTUA financially supported by public and private companies, dealing with quality controls of electrical materials, electric tests and tests of electromagnetic compatibility of electrical and electronic equipment.

Luis Nuño was born in Valencia, Spain, in 1963. He received the Ingeniero (MS. Eng.) degree in telecommunication engineering from the Polytechnic University of Madrid (UPM), Madrid, Spain, in 1987, and the Doctor Ingeniero (Ph.D.) degree in telecommunication engineering from the Polytechnic University of Valencia (UPV), Valencia, Spain, in 1993.He was with the Grupo de Radiación of the UPM during 1986–1988 and with the enterprise Ingeniería de Radiofrecuencias (IRSA) during 1988–1990. Since 1990, he has been with the Department of Communications in the UPV, where he became a Teaching Assistant in 1990, an Associate Professor in 1996, and a Professor in 2002. On the other hand, he was responsible for the EMC Laboratory (ICEM), UPV, since its creation in 1992 until April 2005, when he became the Head of ICEM R&D. From 2006 to 2015, he was the Chairman of IEC/SC 77B: Electromagnetic Compatibility. High Frequency Phenomena. His current research interests include numerical methods in electromagnetism, electromagnetic compatibility and industrial applications of microwaves. Dr. Nuño is a member of IEC/ACEC and the Chairman of the Spanish National Committee on EMC (AEN/CTN 208), since 2011.

Panagiotis Papastamatis was born in 1986 in Athens where he completed his school education. He graduated from the Department of Electrical and Computer Engineering of the National Technical University of Athens in 2013 with the grade «Very Good». Currently he is working on his PhD in the field of Electromagnetic Immunity. He speaks English (Certificate of Proficiency in English, 2001) and German (Abitur-German School of Athens graduate) fluently. Career: During his military service (2013-2014) he worked for the Greek Ministry of Defense in the field of Information Systems. Since 2015 he is working as an EMC test engineer and a laboratory assistant at the High Voltage Laboratory of the National Technical University of Athens. Author Activities: He is the author of 1 diploma thesis entitled "Electromagnetic Compatibility Tests on Telecommunication Lines" and 1 paper in conference proceedings. Other Activities: He participates in research activities of the NTUA, financially supported by public and private companies, regarding electrical tests and tests of Electromagnetic Compatibility of electrical and electronic equipment.

Enrique Pascual-Gil was born in Madrid, Spain, in 1964. He received the M.Sc. degree in aeronautic and computer sciences from Universidad Politécnica de Madrid, Madrid, Spain, in 1985 and 1997, respectively.He has been working more than twenty five years as an EMC and MW Engineer in different areas. He has been the Head of Computational Electromagnetic at Airbus Military since 2008, being responsible for the electromagnetic simulation team dedicated to numerical modeling of all electromagnetic aspects of an aircraft, an Airbus EMC Expert since 2011. He has participated as Technical Manager in various national and international technology projects. His current interest focuses on the application of numerical methods to solve more complex cases in order to achieve a complete Virtual Testing for aircraft certification.

Borja Plaza Gallardo was born in Madrid, Spain, in 1980. He received the M.Sc. degree in industrial engineering in 2005 from the University Carlos III of Madrid, Spain. He has worked more than 10 years in different areas as Engineer at INTA. He is currently working for the Detectability and Electronic Warfare Laboratory. His research interests include computational electromagnetic applied to electromagnetic compatibility, antenna and radar cross section problems.

Marc Pous was born in Barcelona, in 1983. He received the M.Sc. degree in telecommunications engineering and Ph.D. degree from the Universitat Politécnica de Catalunya (UPC), Barcelona, Spain, in 2009 and 2015, respectively.From 2003 to 2006, he was with the LGAI Technological Centre, Electromagnetic Compatibility Department. In 2006, he joined Electromagnetic Compatibility Group at the Universitat Politécnica de Catalunya (GCEM-UPC), where he has been participating in international and national research projects related with automotive, aerospace, railway and medical industries. His research is focused on the development of time-domain measurement techniques to capture interferences that are not properly measured following the harmonized EMC standards.

David Poyatos was born in Madrid (Spain) in 1972. He became Telecommunication Engineer (MSc.) by the Universidad Politecnica de Madrid (UPM) in 1998. He joined INTA, the Spanish National Institute of Aerospace Technology, in 1997 as a junior research assistant in the Electromagnetic Compatibility Area. In 2000 he became R&D+i engineer for the Detectability and Electronic Warfare Lab, participating in numerous national and international projects for the aerospace and defence sectors. He is currently the head of the Radiofrequency Area at INTA. He has also been officially appointed as the Spanish representative in several NATO scientific research groups. His current research activities and interests are related to the analysis, measurement and exploitation of radar signature, the electromagnetic characterization of materials, the use of bio-inspired algorithms in electromagnetics, the theoretical and practical study of radar reflectivity and the application of numerical methods to solve electromagnetic problems, such as, the simulation of electromagnetic compatibility (EMC) tests. He has authored more than 35 publications in books, journals and conferences, all related with the aforementioned areas of expertise.

Pere J. Riu received the M.Sc. and Ph.D. degrees from the Universitat Politecnica de Catalunya (UPC), Barcelona, Spain, in 1986 and 1991, respectively. He is currently a Full Professor of electronics in the Department of Electronic Engineering, UPC, where he has occupied several academic positions since 1986. His research interests include electromagnetic compatibility, computational electromagnetics, interaction of EMF with biological tissues, and biomedical instrumentation design with emphasis on electrical bioimpedance techniques, including EIT. He has published more than 60 peer-reviewed papers on these topics. He was a Visiting Associate Professor in the Department of Bioengineering, University of Pennsylvania, Philadelphia, PA, USA, in 1997.Prof. Riu is currently a member of the Committee on Man and Radiation, COMAR (IEEE-BMES), and was the President of the International Society for Electrical Bio-Impedance until 2010.

Amelia Rubio Bretones was born in Granada, Spain. She received the Ph.D. degree in physics (cum laude) from the University of Granada, Granada, Spain in 1988.Since 1985, she has been with the Department of Electromagnetism, University of Granada, first as an Assistant Professor, and then in 1989 as an Associate Professor, and finally, since the year 2000 as a Full Professor. On several occasions, she was a Visiting Scientist at the Delft University of Technology, Delft, the Netherlands, the Eindhoven University of Technology, Eindhoven, the Netherlands, and the Pennsylvania State University, State College, PA, USA. Her research interest focuses on the field of numerical techniques for applied electromagnetics with an emphasis on time-domain techniques such as finite difference time domain, the application of the method of moments in the time domain for antenna and scattering problems, and hybrid techniques.

Ferran Silva received the M.Sc. and Ph.D. degrees from the Universitat Politécnica de Catalunya (UPC), Barcelona, Spain, in 1989 and 1997, respectively. He is currently an Associate Professor of electronics in the Department of Electronic Engineering, UPC. Since 2000, he is the Director of the Electromagnetic Compatibility Group in the same university (GCEM-UPC). His research interests include electromagnetic compatibility (EMC) in near field and time domain, with application to transport, medical, and industrial areas. He is the author and co-author of more than 90 publications in journals and conferences and contributed to the Wiley Encyclopedia of Biomedical Engineering. He has participated in more than 15 research projects related to EMC.Dr. Silva is a senior member of the IEEE EMC Society and the Head of the IEEE EMCS Spanish chapter. He is also member of the Spanish standardization committees SCTC77-210 and the CTN208 SCCISPR210A. He served as Chairman of the EMC Europe 2006 International Symposium. Since then, he is on the Board of Chairmen of EMC Europe.

Marios Sophocleous received his Mechanical Engineering Degree from the Faculty of Engineering & the Environment at the University of Southampton, UK in 2011 focusing on Mechatronics and Sustainable Energies. During his studies he won two scholarships for being ranked in the top 10 students of the year and also won the Jim Graham Prize for the Best Experimental Project of the year for his 3rd year project. His project focused on the design of screen printed reference electrodes for pH sensors. He received his Ph.D. in 2016 at the University of Southampton, UK. His Ph.D. work was mainly on the design of sensor arrays to be placed underground for soil condition monitoring in an attempt to model the relationships between soil parameters and vegetation growth and diversity. During his PhD studies, he also acted as an innovation consultant on innovation techniques and IP management. He is now a Special Scientist at the Holistic Electronics Research Laboratory, University of Cyprus.

Sotirios T. Spantideas was born in Athens in 1985. He received the diploma in Electrical Engineering from the University of Patras in 2010 and his M.Sc. in Electrophysics from the University of KTH Stockholm in 2013. As of 2014 he is a PhD student in the National Technical University of Athens, and a member of the Wireless and Long Distance Communications Laboratory. His scientific interests include Electromagnetic Compatibility & Interference and Electromagnetic Cleanliness in space missions.

Margarita Tecpoyotl Torres received the Mathematician degree from the University of Puebla, Mexico, in 1991. From this University, she was also graduated as Electronic Engineer in 1993. She received the M.Sc. and Ph.D. degrees in Electronics from National Institute of Astrophysics, Optics and Electronics, INAOE, México, in 1997 and 1999, respectively. Dr. Tecpoyotl works, since 1999, at CIICAp of Autonomous University of Morelos, Mexico, where she is currently titular professor. Her main research interest includes Electronic devices, MEMS, Antenna design, Entrepreneurship and Innovation. She has currently four patents titles from IMPI, and four copyrights. She holds the status of National Researcher (member of SNI), in Mexico since 1999.

Rafael Trallero Vela was born in Madrid, Spain, in 1974.He received the degree in aeronautical engineering from the Universidad Politécnica de Madrid, Madrid, Spain, in 2001 and a Doctoral degree (cum laude) from the Universidad Alfonso X El Sabio, Madrid, Spain, in 2013, in the field of fatigue of composites used on current UAVs.Since 2001, he has been with INTA in the Aerospace Programs Test Center for the qualification of civil and military aerospace structural components. From 2006 to 2009, he was an Associate Professor at UAX University and since 2010, a Full Professor of aerospace engineering. He has published several publications and participated in research programs over the years in different areas of industrial engineering.

Stylianos Tsatalas was born in Athens in 1973. He graduated from University of Patras and he holds a Physics BSc degree and MSc in Electronics. Currently he is Managing Director of EMTECH SPACE P.C (CY) Ltd., a company dealing with EMC consulting and measurement systems in various fields, especially for the SPace sector. The company also provides consulting services in the field of certification of electronic and electrical equipment.

Rafael Vargas-Bernal was born in Irapuato, Guanajuato, México in September 30th, 1972. He received the B. E. degree in Electronics in October 6th, 1995 from the University of Guanajuato. After, the M. S. degree in Electronics from INAOE in September 4th, 1997. Lastly, the Sc. D. degree in Electronics from the same institute in July 19th, 2000. During October 2000 - September 2001, he worked as professor of the University of Guanajuato in the Department of Communications and Electronics Engineering. Since January 2002 to December 2009, he has been with the Department of Electronics Engineering of ITESI as Professor-Researcher. From August 2005 until today, he is also working in the Department of Materials Science in the area of Advanced Materials. He is technical reviewer of standards on microtechnology and MEMS in SEMI from November 2006 to the date. Moreover, he is reviewer of the magazine IEEE Latin America Transactions of IEEE from August 2003, Sensors and Actuators B: Chemical of Elsevier from February 2010, IEEE Sensors Journal from October 2012, Nanoscale of RSC November 2012, and RSC Advances published by RSC from April 2013 until today.

Index

F

G

H

I

J

L

M

T

U

W

www.ingramcontent.com/pod-product-compliance
Lightning Source LLC
LaVergne TN
LVHW080903300926
849162LV00003B/535